MW01469894

Autonomous Horizons

The Way Forward

A vision for Air Force senior leaders of the potential for autonomous systems, and a general framework for the science and technology community to advance the state of the art

DR. GREG L. ZACHARIAS
CHIEF SCIENTIST OF THE UNITED STATES AIR FORCE 2015–2018

The second volume in a series introduced by:
Autonomous Horizons: Autonomy in the Air Force – A Path to the Future,
Volume 1: Human Autonomy Teaming (AF/ST TR 15-01)

March 2019

Air University Press
Curtis E. LeMay Center for Doctrine Development and Education
Maxwell AFB, Alabama

Chief of Staff, US Air Force
Gen David L. Goldfein

Commander, Air Education and Training Command
Lt Gen Steven L. Kwast

Commander and President, Air University
Lt Gen Anthony J. Cotton

Commander, Curtis E. LeMay Center for Doctrine Development and Education
Maj Gen Michael D. Rothstein

Director, Air University Press
Lt Col Darin M. Gregg

Project Editor
Donna Budjenska

Cover and Book Design and Illustrations
Daniel Armstrong

Composition and Prepress Production
Nedra Looney

Print Preparation and Distribution
Diane Clark
MSgt Ericka Gilliam

Air University Press
600 Chennault Circle, Bldg 1405
Maxwell AFB, AL 36112-6026
http://www.airuniversity.af.edu/AUPress/

Facebook:
https://www.facebook.com/AirUnivPress
and
Twitter: https://twitter.com/aupress

Library of Congress Cataloging-in-Publication Data

Names: Zacharias, Greg, author. | Air University (U.S.). Press, publisher. | United States. Department of Defense. United States Air Force.
Title: Autonomous horizons : the way forward / by Dr. Greg L. Zacharias. Description: First edition. | Maxwell Air Force Base, AL : AU Press, 2019. "Chief Scientist for the United States Air Force." | "January 2019." |Includes bibliographical references.
Identifiers: LCCN 2018061682 | ISBN 9781585662876
Subjects: LCSH: Aeronautics, Military—Research—United States. | United States. Air Force—Automation. | Artificial intelligence—Military applications—United States. | Intelligent control systems. | Autonomic computing—United States.
Classification: LCC UG643 .Z33 2019 | DDC 358.407—dc23 | SUDOC D 301.26/6: AU 8/2
LC record available at https://lccn.loc.gov/2018061682

Published by Air University Press in March 2019

Disclaimer

Autonomous Horizons: The Way Forward is a product of the Office of the US Air Force Chief Scientist (AF/ST). Opinions, conclusions, and recommendations expressed or implied within are solely those of the author and do not necessarily represent the views of Air University, the United States Air Force, the Department of Defense, or any other US government agency. Cleared for public release, distribution unlimited.

AIR UNIVERSITY PRESS

Contents

Illustrations

Tables

Abbreviations and Acronyms

A2A	air to air
A2/AD	antiaccess/area denial
ACM	Association of Computing Machinery
ACOI	Autonomy Community of Interest
ACT	Autonomy Capabilities Team
ACT-R	Adaptive Control of Thought-Rational
AESA	active electronically-scanned array
AFFOC	Air Force Future Operating Concept
AFIT	Air Force Institute of Technology
AFMC	Air Force Materiel Command
AFOSR	Air Force Office of Scientific Research
AFRL	Air Force Research Laboratory
AGI	artificial general intelligence
AI	artificial intelligence
ANN	artificial neural network
AOC	air operations center
APU	auxiliary power unit
AS	autonomous system
ASoS	agile system of systems
ATO	air tasking order
AUAV	autonomous unmanned air vehicle
Auto-GCAS	Auto Ground-Collision Avoidance System
AWACS	airborne warning and control system
BBN	Bayesian belief network
BD	big data
BDA	battle damage assessment
BDI	Belief-Desire-Intention
BDP	big data platform
BICA	biologically inspired cognitive architecture
BMC2	battle management C2
BSD	Berkley Software Distribution
C2	command and control
CBR	case-based reasoning
CCD	concealment, camouflage, and deception

CE	collaborative environment
CEP	complex event processing
CF	certainty factor
CLARION	Connectionist Learning with Adaptive Rule Induction ON-line
COA	course of action
CoAP	constrained application protocol
COI	community of interest
COTS	commercial-off-the-shelf
CPEF	Continuous Planning and Execution Framework
CPT	conditional probability table
CPU	central processing unit
CRISP-DM	CRoss-Industry Standard Process for Data Mining
CSAR	combat search and rescue
D2D	data to decisions
DARPA	Defense Advanced Research Projects Agency
DCGS	Distributed Common Ground System
DDOS	distributed denial of service
DEAD	destruction of enemy air defenses
DEFT	Deep Exploration and Filtering of Text
DISA	Defense Information Systems Agency
DL	deep learning
DM	decision making
DNN	deep neural network
DOD	Department of Defense
DSB	Defense Science Board
DSP	Defense Support Program
EA/EW	electronic attack/electronic warfare
EO	electro-optic
EO/IR	electro-optic/infrared
EPIC	Executive-Process/Interactive Control
ES	expert system
EW	electronic warfare
EW/EA	electronic warfare/electronic attack
FAQs	frequently asked questions

FCS	flight control subsystem
fMRI	functional magnetic resonance imaging
FMV	full-motion video
FSA	fast simulated annealing
GA	genetic algorithm
GAO	Government Accountability Office
GCAS	Ground-Collision Avoidance System
GIISR	global integrated ISR
GNU	GNU's Not Unix!
GOMS	goal, operator, method, selection
GOTS	government-off-the-shelf
GPS	General Problem Solver
GPS	Global Positioning System
GPU	graphics processing unit
GSA	generalized simulated annealing
GUI	graphical user interface
HCI	human-computer interaction
HCI	human-computer interface
HFE	human factors engineering
HFES	Human Factors and Ergonomics Society
HIPF	Human Information Processing Framework
HMI	human-machine interface
HPC	high-performance computing
HSI	human-systems integration
HSI	human-system interaction
I&W	indications and warnings
IADS	integrated air defense system
IDS	intrusion detection system
IEEE	Institute of Electrical and Electronic Engineers
IIT	integrated information theory
INTEL	(operational) intelligence
IoT	Internet of Things
IR	infrared
ISR	intelligence, surveillance, and reconnaissance
IT	information technology

xiv | AUTONOMOUS HORIZONS

JDL	Joint Director Laboratories
JFACC	Joint Forces Air Component Commander
JPADS	Joint Precision Air Drop System
JSTARS	Joint Surveillance Target Attack Radar System

KP	Knowledge Platform
KSA	Knowledge, Skill, and Ability

LISP	list processing
LRSB	Long-Range Strike Bomber
LVQ	learning vector quantization
LWM2M	lightweight machine-to-machine (protocol)

M&S	modeling and simulation
MAD-E	Multi-Source Analytic Development and Evaluation
MAS	multiagent system
MDC2	multidomain command and control
MDOC	multidomain operations center
MDP	Markov decision problem
MDSA	multidomain situation awareness
MDSC	multidomain situated consciousness
MEC	mission effect chain
MHP	Model Human Processor
MI-ISR	multi-intelligence intelligence, surveillance, and reconnaissance
ML	machine learning
ML/PR	machine learning and pattern recognition
MQTT	Message Queuing Telemetry Transport (protocol)
MTL	medial temporal lobe

NDIA	National Defense Industrial Association
NEAT	neuroevolution of augmenting topology
NLP	natural language processing
NRC	National Research Council
NSA	National Security Agency
NTSB	National Transportation Safety Board

ONA	Office of Net Assessment
OSA	open systems architecture
OSD	Office of the Secretary of Defense

P2F2T2EA4	predict, prescribe, find, fix, track, target, engage, and assess anything, anytime, anywhere in any domain
PCF	Perceptual-Cycle Framework
PED	processing, exploitation, and dissemination
PID	proportional-integral-derivative
PME	professional military education
POMDP	partially observable Markov decision problem
PRM	probabilistic relational model
QA	question answering
R&D	research and development
RDDL	Relational Dynamic Influence Diagram Language
ROE	rules of engagement
RPA	remotely piloted aircraft
RPD	recognition primed decision-making
RPM	revolutions per minute
S&T	science and technology
SA	situation awareness
SAA	sense-assess-augment
SADM	situation assessment and decision-making
SAE	Society of Automotive Engineers
SCADA	supervisory control and data acquisition
SDB	Small Diameter Bomb
SDPE	Strategic Development Capabilities and Experimentation
SEAD	suppression of enemy air defenses
SIGINT	signals intelligence
SIPE	System for Interactive Planning and Execution
SME	subject matter expert
SOM	self-organizing (feature) map
SOM	Society of Mind
SoS	system of systems
SQL	Structured Query Language
SVM	support vector machine
SWAP	size, weight, and power
T&E	test and evaluation

TB	terabyte
TCP	Transmission Control Protocol
TIE	technical integration experiment
TPED	tasking, processing, exploitation, and dissemination
TRL	Technology Readiness Level
TSA	team situation awareness
TTP	tactics, techniques, and procedures
TWEANN	topology and weight-evolving ANN
UAV	unmanned aerial vehicle
UCT	upper confidence bounds applied to trees
UDP	User Datagram Protocol
UI	user interface
UUV	unmanned <undersea/underwater> vehicle
UXV	unmanned <air/ground/sea/undersea> vehicle
V&V	verification and validation
VAST	Visual Analytics Science and Technology
XAI	eXplainable artificial intelligence
ZB	zettabyte

Foreword

THE SECRETARY OF THE AIR FORCE
CHIEF OF STAFF, UNITED STATES AIR FORCE
WASHINGTON DC

AUG 3 0 2018

Rapid advances in autonomous system (AS) development and artificial intelligence (AI) research will change how we as an Air Force work with machines in the future. The payoff will be considerable, including significant protection for our Airmen, greater employment effectiveness, and unlimited opportunities for novel and disruptive concepts of operations. The commercial world already recognizes this potential and has pounced on it; for our military, the pace is slower, but the promise is just as transformational. And the time is right: we can take advantage of the dizzying pace of commercial advances in AS and AI applications, as well as explosive gains in the underlying computational infrastructure growth in computational power, memory, networking, and data availability.

Autonomous Horizons: The Way Forward identifies issues and makes recommendations for the Air Force to take full advantage of this transformational technology. Four broad recommendations are made:

- Embrace a common technical framework and technology set to bring together the many disparate communities working different but key aspects of the problem
- Work on a common set of "challenge problems", both basic and applied, to drive advances in far-term opportunities, as well as near-term capability requirements
- Create a product-focused organizational structure incorporating an iterative development process of rapid prototyping and experimentation, for faster fielding of these systems
- Transition from a legacy platform-centric service to an information-centric enterprise, powered by proliferating autonomous systems that enable resilient human-systems operations

Autonomous Horizons: The Way Forward was created as a collaborative effort between the AF Chief Scientist's office (AF/ST) and the Air Force Research Laboratory (AFRL). Because the recommendations span technical, organizational and cultural issues, the report is written for both the technical/acquisition community, to more rapidly develop and transition this technology to the warfighter, and for senior leadership, to envision future opportunities for transforming the Air Force and how we do business.

Autonomous systems are on the horizon, whether we develop them or not. We have a unique opportunity to transform the Air Force from an air platform-centric service, to a truly multi-domain and knowledge-centric organization, by delivering autonomous systems to the warfighter, so that every mission in air, space, and cyber will be improved, and not just incrementally, but multiplicatively. We encourage all Airmen to consider how their missions can be enabled, or better, transformed by this technology.

Heather Wilson
Secretary of the Air Force

David L. Goldfein
General, USAF
Chief of Staff

Acknowledgments

This report would not have been possible without the encouragement, technical support, and many contributions of the Air Force Research Laboratory (AFRL) and the Air Force Institute of Technology (AFIT). I thank Dr. Morley Stone and Mr. Jack Blackhurst at AFRL Headquarters for their encouragement and support from the very beginning and throughout this effort. I also thank the key technical leaders at Wright–Patterson AFB, Ohio, for their dedication to the effort, their insight into many of the issues, and their contributions throughout the report, particularly Dr. Steven "Cap" Rogers (AFRL), Prof. Bert Peterson (AFIT), and Dr. Mike Mendenhall (AFRL). Finally, considerable thanks are due to Ms. Cathy Griffith (AFRL) for dealing with the extensive reference list and citations throughout.

This AFRL/AFIT team was also supported by several other contributors to issues discussions and written sections, including:[1] Dr. "Wink" Bennett; Mr. Jim Burlingame (LeMay Center); Dr. Gloria Calhoun; Col Doug "Cinco" DeMaio (LeMay Center); Dr. Mark Draper; Capt Justin Fletcher; Dr. Kevin Gluck; Mr. Jeff Hukill (LeMay Center); Lt Col Clay Humphreys, PhD; Maj John Imhoff; Ms. Kris Kearns; Mr. Aaron Linn; Dr. Joe Lyons; Maj Lori Mahoney; Capt Bob Mash; Mr. Brian McLean (LeMay Center); Ms. Dusty Moye (LeMay Center); Capt Jeffrey Nishida; Prof. Mark Oxley (AFIT); Dr. Robert Patterson; Mr. Bob Pollick; Mr. Derek "Edge" Steneman; Mr. Drew Strohshine; Dr. Mike Talbert; Capt Khoa Tang; Maj Albert Tao; Mr. Zachary "Zen" Wallace; Dr. Mike Vidulich; and Mr. Ed Zelnio. I thank them all for their many contributions throughout the report.

The publication of this report was made possible by the support of the Air University Press, and particularly the editorial ministrations of Ms. Donna Budjenska, who brought a professional polish to the effort. Finally, I wish to thank my AF/ST staff for their support throughout the effort and their encouragement to push to the finish line on this document: my military assistants, Col Anne Clark (at the beginning) and Col Michelle Ewy (at the end) and my executive assistant, Ms. Penny Ellis (from beginning to end). It was very much a team effort.

---

1. All members of various directorates of AFRL, except as indicated.

Executive Summary

As the Air Force realigns itself from a counterterrorism focus to dealing with near-peer rivalries with potential existential consequences, a "business as usual" approach to systems development will no longer suffice: we will not be able to continue with incremental advances on concepts developed decades ago. Instead, we need new technologies, affording us new capabilities, and new operational concepts to employ them. Fortunately, there now exists a broad and deep technology push in the information sciences, particularly in the area of autonomous system (AS) development and its associated foundational technology, artificial intelligence (AI). Our knowledge of cognition and neurophysiology—the basis of what makes us "smart," most of the time—grows at a dizzying pace, while our ability to build autonomous systems—like self-driving cars and game-playing robots—continues to make front-page news, as new AI algorithms and learning techniques are developed and employed in novel ways. And these advances are compounded by explosive gains in the underlying computational infrastructure afforded by Moore's law growth in computational power, memory, networking, and data availability.

Our goal here is twofold: to provide a vision for Air Force senior leaders of the potential of autonomous systems and how they can be transformative to warfighting at all levels and to provide the science and technology community a general framework and roadmap for advancing the state of the art while supporting its transition to existing and to-be acquired systems. Like others, we believe the payoff from the employment of these systems will be considerable, simply because the individual capabilities of these autonomous systems will afford us greater degrees of freedom in employment and opportunities for novel concepts of operations. But this is a traditional view. A potentially more far-reaching payoff will come from becoming more information-centric and aided by proliferating autonomous systems, so that we can leave our legacy platform-centric way of thinking behind and become an enterprise that is service oriented, ubiquitously networked, and information intensive.

Our approach in this document is to first lay out what we need in the way of AS "behaviors": that is, no matter what the underlying technical means, what are the resultant behaviors of these systems across the key dimensions of proficiency, trust, and flexibility? We then focus on architectural approaches that have the potential of bringing together several different communities working on the problem and then discuss enabling technologies that could bring these architectures to life. We close with recommendations that are not only technical but also that touch on the kinds of problem sets we should be addressing, the developmental processes and organizational structures needed to attack them, and the broader structuring of a knowledge platform that enables the vision we have put forth.

Our recommendations cover six specific areas, summarized as follows.[1]

R1. Behavioral Objectives

These are basically generalized design requirements specifying how we want an AS to behave, in terms of proficiency, trustworthiness, and flexibility.

- *Recommendation 1a*: ASs should be designed to ensure *proficiency* in the given environment, tasks, and teammates envisioned during operations. Desired *properties of proficiency* include situated agency, a capacity for adaptive cognition, an allowance for multiagent emergence, and an ability to learn from experience.

- *Recommendation 1b*: ASs should be designed to ensure trust when operated by or teamed with their human counterparts. Desired *tenets of trust* include cognitive congruence and/or transparency of decision making, situation awareness, design that enables natural human-system interaction, and a capability for effective human-system teaming and training.

- *Recommendation 1c*: ASs should be designed to achieve *proficiency* and *trust* in a fashion that drives behavioral *flexibility* across tasks, peers, and cognitive approaches. Desired *principles of flexibility* for an AS include an ability to *change its task or goal* depending on the requirements of the overall mission and the situation it faces.

---

1. Specific details are to be found in the body of the report and the accompanying appendices.

It should be able to *take on a subordinate, peer, or supervisory role* and change that role with humans or other autonomous systems within the organization. And it should be able to *change how it carries out a task*, both in the short term in response to a changing situation and over the long term with experience and learning.

R2. Architectures and Technologies

This covers unifying frameworks and architectures that will support cross-disciplinary research and development, along with the technology investments needed to support desired functionalities within an architecture.

- *Recommendation 2a*: Develop one or more *common AS architectures* that can subsume multiple frameworks currently used across disparate communities. Architectures should, at a minimum, provide for "end-to-end" functionality, in terms of providing the AS with a sensory ability to pick up key aspects of its environment; a cognitive ability to make assessments, plans, and decisions to achieve desired goals; and a motor ability to act on its environment, if called upon. The architecture should be functionally structured to enable extensibility and reuse, make no commitment on symbolic versus subsymbolic processing for component functions, incorporate memory and learning, and support human-teammate interaction as needed. Whatever the form, an architecture should be extensible to tasks assigned, peer relationships engaged in, and cognitive approaches used. A key metric of an architecture's utility will be its capability of bridging the conceptual and functional gaps across disparate communities working autonomy issues.

- *Recommendation 2b*: Pursue the *development of enabling technologies* that provide the needed functionality at the component level. This includes technologies that support not only the basic "see/think/do" functions but also those that enable effective human-computer interfaces (HCI), learning/adaptation, and knowledge-base management, both of a general purpose and of a domain-specific nature. The nature of technology development should range from basic research to exploratory development to early prototyping, depending on the maturity of the specific technology and its envisioned application.

- *Recommendation 2c*: Develop and promulgate a *multitiered hardware and multilayered software architecture* to support AS development, validation, operation, and modification, where each tier provides for physical structuring across distinct hardware implementations/hosts for given high- and low-level functions and each layer provides distinct software implementations of similar functions. A variety of complex architectural patterns may be needed to take full advantage of emerging technology trends, particularly in the commercial sector.

R3. Challenge Problems

Addressed here are both domain-independent (or functional) problems such as dynamic replanning and domain-dependent (or mission-oriented) problems, such as multidomain fusion.

- *Recommendation 3a*: Drive basic behavior, architecture, and function development of ASs with an appropriately scoped, scaled, and abstracted set of functionally oriented challenge problems that allow different members of the science and technology (S&T) community to focus down on different contributors to AS behavior. Select the set of challenge problems based on an initially nominated architecture and function set, in a fashion that spans the full set of functionalities represented in the architecture (exhaustiveness) and that minimizes the overlap in functionalities needed to address any two challenge problems (exclusivity).

- *Recommendation 3b*: Select mission-oriented challenge problems with the two objectives of: a) addressing current or future operational gaps that may be well-suited for AS application; and b) challenging the S&T community to make significant advances in the science and engineering of AS functionality. Ensure that the challenge problems can be addressed within the context set by the architectures and functions selected earlier, to ensure consistent efforts between the domain-independent and domain-dependent efforts and to avoid "one off" application efforts that end up having little to contribute to other mission-oriented problem sets. Consider both "partial" mission-focused challenge problems as well as "end-to-end" challenge problems. Finally, do not allocate S&T resources to solving operational problems that have close analogs in other sectors, unless the AF-specific attributes make the problem so unique that it can't be solved in an analogous fashion.

R4. Development Processes

This includes processes—in contrast to our traditional waterfall process of requirements specification, milestone satisfaction, and end-state test and evaluation (T&E)—that support innovation, rapid prototyping, and iterative requirements development to support rapid AS development and fielding.

- *Recommendation 4a*: Create an educational and intern-like personnel pipeline to send selected staff to the Air Force Institute of Technology for an introductory autonomy short course, focusing on AI enablers. Individual members would then be embedded into an AI-focused special operations activity: an Autonomy Capabilities Team (ACT) to learn how to apply the skillsets they acquired in addressing USAF autonomy needs. Support this effort over the course of four years to grow AI manpower by an order of magnitude over today's level. Assure retention via a number of special incentive programs. Supplement this cadre with appropriate and long-term support of key extramural researchers.

- *Recommendation 4b*: Use a three-phase framework for iterative selection of challenge problems, for modeling the impact of potential solutions, and for solution development, prototyping, and assessment. Conduct an initial phase of wargaming–based assessment with the goal of identifying key challenge problems and AS-based solutions that can address those threats or take advantage of potential opportunities. Provide a deeper assessment of promising AS candidates via formalization of those concepts with quantitative models and simulations (M&S) and parameters of performance. Finally, focus on the design of one or more engineering prototypes of promising AS candidates identified in the M&S studies. Develop and experimentally evaluate a prototype AS that can serve as: a) design prototype for acquisition; and b) a design driver for additional needed S&T.

- *Recommendation 4c*: Through the Air Force Chief Data Officer, acquire space to store USAF air, space, and cyber data so that AI professionals can use it to create autonomy solutions to challenge problems. Establish data curator roles in relevant organizations to manage the data and to create streamlined access and retrieval approaches for data producers and consumers.

- *Recommendation 4d*: Support the movement to cloud-based computing while also leveraging quantum computing as a general computational paradigm that can be exploited to meet embedded and high-performance computing processing demands.

R5. Organizational Structures

This includes organizing around a project (or outcome) focus, rather than, for example, along traditional technical specialty domains.

- *Recommendation 5*: Establish the ACT within the Air Force Research Laboratory (AFRL), incorporating a "flatarchy" business model to bring 6.1–6.4 experts into a single product-focused organization to develop the science of autonomous systems while delivering capabilities to the warfighter. Collaborate with Air Force Office of Scientific Research and other key AFRL Technical Directorates, and coordinate with USAF organizations outside AFRL, including the DOD Autonomy Community of Interest (COI), AFWERX, and other offices that can facilitate technology transition to the warfighter. Within the ACT, incorporate product-focused business processes based on a Skunk Works–like set of "guiding rules" and facilitate the move towards an information-centric business platform model for the future Air Force.

R6. Knowledge Platform

This provides us with a holistic means of integrating across AS behavioral principles, architectures/technologies, challenge problems, developmental processes, and organizational structures.

- *Recommendation 6*: Develop a Knowledge Platform (KP) centered on combining an information technology (IT) platform approach, with a platform business model. A KP designed for the multidomain operating Air Force should monopolize the connection of observation agents with knowledge creation agents and with warfighting effects agents, which can be either human or machine-based agents (ASs). The KP provides the ecosystem necessary to create capabilities, and those capabilities are used to create combat effects. This ecosystem will come to fruition by exploiting the three behavioral principles of autonomy; the architectures and technologies that enable those behaviors; the driving challenge problems; the developmental processes across people, architectures/applications, data, and computational infrastructures; and, finally, the organizational structures that need to be in place to advance the technology, exploit it, and deliver capability. This approach will provide us with the means of transitioning the USAF from the traditional tools-based approach that solves a small number of problems to a Knowledge Platform approach applicable to a far greater set of problems.

Summary

In summary, our recommendations for AS development and application cover:

- The behaviors these systems must have if they are to be proficient at what they do, trusted by their human counterparts, and flexible in dealing with the unexpected

- The unifying frameworks, architectures, and technologies we need to bridge across not only insular S&T communities but also operational stovepipes and domains

- The focused challenge problems, both foundational and operational, needed to challenge the S&T community while providing operational advantages that go far beyond our traditional platform-centric approach to modernization

- New processes for dealing with people, systems, data, and computational infrastructures that will accelerate innovation, rapid prototyping, experimentation, and fielding

- A new organizational structure, the Autonomy Capabilities Team, that brings together technical specialties into a single organization focused on innovative product development, with outreach to other organizations and communities, as needed

- A Knowledge Platform for holistically integrating across AS behavioral principles, architectures/ technologies, challenge problems, developmental processes, and organizational structures

The AFRL, and specifically the ACT, cannot simply limit its attention to the research space of autonomous systems—nor can it simply perpetuate the model of applying modern AI and AS technology to provide incremental mission capability improvement in one-off demonstrations. Challenge problems must be chosen to advance the Knowledge Platform's ability to provide, in an agile fashion, ASs that exhibit proficient, trustworthy, and flexible behaviors, in transformational applications. In addition to project-focused efforts, the ACT can serve to prioritize and coordinate AFRL's entire autonomy S&T portfolio—synchronizing efforts to maximize investment impact—bringing AS capabilities to mission challenges at scale, and in a timely fashion, all while "sharing the wealth" of new architectures, technologies, and processes across the S&T directorates. Finally, when successful, the ACT can serve as an "existence proof" of how AFRL can transform itself from its legacy of a discipline-focused organization to one that is more cross disciplinary and project oriented, solving transformative, USAF enterprise–wide problems.

We have a unique opportunity to transform the Air Force from an air platform–centric service, where space and cyber often take a back seat, to a truly multidomain and knowledge-centric organization. By delivering autonomous systems to the warfighter by way of a Knowledge Platform, every mission in air, space, and cyber will be improved—and not just incrementally, but multiplicatively. We will become an enterprise that is service oriented, ubiquitously networked, and information intensive. In short:

An agile, information-centric enterprise making timely decisions executed via friction-free access to exquisitely effective peripherals.

Chapter 1

Introduction

As the Air Force realigns itself from a counterterrorism focus to dealing with near-peer rivalries with potential existential consequences, a business-as-usual approach to systems development will no longer suffice. We will not be able to continue with incremental advances on concepts developed decades ago. Instead, we need new technologies, affording us new capabilities, and new operational concepts to employ them. Fortunately, there now exists a broad and deep technology push in the information sciences, particularly in the area of autonomous system (AS) development and its associated foundational technology, artificial intelligence (AI). Our knowledge of cognition and neurophysiology—the basis of what makes us "smart," most of the time—grows at a dizzying pace, while our ability to build ASs—like self-driving cars and game-playing robots—continues to make front-page news, as new AI algorithms and learning techniques are developed and employed in novel ways. And these advances are compounded by explosive gains in the underlying computational infrastructure afforded by Moore's law growth in computational power, memory, networking, and data availability (Moore 1965).

Our goal here is twofold: to provide a vision for USAF senior leaders of the potential of ASs and how they can be transformative to warfighting at all levels, and to provide the science and technology (S&T) community a general framework and roadmap for advancing the state of the art *while* supporting its transition to existing and to-be-acquired systems. Like others, we believe the payoff from the employment of these systems will be considerable, simply because the individual capabilities of these ASs will afford us greater degrees of freedom in employment and opportunities for novel concepts of operations. But this is a traditional view. A potentially more far-reaching payoff will come from becoming more information-centric as we are aided by proliferating ASs, enabling us to leave our legacy platform-centric way of thinking behind and becoming an enterprise that is service oriented, ubiquitously networked, and information intensive. Our vision is:

> *An agile, information-centric enterprise making timely decisions executed via friction-free access to exquisitely effective peripherals.*

Our approach in this document is to first lay out what we need in the way of AS "behaviors": that is, no matter what the underlying technical means, determine the resultant behaviors of these systems across key dimensions of proficiency, trust, and flexibility. We then focus on architectural approaches that have the potential of bringing together several different communities working on the problem and then discuss enabling technologies that could bring these architectures to life. We close with recommendations that are not only technical but also touch on the kinds of problem sets we should be addressing, the developmental processes and organizational structures needed to attack them, and the broader structuring of a knowledge platform that addresses the vision we have put forth.

The remainder of this chapter focuses on the potential benefits of ASs (section 1.1), attempts to define them and summarizes past studies in this area (1.2), identifies operational challenges (1.3) and developmental opportunities (1.4), and presents a future vision of where we should be going (1.5). This chapter closes with a chapter-by-chapter outline (1.6).

1.1 Motivation and Benefits

Over the last decade the United States has witnessed a renewal of a Great Power Rivalry (Thompson 1999), with Russia in the Eurasian continent and China in the Pacific, both of which are making investments and technological progress in weapons systems that may not only provide parity with our own systems (e.g., via stealthy air platforms and precision-guided munitions) but also may counter existing US advantages (e.g., via advanced integrated air defenses [IADS] and antisatellite weapons). A primary finding of the Office of the Secretary of Defense (OSD) is that the United States is facing a diminishing gap in relative technology advantage, due to a number of potential factors, including a multidecade focus on counterinsurgency warfare, a relative lack of investment in game-changing technologies, and the rapid and global diffusion of technology outside of the defense community—a community which has had a long heritage of innovation. Compounding these advancements by our adversaries is our own adherence to a linear/incremental development approach to next-generation systems, which has led to increasingly expensive weapons-systems costs and development/fielding times.

In 2016, Secretary of Defense Carter proposed to address these shortcomings via a threefold approach (Carter 2016):

- Drive smart and essential technological innovation

- Update and refine warfighting strategies, operational concepts, and tactics

- Build the force of the future and reform the Department of Defense (DOD) enterprise

Deputy Secretary of Defense Work expanded upon this with a "framework-to-solution space" bringing together technology, concepts, and culture as illustrated in figure 1.1, under the rubric of the Third Offset Strategy (Pellerin 2016). It is as much a process as it is a product, with wargaming, simulation, experimentation, and analysis iteratively generating solutions to provide us with the operational, and not just technological, advantage.

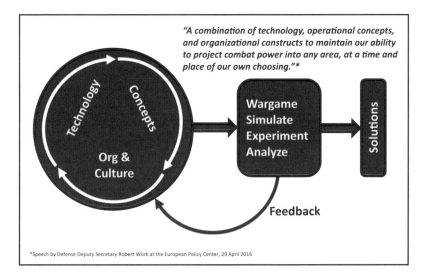

Figure 1.1. DOD Framework-to-Solution space under the Third Offset Strategy

Secretary Work also gave as an example the following Third Offset "building blocks" that could support the transformation of all the services (illustrative examples included) (Work 2015):

- Autonomous deep-learning systems
  - Coherence out of chaos: Analyzes overhead constellation data to queue human analysts
- Human-machine collaboration
  - F-35 helmet: portrayal of 360 degrees on heads-up display
- Assisted human operations
  - Wearable electronics, heads-up displays, and exoskeletons
- Human-machine combat teaming

o Army's Apache and Gray Eagle unmanned aerial vehicles (UAV), and Navy's P-8 aircraft and Triton UAV

- Network-enabled semi-autonomous weapons

o Air Force's Small Diameter Bomb (SDB)

Four of the above five items clearly identify the importance of AS, human-machine teaming, and machine-assisted human augmentation. Our focus in this report is on ASs[1]—how they interact with humans and how they can augment human capabilities.

In parallel with the OSD efforts to transform the department, the Air Force developed the "Air Force Future Operating Concept" (AFFOC; US Air Force 2015) to envision how the future USAF in 2035 will conduct its five core missions across air, space, and cyber domains.[2] In the document are included many illustrative vignettes, but a key product is the generation of 19 "implications" for future capabilities. These are illustrated in figure 1.2. many illustrative vignettes, but a key product is the generation of 19 "implications" for future capabilities. These are illustrated in figure 1.2.

Highlighted in red are those implications that could benefit from ASs, acting independently or in concert with humans. Clearly, autonomy can be an enabler across a broad swath of future functionality.

There have been several studies addressing the use and development of autonomy for DOD systems over the last decade, and we summarize some of the findings in the next section; one of the more recent studies supported by the Intelligence Advanced Research Projects Activity (IARPA) and conducted at Harvard's Belfer Center for Science and International Affairs highlights its "transformative" potential—along with that of AI and machine learning (ML)—equating it to earlier revolutionary technologies including aircraft, nuclear weapons, computers, and biotechnology (Allen and Chan 2017). But it is clear our adversaries have come to the same conclusions regarding the potential for AS development.

China opened a new AI "Development Planning Office" in July 2017[3] with a multibillion-dollar budget (Mozur 2017) in which several goals are laid out,

---

1. Which we will define at greater length in the next section. For now, we can think of an AS as an agent that can accomplish a task on its own, once told what to do, but not necessarily how to do it.

2. These are: adaptive domain control, global integrated ISR, rapid global mobility, global precision strike, and multidomain command and control.

3. As announced at "Guówùyuàn guānyú yìnfā xīn yīdài réngōng zhìnéng fāzhan guīhuà de tōngzhī (guó fā [2017] 35 hào)_zhèngfu xìnxī gōngkāi zhuānlán" 国务院关于印发新一代人工智能发展规划的通知（国发〔2017〕35号）_政府信息公开专栏 [Notice of the

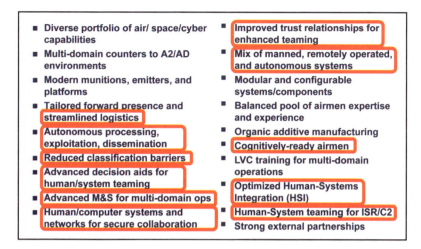

Figure 1.2. Air Force Future Operating Concept—19 implications

including international parity with the United States by 2020 and world dominance by 2030, targeting a $150B market by then. Social stability—that is, maintenance of the Communist Party's power—is a major emphasis through internal monitoring and security measures (Larson 2018), but so is national security. A whole-of-government approach is outlined, including encouragement of the domestic AI enterprises to "go global" and provide facilities and services for the powerful AI enterprises to carry out overseas mergers and acquisitions, equity investment, venture capital investment, and establishment of overseas research-and-development (R&D) centers, as well as encouragement of foreign AI enterprises and research institutes to set up R&D centers in China.[4]

It is noteworthy that this is already happening with significant Chinese-backed venture-capital investments in Silicon Valley and with Google opening an AI lab in China in December 2017.[5] China is also making large investments in specialized AI chips to speed computations (Yuan 2018) and has access to large datasets needed for training the current generation of artificial neural

State Council on Printing and Distributing a New Generation of Artificial Intelligence Development Plan (Guo Fa [2017] No. 35, Government Information Disclosure Column), via Google Translate], accessed 12 October 2018, http://www.gov.cn/zhengce/content/2017 -07/20/content_5211996.htm.

4. Ibid.

5. Jon Russell, "Google Is Opening a China-Based Research Lab Focused on Artificial Intelligence," TechCrunch, accessed 12 October 2018, https://techcrunch.com/2017/12/12 /google-opening-an-office-focused-on-artificial-intelligence-in-china/.https://techcrunch .com/2017/12/12/google-opening-an-office-focused-on-artificial-intelligence-in-china/.

networks (ANN) behind many of the recent successes in AI. As noted by (Schenechner et al. 2017), Yitu Tech, a Shanghai-based AI startup, won a 2016 face recognition contest hosted by the US intelligence community partly, if not totally, because the company "works on behalf of multiple security agencies in China and has access to a database of 1.8 billion photos of faces."

Although we do not have the same insight into the direction and investments planned for Russia, it is clear what President Putin thinks about this area. According to CNBC, on 15 June 2017 Putin noted, "The one who becomes the leader in this [AI] sphere will be the ruler of the world" (CNBC 2018). And that, when drones fight future wars, "when one party's drones are destroyed by drones of another, it will have no other choice but to surrender" (CNBC 2018).

It is therefore appropriate that we move forward briskly in the development of the AI technology and the ASs that incorporate it. The Belfer Study cited above recommends a whole-of-government approach (although not nearly as encompassing as the Chinese Plan outlined above), including funding activities that ensure technological leadership for national security, supporting dual-use technologies, and "managing catastrophic risks" via treaty restrictions, development of "fail-safe" systems, and so forth (Allen and Chan 2017). Our goal with this study is much narrower, with a focus on the performance and architectural aspects of these systems, including recommendations on challenge problems, development processes, and organizational structures.

Complementing this capability-pull driven by national security concerns, by both us and our adversaries, is a concurrent technology-push, driven mainly by the commercial sector. What we have seen over the last few years are the dramatic achievements of autonomy and AI, in "challenge" applications that have yet to reach their full commercial potential: Self-driving cars like those produced by Waymo (2018) that embody the "see/think/do" paradigm of robotics and promise an order-of-magnitude reduction in traffic accidents (Bertonocello and Wee 2015); predictive analytics for logistics management like GE's Predix platform that can provide insight across the entire lifecycle of industrial assets;[6] voice-recognition software that understands human speech at a 95-percent accuracy level like Google's Assistant;[7] facial-recognition programs that boast of 99 percent database-matching accuracy and that are becoming omnipresent, such as Face++'s video surveillance-and-recognition software (Knight 2017a); language-translation programs that are almost rival-

---

6. "Predix Platform," GE Digital, accessed 12 October 2018, https://www.ge.com/digital/iiot-platform.

7. iamcarolfierce13, "Google Improves Voice Recognition, Hits 95% Accuracy," *Android-Headlines.Com* (blog), 2 June 2017, https://www.androidheadlines.com/2017/06/google-improves-voice-recognition-hits-95-accuracy.html.

ing expert human translators, like Google Translate (Turner 2017); and game-playing programs that beat the best human players in the world, like IBM's Deep Blue in chess[8] and IBM's Watson in *Jeopardy!* (Markoff 2011; Ferrucci et al. 2010), and Deep Mind's AlphaGo and AlphaGo Zero in Go (Moyer 2016; Silver et al. 2016; Silver et al. 2017).

Behind these application achievements are several "converging" technologies, one of which is robotics-influenced agency that includes improved sensors, such as "smart" hybrid sensors being developed for self-driving cars (Condliffe 2017); onboard computational resources that continue to drop in price due to Moore's law (Moore 1965); motors and effectors that proliferate to provide greater interaction with the "real world"; and networks and communication protocols that enable novel paradigms such as "fleet learning" across large numbers of autonomous or semi-autonomous vehicles, such as used by Tesla (Frommer 2016), taking advantage of Metcalfe's law (Hendler and Golbeck 1997). Another enabling technology, of course, is the exploding field of ANNs, particularly in the development of many-layered networks ("deep networks"), associated ML algorithms that allow deep networks to be "trained" on sample datasets, and large datasets ("big data") that facilitate that training—all of which will be discussed at greater length in chapter 4. These ANN advances in algorithms and data, in turn, have been made possible by the continuing growth in *computational power* and *memory*, particularly by the proliferation of graphics processing units (GPU)[9] that provide the kind of parallel computational architectures called for by the parallelism of the ANN algorithms themselves and the exponential growth of cloud-based storage, expected to double this year over last year to exceed a ZB[10] in size (Puranik 2017).

We discuss these technologies at greater length below in section 1.4 and later in chapter 4. The key point, however, is that we are in the midst of a paradigm-shifting approach to bringing "intelligence" to computing, with a convergence of techniques, technologies, and infrastructure. We must take advantage of it if we, as an Air Force, are to move from the industrial age to the information age.

Clearly, there will be far-reaching implications for Air Force capabilities with the development and application of ASs that embrace advanced AI tech-

---

8. "IBM Archives: Deep Blue," TS200, 23 January 2003, https://www-03.ibm.com/ibm/history/exhibits/vintage/vintage_4506VV1001.html.

9. Originally developed for the videogaming industry; see, for example, Graham Singer, "The History of the Modern Graphics Processor," TechSpot, accessed 15 October 2018, https://www.techspot.com/article/650-history-of-the-gpu/.

10. A zettabyte (ZB) is 1 billion terabytes (TB), where a TB is roughly the size of a modern laptop hard drive.

nologies. Major benefits, at a generic or functional level, include extending and complementing human performance and proficiency, and not necessarily simply replacing humans, in functional areas such as (Endsley 2015c):

- Extending human reach (e.g., operating in riskier areas)

- Operating more quickly (e.g., reacting to cyberattacks)

- Permitting delegation of functions and reduction of manpower (e.g., information fusion, intelligent information flow, assistance in planning/ replanning)

- Providing operations with denied or degraded communications links

- Expanding into new types of operations (e.g., swarms)

- Synchronizing activities of platforms, software, and operators over wider scopes and ranges (e.g., manned-unmanned aircraft teaming)

Figure 1.3 illustrates more specific areas that could benefit from the introduction of ASs, with the top row illustrating platform-centric applications ("autonomy in motion" as coined by Defense Science Board [DSB] Chair Dr. Fields during a recent DSB study on autonomy [DSB 2016]), and the bottom row illustrating more pervasive and enterprise-centric applications ("autonomy at rest").

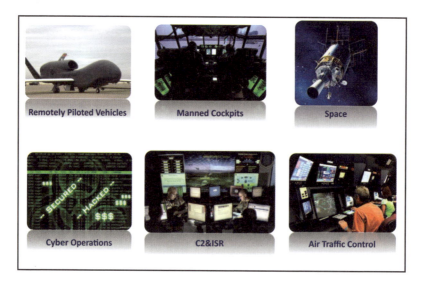

Figure 1.3. Autonomy could transform many Air Force missions

Regarding the top row of applications, we are all well aware of the fact that current-generation remotely piloted vehicles are in no sense autonomous. They could greatly benefit from more onboard awareness, platform self-tasking, and autonomous payload processing. Manned cockpits are already seeing benefits from greater levels of automation in sensor fusion, guidance, navigation, and control and subsystem management, but true autonomous decision-aiding, as envisioned by the Pilot's Associate program of the 1990s (Banks and Lizza 1991) and the Intelligent Autopilot System under way now (Baomar and Bentley 2016), is still a long way off. Finally, our orbiting space platforms are very much situationally *unaware* of much beyond their mission focus (for example, nuclear detonation detections via Defense Support Program satellites) (USAF Space Command 2015); more situation awareness (SA) and autonomous capabilities would not only provide a capability for self-defense in an increasingly contested environment but would also provide needed flexibility for mission success under changing operating conditions.

Regarding the bottom row of applications in figure 1.3, cyber defenses demand sub-millisecond response times; bringing the cyber operator out of the loop via autonomy would not only increase speed of response but also support dealing with the complexity of dealing with multiple attack surfaces subject to multiple types and numbers of attacks (Corey et al. 2016). The fundamental information management tasks of the current air operations center (AOC) and the future multidomain operations center—in both intelligence, surveillance, and reconnaissance (ISR) and command and control (C2)—are currently accomplished via a manually intensive, procedure-driven cycle that is not designed to take advantage of the full datasets available to it, is stovepiped by its organizational structures, and is slowed by manual interventions. The AOC could benefit enormously with a redesign built around autonomous processing capabilities. Finally, we show an air traffic control center as an exemplar of many of the support functions needing manual oversight and supervision, functions which, with the introduction of ASs, could not only reduce manning requirements but also improve system performance and flexibility through rapid adaptivity to changing conditions.

Other applications of autonomy have been identified by several other studies, which we summarize in section 6.3 and review in appendix A. As noted above, the AFFOC also presents several vignettes that incorporate a variety of AS concepts, although for a vision that purports to look out 20 years, it is, in our opinion, a conservative view of what could be accomplished with autonomous capabilities in that time frame. For an example of a more forward-looking vignette, see appendix H, which includes both "at rest" and "in motion" AS assets and progresses from an ISR mission to an area-defense mission and, finally, to a humanitarian recovery mission. The vignette displays key AS task, peer, and cognitive flexibilities that we will discuss at greater length below and later in this chapter.

To appreciate the potential for the introduction of ASs in future engagements, also consider the following vignette illustrated in figure 1.4. Here, an antiaccess/area-denial (A2/AD) environment precludes the ability of Blue (our forces) to operate with impunity in Red's air, ground, maritime, and cyber domains. This can be attributed to Red's advances in, for example, advanced active and passive sensing (e.g., radar and electro-optic [EO]), digital signal-processing, networking, C2, and cyber capabilities, which will constrain Blue to operate in more permissive environments hundreds of miles from Red. How can Blue use ASs, for example, in the areas of ISR, battle management C2 (BMC2), electronic warfare/cyber (EW/Cyber), and strike, to "push back the bubble" of A2/AD-defended domains to operate freely or with limited adversary impact in all domains?

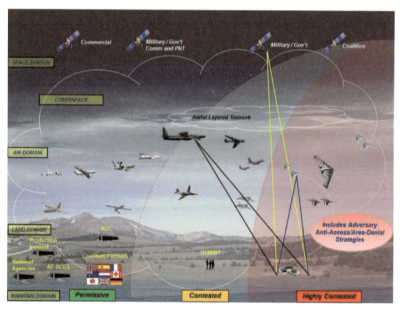

Figure 1.4. Antiaccess/Area-Denial (A2/AD) operational space

The advances of Red IADS and continued improvements in surface-to-air missiles (SAM) are daunting challenges. Blue can counter those advances by leveraging disruptive advances in multidomain "sense-making," combining intelligence from air, space, and cyberspace assets; fusing different data feeds;

detecting critical "events" and "signals" of key activities; and assessing the situation in real time to manage the multidomain infosphere while calling for collection assets when key information is missing or there exists a need for, for example, battle damage assessment (BDA). Advances in understanding quickly, proficiently, and flexibly and being able to respond quickly to changes in the battlespace will enable Blue to selectively shrink and/or eliminate Red's previously imposed A2/AD zone. Blue's ability, teaming with AS partners, to rapidly collect, update, replan, and disseminate changes to Blue forces operating in the previously imposed A2/AD zone will change the balance of outcomes.

A successful strike package, mixing manned and unmanned autonomous platforms (e.g., a "Loyal Wingman" [Kearns 2016]), could take advantage of autonomously coordinated attacks designed to dynamically interfere with, say, Red long-range passive sensors by integrating cyber and EW attacks. As the package closes in, ASs organic to the strike package can provide real-time updates on threat information and propose (and manage) targeted cyber/ EW/kinetic attacks against long-range SAM search-and-acquisition radars, updating attack vectors as these Red assets are nullified. These coordinated attacks would then have served to push the A2/AD bubble back to a point where integrated suppression of enemy air defenses (SEAD)/destruction of enemy air defenses (DEAD) could be conducted using strike assets and cruise missiles to neutralize the Red IADS. A key to all of these capabilities is ensuring high levels of flexibility in our AS designs, across different command structures, mission taskings, and solution approaches, in a fashion similar to that embraced by our most experienced and proficient operators today. This is illustrated in the graphic of figure 1.5, in which a number of ASs (e.g., ISR assets that are also jammers, long-range strike assets that can retask themselves depending on how the mission evolves, etc.) behave *flexibly* in terms of dealing with different taskings, coordinating with their peers to meet overall mission intent, and taking on different approaches to accomplish tasks within a given peer structure. We summarize these flexibilities as *task*, *peer*, and *cognitive flexibility* and will discuss these general requirements at greater length in section 1.3.

ISR to Global Strike mission initiated...

ISR assets become jammers disrupting command links & IADS
while target is destroyed with long range strike assets,

relationships between manned/unmanned assets remain fluid
systems operate with adaptability to
dynamic environment and new
unforeseen threats

Task Flexibility – Autonomy completes a *range of tasks* to achieve combat effects
Peer Flexibility – Autonomy *negotiates peer relationships* to meet intent
Cognitive Flexibility – Autonomy uses *various cognitive models* for task completion

Figure 1.5. Autonomy behaviors and task, peer, and cognitive flexibilities

1.2 Definitions and Summary of Past Studies

The definitions for the terms *autonomy* and *autonomous systems* have undergone continuous refinement as the research and engineering communities further their understanding of the concepts and applications of the evolving technology. Early beginnings of autonomy are rooted in automation, referring to mechanical operation of equipment, a process, or a system with minimal human interference and interaction, completing well-defined tasks traditionally accomplished manually (e.g., an autoland system performing a "hands off" landing in a modern jetliner). As automation systems have become more complex and adept at dealing with more complex situations, the term autonomous has often been applied to such systems, but this is an incorrect intertwining of terms. Autonomy is defined in the Merriam-Webster dictionary as "the quality or state of being self-governing; the state of existing or acting separately from others."[11] In practical terms, for example, working with colleagues or subordinates, the term refers to a degree or level of freedom and discretion to an employee over the conduct of his/her job; note the distinction from the dictionary definition, in which complete freedom and separate action is not presumed, since such factors as task assignments, performance

_____

11. *Merriam-Webster*, 2016, s.v. "autonomy," https://www.merriam-webster.com/dictionary/autonomy.

quality, timeliness, and trust are all explicitly stated or implicitly assumed in a workplace environment.

When we turn our attention to describing ASs, initial attempts were made to have autonomy anchor one end of a scale of machine-aided sophistication, ranging from no autonomy (complete human decision making or manual control), to automatic (where the system takes on some limited range of well-defined tasks), to fully autonomous (where the system makes all decisions and/or actions, often ignoring the human) (Parasuraman and Miller 2004). Because of the difficulty in defining different levels that could be of use in designing or analyzing these systems, and because of the complexity of interactions that can arise between human and machine (e.g., humans often adapt to automation in unexpected ways), a more subtle approach is evolving in the community. In 2012, the DSB task force recommended that the DOD abandon the notion of defining levels of autonomy and replace it with a framework that (DSB 2012):

- Focuses design decisions on the explicit allocation of cognitive functions and responsibilities between the human and computer to achieve specific capabilities
- Recognizes that these allocations may vary by mission phase as well as echelon (or place in the organization chart)
- Makes visible the high-level system trades inherent in the design of autonomous capabilities

In 2013, the Air Force Research Laboratory (AFRL) defined autonomy as "systems which have a set of intelligence-based capabilities that allow it to respond to situations that were not pre-programmed or anticipated in the design (i.e., decision-based responses). Autonomous systems have a degree of self-government and self-directed behavior (with the human's proxy for decisions)" (Masiello 2013).

The Autonomy Community of Interest (ACOI) also adopted this definition (DOD R&E 2015). More recently, in 2016, a second DOD autonomy task force noted that an AS "must have the capability to independently compose and select among different courses of action to accomplish goals based on its knowledge and understanding of the world, itself, and the situation" (DSB 2016).

These last three definitions are certainly driving in the right direction when we think in terms of operational desires when interacting with ASs. As when working with peers or subordinates in an interdependent task-focused environment, we would like to be able to "tell" ASs[12]

- WHAT to do, in terms of the task goals/objectives and the context of the overall mission

- WHAT NOT to do, specified as operating constraints or rules of engagement (ROE)

- And perhaps some of the WHYs, specified as the supervisor's objectives or commander's intent

- But NOT be required to provide details on HOW to do it, that is, the methods, nor HOW to deal with "unplanned for" conditions, such as when dealing with system failures or an adversary's actions

Referring to the earlier "definitions," we see that being able to interact in this fashion calls on the capability for task sharing and allocation highlighted in the 2012 DSB definitions above, the adaptability and self-direction called out in the AFRL definition, and the situation understanding and course-of-action composition called for in the 2016 DSB study. All of these are clearly relevant attributes of autonomous behavior.

In this report we will not attempt a formal definition of autonomy. Rather, we propose initially that we accept the above working definition as a placeholder—that is, the (WHAT, WHAT-NOT, WHY, and NOT-HOW) interaction dimensions and expectations of working with ASs. In the next section, instead of a definition, we will propose three sets of attributes we believe to be central to autonomous behavior as envisioned in the context of a military organization and mission. But first, to provide additional context, we will summarize some of the key findings and recommendations of several prior studies in this area.

1.2.1 Summary of Past Studies

In appendix A we review seven past studies of ASs for military applications and summarize and categorize major findings and recommendations across the studies. We have structured the study findings into five categories:

Behavioral objectives: These are basically generalized design requirements specifying how we want an AS to behave, in terms of proficiency, operator interactions, and so forth. These requirements, which take up the bulk of the study findings, can be broken down into two subcategories. The first deals with the performance of the AS itself, such as ensuring task proficiency, that

---

12. And certainly, we would like reciprocity in the relationship, so that ASs can "tell" us relevant information when appropriate. We discuss this further in chapter 3.

behaviors are directable and predictable, and that the AS can accomplish tasks with adequate flexibility and adaptivity. The second focuses on human-system teaming, including the desirability of being able to set mutual goals, to maintain adequate mutual shared awareness of the team and the adversary, and to communicate and coordinate effectively.

Architectures and technologies: This covers unifying frameworks and architectures that will support cross-disciplinary research and development, along with the technology investments needed to support desired functionalities within an architecture. These were covered in only one of the studies that we reviewed, with recommendations in the areas of hybrid frameworks (e.g., classical algorithmic approaches combined with more contemporary deep-learning network approaches) and biomimetic architectures that are inspired by animal anatomy and physiology.

Challenge problems: Addressed here are domain-independent (or functional) problems—like dynamic replanning—and domain-dependent (or mission-oriented) problems—like multidomain fusion across air, space, and cyber. The domain-*independent* problems cover a range of general functional areas like collection/sensing/fusion of information, decision-aiding (with a human) and decision-making (autonomous) subsystems, and operation in adversarial environments that demand improvisation. The domain-*dependent* problems range from generic concepts like autonomous swarms and fractionated platforms to specific operations in air (including ISR, air operations planning, EW, and logistics), space (including fractionated platforms and embedded health diagnostics), and cyber (defensive operations, offensive operations, and network resiliency).

Development processes: This includes processes that support innovation, rapid prototyping, and iterative requirements development to support rapid AS development and fielding. A broad span of recommendations is put forward covering the following: (1) the need to actively track adversarial AS capabilities; (2) the importance of human-capital recruiting, retention, and alliances in the AS and AS-related specialties; (3) continued support of basic and applied research in a broad area of underlying technologies (not just deep learning [DL]), coordinated across research communities, and informed by operational experience and requirements; (4) support of advanced systems development that separates the development of platforms from the autonomy software that governs them; (5) establishing processes for upgrading legacy systems with new AS capabilities; and (6) recognizing the difficulty of conducting test and evaluation of these systems and the need to establish a research program in this area.

Organizational structures: This includes organizing around a project (or outcome) focus, rather than, say, along traditional technical specialty domains. In our survey of past studies summarized in appendix A, we found no recommendations in this specific area but include it to highlight the lack of the research community's concern in this area.

The above four sets or categories of findings associated with previous studies, along with the fifth set called out regarding organizational structures, will form the framework for our recommendations later in chapter 6. But first, we will focus much of this report on the first two areas noted above, behaviors (in chapters 2 and 3) and architectures and technologies (in chapters 4 and 5), since these design requirements and potential solutions will influence the other three sets of recommendations. In the next section, we outline the behavioral objectives driven by the operational challenges and introduce the need for AS proficiency, trust, and flexibility.

1.3 Operational Challenges and Behavioral Implications

Endsley (2015c) presents in *Autonomous Horizons*, volume I, a summary of challenges in dealing with automation and autonomous systems, including:

- Difficulties in creating autonomy software that is robust enough to function without human intervention and oversight. After years of dealing with hand-coded (and hard-coded) software, this is starting to be addressed via the resurgence of AI applications, discussed later in this chapter and in chapter 4.

- The lowering of human SA that occurs when using automation/autonomy, leading to out-of-the-loop performance decrements. Dramatic examples occurred with the introduction of highly automated commercial cockpits in the 1990s, with many lessons learned by the aviation community (Wiener 1989; Sarter and Woods 1995); the automobile industry is now in the process of relearning some of these same design issues, with the introduction of "autonomous" or "autopilot" modes (Casner et al. 2016; Mitchell 2018).

- Increases in cognitive workload required to interact with the greater complexity associated with automation/autonomy. Reducing one kind of workload (e.g., task workload) via the introduction of an autonomous "workmate" may introduce another (e.g., management workload).

- Increased time (for the human) to make decisions when decision aids are provided, often without the desired increase in decision accuracy. A simple advisory by a decision aid, without the underlying rationale and/ or an estimate of the uncertainty in the answer, may add a cognitive load and extra decision time to evaluate the "goodness" of the advice.

- Challenges with developing a level of trust that is appropriately calibrated to the reliability and functionality of the AS in various circumstances. *Overtrust* by the operator can lead to AS misuse and incidents and accidents, because the AS is not proficient enough with the task assigned; *undertrust* by the operator can lead to nonuse, even when the AS could be a useful aide. Appropriate levels of trust are always called for, whether high or low.

Note that the first issue above focuses on the unaided *proficiency* of an AS, while the remaining four issues focus on human-system interaction and *trust* in that interaction. This is appropriate, since Endsley notes, "Given that it is unlikely that autonomy in the foreseeable future will work perfectly for all functions and operations, and that airman interaction with autonomy will continue to be needed at some level, each of these factors works to create the need for a new approach to the design of autonomous systems that will allow them to serve as an effective teammate with the airmen who depend on them to do their jobs" (Endsley 2015c).

When we look more deeply into issues that limit automatic or AS *proficiency,* we can identify four classes of shortcomings:[13]

- Inadequate "loop closures" with the environment, in terms of sensing the key aspects of the environment, reasoning about it, and then acting on it with appropriate effectors (Wiener 1948). Thermostatic control of room temperature is a classic example. Initially designed to sense the current temperature and open or close a heating switch in response, thermostats have evolved to now include the time (for multiple time-dependent set points), to the Nest that detects when people are in the room,[14] so that temperature control becomes more context sensitive (and remotely accessible via the internet). As we noted in the previous section, we expect this trend to continue as sensors, computational power/memory, effectors, and networking drop in price, a trend we have seen embodied in the exponential growth of self-driving cars, for example.

---

13. We include automatic systems because of the human-factors community's long history of dealing with automation and its experience in automation's many shortcomings, as pointed out by Endsley.

14. "A Better Thermostat," Carnegie Mellon University, accessed October 19, 2018, https://www.cmu.edu/homepage/environment/2012/winter/the-nest-thermostat.shtml.

- Single-strand and often brittle approaches to "reasoning" about a problem, including how to deal with multiple, possibly conflicting, feeds of sensor information, "understanding the meaning" of the sensed data in the context of the AS's tasking or goal objectives, and deciding on an appropriate action (Domingos 2015). At each one of these stages of processing, it has been the tradition to embrace a single "best" approach to processing, based on the designer's knowledge or experience. But experience shows that it is impossible to anticipate all potential operating scenarios, so that what may be best in the anticipated design envelope/situation may not be optimal outside that envelope/situation, especially under operating conditions where an adversary is deliberately trying to degrade system performance.

- A focus on *single-agent approaches* to autonomy, under the assumption that developing a single agent that "does it all" is easier and results in more predictable behaviors, vice dealing with multiple interacting agents that are simpler and possibly more limited in their repertoire, but whose collective behavior may be harder to anticipate (Mataric 1993). *Multiagent approaches* also incur communications limitations, both on a syntactic level as well as a semantic level—especially if the agents are fundamentally different in their internals.

- A lack of a capability for learning over time and with experience or, over a multiagent ensemble, over the ensemble, as we noted with "fleet learning" in the previous section.[15] We as humans do this as individuals via experience and as part of a larger culture via education. In contrast, most of our automation and fledging efforts in autonomy are hand coded and fixed for the life of the system. It has only been recently with the collection of "big data" and the development of DL neural networks operating on that data (Najafabadi et al. 2015; also see chapter 4) that we are beginning to see evolving systems that learn over time and experience.

In chapter 2, we address these issues and outline four key *properties* for *proficiency* in ASs. While the list may not be exhaustive, we believe these to be necessary for ASs to realize their full potential in future defense systems. These properties are:

---

15. That is, when one agent learns something, it transmits that "learning instance" to all the others, as Tesla autos do now: "The whole Tesla fleet operates as a network. When one car learns something, they all learn it." Elon Musk quoted by Fehrenbacher (2015); http://fortune.com/2015/10/16/how-tesla-autopilot-learns/.

- *Situated Agency.* Embedding the AS within the environment, with component abilities to sense or measure the environment, assess the situation, reason about it, make decisions to reach a goal, and then act on the environment, to form a closed loop of "seeing/thinking/doing," iteratively and interactively.

- *Adaptive Cognition.* A capability to use several different modes of "thinking" about the problem (i.e., assessing, reasoning, and decision making), from low-level rules to high-level reasoning and planning, depending on the difficulty of the problem, with sufficient flexibility for dealing with unexpected situations.

- *Multiagent Emergence.* An ability to interact with other ASs via communications and distributed function allocations (e.g., sensing, assessing, decision making, etc.), either directly or through a C2 network, in a manner that can give rise to emergent behavior of the group, in a fashion not necessarily contemplated in the original AS design.

- *Experiential Learning.* A capability to "learn" new behaviors over time and experience, by modifying internal structures of the AS or parameters within those structures, based on an ability to self-assess performance via one or more performance metrics (e.g., task optimality, error robustness, etc.), and an ability to optimize that performance via appropriate structural/parametric adjustments over time.

In chapter 2, we describe these *properties for proficiency* at greater length. But as noted by many, we can anticipate significant human-system interactions over the foreseeable future and, therefore, need to address human-system interactions (HSI) and associated trust issues (Dahm 2011; DSB 2012; DSB 2016; Endsley 2015c). Many HSI design issues will be similar to those that arise when dealing with traditional non-ASs, such as dealing with the "ilities" (e.g., usability, maintainability, etc.; deWeck et al. 2011), and with the issues raised by Endsley (2015c) under the second, third, and fourth bullets at the top of this section. Of particular concern here, however, are the trust issues identified in Endsley's (2015c) last bullet—issues that arise particularly because of the properties that are unique to, or that strongly characterize, ASs, including:

- *Lack of analogical "thinking" by the AS.* When the AS approaches and/or solves a problem in a fashion that is not at all like a human would attack the problem, trust can become an issue because of human concern that the approach may be faulty or unvalidated.

- *Low transparency and traceability in the AS solution.* Lacking an ability to "explain" itself, in terms of assumptions held, data under consideration, reasoning methods used, and so forth, it is difficult for the AS to justify its solution set and thus engender human trust.

- *Lack of self-awareness or environmental awareness by the system.* In the former, this might include AS health and component failure modes, while in the latter, this might include environmental stressors or adversary attacks. Either may unknowingly affect performance and proficiency and overstate the confidence in an AS-based solution made outside of its nominal "operating envelope."

- *Low mutual understanding of common goals.* When a human and AS are working together on a common task, a lack of understanding of the common goals, task constraints, roles, and so forth can lead to a lack of trust on the part of the human in terms of the system's anticipated proficiency over the course of task execution.

- *Non-natural communications interfaces.* The lack of conventional bidirectional multichannel communications between human and system (e.g., verbal/semantic, verbal/tonal, facial expressions, body language, etc.) not only reduces communications data rates but also reduces the opportunity to convey nuances associated with operations by well-practiced and trusting human-only teams.

- *Lack of applicable training and exercises.* Lack of common training and practice together reduce the opportunities for the human to better understand the system's capabilities and limitations and how it goes about problem solving and thus diminishes opportunities for understanding a system's "trust envelope," that is, where it can be trusted and where it cannot.

In chapter 3, we address these issues and outline four key tenets of trust for ASs. As with the proficiencies listed above, the list may not be exhaustive, but we believe these to be necessary for ASs to realize their full potential in future defense systems. These tenets are:

- *Cognitive Congruence and Transparency.* At a high level, the AS operates congruently with the way humans parse the problem, so that the system approaches and resolves a problem in a manner analogous to the way a proficient human does. Whether or not this is achievable, there should be some means for transparency or traceability in the system's solution, so that the human can understand the rationale for a given system decision or action.

- *Situation Awareness.* Employing sensory and reasoning mechanisms in a manner that supports SA of both the system's internal health and component status and of the system's external environment, including the ambient situation, friendly teammates, adversarial actors, and so forth. Using this awareness for anticipating proficiency increments/decrements within a nominal system's operating envelope to support confidence in estimates of future decisions and actions.

- *Human-Systems Integration.* Ensuring good HSI design to provide natural (to the human) interfaces that support high-bandwidth communications if needed, subtleties in qualifications of those communications, and ranges of queries/interactions to support not only tactical task performance but also more operational issues dealing with goal management and role allocation (in teams).

- *Human-System Teaming and Training.* Adapting or morphing human-system team training programs and curricula to account for the special capabilities (and associated limitations) of humans teaming with ASs. Conducting extensive training so that the team members can develop mutual mental models of each other, for nominal and compromised behavior, across a range of missions, threats, environments, and users.

As a final note, we wish to introduce the broader notion of *flexibility*, dependent on both proficiency and trust and that often characterizes our notion of autonomy when dealing with human agents. Several studies, including three reviewed in appendix A (Klein et al. 2005; Dahm 2011; DSB 2012) and Endsley (2015c), call for flexible autonomy in terms of the scope of tasks taken on by the AS; the outcomes that could ensue in pursuing or failing at those tasks; the levels of interaction with other agents, particularly humans; and the control granularity of an AS and its design for ensuring robustness to uncertainty and changing conditions. We take a slightly simpler approach here, generalizing upon Scharre's (2015) "dimensions of autonomy," and focus on three aspects of AS flexibility associated with task scope, peer relations, and solution approaches. These are driven by our working definition of autonomy given in the previous section and by a strong consideration of how we, as humans, expect flexibility in the behavior of other humans, whom we consider autonomous in some fashion. Specifically, we introduce three *principles of flexibility*:

- *Task Flexibility.* An AS should be able to change its task or goal depending on the requirements of the overall mission and the situation it faces. Humans are not *optimizers* designed for only accomplishing one task, even if they are experts in one (e.g., Olympic athletes or world-class chess players); rather, they are *sufficers* in many tasks, flexibly changing from one to another as the need arises. Humans can accomplish multiple tasks, serially and in parallel, dynamically changing priorities over time, shedding tasks and taking on new ones, depending on the situation and motivation (rewards). We believe the same task flexibility needs to be embodied in ASs and that this capability is enabled by situated agency: sensing the environment, assessing the situation, deciding on a course of action to accomplish its tasking, and acting on that course of action, all the while closing the loop by monitoring the outcome and communicating with the other agents in its team.[16]

- *Peer Flexibility.* An AS should be able to take on a subordinate, peer, or supervisory role, depending on the situation and the other agents, human or machine, populating the environment. Humans accomplish this type of relational flexibility as they move through different roles throughout the day, dynamically changing their relationships depending on the situation and the peers they are interacting with. We believe the same peer flexibility needs to be embodied in an AS, changing its relationship role with humans or other autonomous systems within the organization as the task or environment demands. An AS should participate in the negotiation that results in the accepted peer-relation change, requiring the autonomous system to *understand* the meaning of the new peer relationship to respond acceptably. This capability is enabled by situated agency providing environmental and task awareness, an understanding of its peer population (humans and machines), and learning over time to develop proficiency.[17]

- *Cognitive Flexibility.* An AS should be able to change how it carries out a task, both in the short term in response to a changing situation and over the long term with experience and learning. Humans accomplish tasks in multiple different ways, using visualization, verbalization, rote memory, solutions from first principles, and so forth (Davidson et al. 1995; Casakin

16. Examples include an AS capable of changing its role from ISR to jamming; from close air support to search and rescue; from route navigation to targeting; from data compression to image super-resolution; and from ISR analysis to tasking, processing, exploitation, and dissemination (TPED).

17. The F-16 Auto Ground Collision Avoidance (Auto-GCAS) is an example system. It is subordinate to the pilot until a potential ground collision is detected, and if the pilot does not respond to system warnings, Auto-GCAS then takes over (now acting as the pilot-in-control) to put the aircraft on a safe trajectory. Once the pilot takes over control, Auto-GCAS becomes a subordinate team member again.

and Goldschmidt 1999; Dixon 2011). They also change their approaches as they become more expert in tasks, with learning and skill acquisition over time. Finally, they may employ parallel approaches to problem solutions (e.g. Minsky's Society of Mind [Minsky 1988]), and they may also act consciously or unconsciously (Kahneman 2011b). We believe that an AS should embrace this type of *cognitive* flexibility in addressing a problem by bringing to bear a variety of techniques to assess and then decide, selecting those techniques based on the current situation, past experience with given methods, the need to trade optimality versus timeliness, and so forth. In the long term, the AS can also learn new "solution methods" over time, assessing and readjusting a technique's contribution to task performance for a given situation and mission tasking.[18]

Figure 1.6 attempts to show the relationships among the three major behavioral dimensions needed in ASs, linking together *properties for proficiency, tenets of trust,* and *principles of flexibility.* On the left, we have the four properties for proficiency discussed earlier and covered in more detail later in chapter 2. On the right, we have the four tenets of trust, also discussed above and covered in more detail in chapter 3. In the middle, we have the three principles of flexibility, which depend both on proficiency and trust. Peer flexibility is driven mainly by an awareness of the environment and the other agents populating it, subserved by situated agency; an ability to communicate with and team within its multiagent environment; and learning over time and experience. Task flexibility depends on a knowledge of the environment and tasking, subserved by situated agency, and learning over time and experience, to support the development of a broad range of skills, including multitask management. Finally, cognitive flexibility depends on a knowledge of the problem being addressed, subserved by situated agency; adaptive cognition, which brings to bear one or more appropriate problem-solving methods; and learning over time and experience to develop meta-problem-solving skills. We discuss these properties for proficiency at greater length in chapter 2. On the right-hand side of the figure we show how the four tenets of trust contribute to overall trust in the AS and how, in turn, increasing trust levels can support additional flexibility as other agents (human and machine) allow for a greater span of task assignments, peer relationships, and problem-solving approaches engaged in on the part of the AS.

---

18. Two well-known examples in the AI community are IBM's Watson (Ferrucci et al. 2010) and Deep Mind's AlphaGo (Silver et al. 2016), both employing multiple approaches to problem solving—Watson with hundreds of question-answering "experts" and AlphaGo with a complementary combination of tree searching and specialized deep neural networks. We discuss these at greater length in chapter 4.

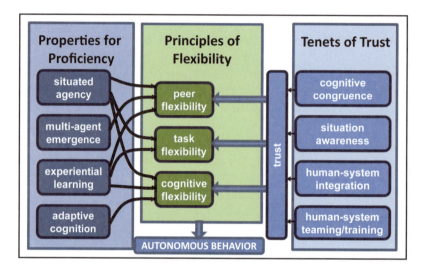

Figure 1.6. Relationships among AS proficiency, trust, and flexibility

1.4 Development Challenges and Opportunities

There exists a broad range of challenges to the development and use of autonomy in DOD systems in general and Air Force systems specifically. We discuss here four of these major challenges and the opportunities we have in dealing with them, including:

- The difficulty of defining what the term autonomy even means and the opportunity that is afforded us when we think in terms of behavioral requirements across the dimensions of *proficiency*, *trust*, and *flexibility*.

- The challenge we face in understanding and designing these systems because of the many different S&T communities working on different aspects of the problem, with different objectives, concepts, methods, and languages; and the opportunity afforded us with greater awareness of one another's efforts and results, a rapid proliferation of "best practices"—such as computational modeling, computer simulations, and rigorous statistical analysis—and a common vision instantiated by one or more unifying frameworks that can span across communities, improve collaboration, and accelerate progress.

- The inherent difficulty in attempting, in some fashion, to mimic biologically based animal cognition, with all the uncertainties of how these existing systems operate, learn, replicate, and evolve in a digital software/hardware environment that brings many constraints with it; counterbalanced with rapid advances in AI architectures (e.g., deep neural networks [DNN]), new learning paradigms afforded by new training schemes and "big data," and continuing growth in the software and hardware (e.g., GPUs that make these computational approaches all possible).

- And, finally, the challenge we face as an institution in bringing innovative approaches to designing and developing these systems, which are fundamentally different from the industrial age products the DOD and USAF are accustomed to developing and deploying; and the opportunity we have to change that legacy approach with the introduction of thought-provoking challenge problems, new development processes that embrace prototyping and experimentation, and cross-disciplinary organization structures that bring together the needed skill sets to effectively tackle the S&T and development challenges facing us.

In the remainder of this section, we discuss these four areas at greater length and point to opportunities for overcoming these challenges, which are arising now and will evolve rapidly over the next several years.

We noted earlier in section 1.2 that several past attempts have been made to define the term *autonomy*, with differing degrees of granularity and, frankly, success. For this reason, we chose a working definition based on the (WHAT, WHAT-NOT, WHY, and NOT-HOW) interaction dimensions and expectations of working with autonomous systems, assuming that humans and other ASs would not be operating in a vacuum but, rather, in a network of other agents, both human and machine. However, even a working definition is not sufficiently fine-grained if we are expected to build and use these systems. We were therefore motivated to build on Endsley's (2015c) work and present, in the previous section, three categories of *behavioral requirements* to which designers of ASs should aspire—three sets of attributes we believe to be central to autonomous behavior, as envisioned in the context of a military organization and mission:

- *Properties for Proficiency*, covering situated agency, adaptive cognition, multiagent emergence, and experiential learning

- *Tenets of Trust*, covering cognitive congruence and transparency, SA, human-systems integration, and human-system teaming/training

- *Principles of Flexibility*, covering task, peer, and cognitive flexibility

So rather than attempt to *define* autonomy or what an AS is, we propose to set requirements on an AS's behavior in a way that supports recursively finer requirements in each of these categories and their components, as discussed further chapters 2 and 3.

A second issue facing the AS-development community is the fact that there is no single such community. Work in this area is distributed across many different communities, ranging from those developing psychological models of human perception, cognition, and behavior—unconstrained by any theories of the underlying neural processes supporting cognition and behavior—to software engineers developing decision algorithms for "autonomous" vehicles, perhaps unaware of how humans operate as autonomous driving agents. Even within the engineering-development community, there are numerous stovepipes of isolated efforts. We should expect this in a competitive commercial environment, but a casual review of ongoing efforts within the DOD—including the DOD ACOI—reflects similar issues associated with low awareness of other efforts; a lack of cross-transference of concepts, frameworks, and technologies; multiple "one-off" prototypes that do not make it through the S&T "valley of death" (see, for example, GAO 2015); and slow overall progress toward the development of fieldable and useful ASs for military applications. There are likely a number of contributing factors to this situation, but a fundamental one has to do with a deficiency of awareness due to the lack of communication between disparate communities (e.g., the cognitive scientist not collaborating with the algorithm designer), and this, in turn, may be due to a lack of even a common language to describe similar autonomous behaviors and mechanisms observed or developed by different communities.[19]

Mitigating against this situation is, we believe, a gradual convergence toward very broad common frameworks that describe the problem of autonomous behavior from radically different viewpoints. Figure 1.7 attempts to illustrate, at a very high level, how six distinct communities (with, alas, different languages) may be converging onto a common understanding of cognition and autonomous behavior, whether human- or machine-based. The six communities are as shown:

- At the top, the robotics and cybernetics communities, which have driven a better understanding of machine-based autonomy and human-system integration

---

19. This becomes readily apparent when members of these disparate communities become part of a project-oriented team and are forced to communicate and collaborate.

- At the bottom, the cognitive psychology and neurosciences communities, which bring us closer to understanding human cognition via concepts, cognitive architectures, models, and simulations

- In the middle, the AI communities, both "hard" and "soft," which continue to provide us with nontraditional computational approaches to perceptual/cognitive problems that only recently were thought to be unaddressable via machines (e.g., the game of Go [Silver et al. 2016])

In chapter 4 we go into greater detail regarding this "convergence" and discuss how each community can contribute to our understanding of what constitutes autonomous behavior; how perception, cognition, and action drive that behavior; and how research and engineering can work in tandem to advance understanding of existing systems (e.g., humans) and support more rapid development of robust, trustworthy, and flexible machine-based systems. We believe the time is right to make the effort to reach out to a diverse set of communities and embrace a cross-disciplinary approach to the engineering of future ASs.

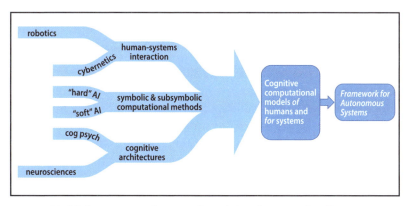

Figure 1.7. Multiple streams of research and development leading to a common framework for ASs

The figure above also indicates the potential for this convergence to yield one or more computational models of how humans perform "cognition" as well how systems might be designed to perform in a like manner.[20] What this requires is some sort of convergence of terms across the different languages used by the different research/engineering communities or at least some

---

20. Or even in an improved manner as some of our "idiot-savant" machines do now, in narrowly restricted tasks.

means of translating from one set of terms into another.[21] We believe this could be facilitated by a common framework that encompasses the basic functions enabling perception, cognition, and autonomous behavior, for both humans and machines. Such a framework would include not only the component functions underlying a full repertoire of behaviors attributed to an autonomous/cognitive system but also the "data flow" (or signals) between these component functions and the "control flow" (or executive function), which serves to orchestrate component function execution, whether explicitly programmed or implicitly driven as an emergent quality of the framework.

We discuss these ideas at greater length in chapter 5, where we present an example framework, albeit engineering focused and with some shortcomings, to illustrate what the community should be considering, to help bridge the conceptual gaps and focus the effort. We have several communities researching different (and oftentimes the same) aspects of human and machine intelligence, and if we are to accelerate their convergence to a common understanding, we need to create a common framework to express common concepts, to test different theories, and to explore different paths toward building ASs with all the behavioral properties we discussed earlier.

We believe that the payoff in striving toward a common AS framework is threefold:

- From a theoretical point of view, it would be particularly appealing to attempt to unify a wide range of research efforts now under way exploring what constitutes autonomous behaviors and an equally diverse number of efforts attempting to build and test such systems. The unification of some of these efforts under a common "architectural umbrella" would drive the research and the engineering and perhaps support faster development along both fronts, with lasting tenets that might eventually serve as the basis of a "science of autonomy."

- From a practical point of view, common architectures can provide insight into best practices in terms of explaining observed behaviors and/or supporting the development of new ASs across different domains and for different applications. Common architectures can also make possible the "reuse" of solutions in one domain that have already been developed in another. At the very least, they can encourage, through the introduction of common constructs and nomenclature, greater communication

---

21. In appendix B, we provide a brief set of "Frequently Asked Questions" developed by the AFRL and aimed at the engineering/operational community, which begins to address some of these issues, but we recognize that it is only a start at attempting to bridge the communications gaps across S&T communities.

among different groups pursuing what, on the surface, may appear to be significantly disjointed aspects of the same or closely related problem.

- And in terms of the opportunity, we believe that the time is ripe for attempting to develop one or more unifying architectural views, because of the long-term and now rapidly accelerating "convergence" across multiple R&D communities focusing on many different aspects of what may very well be a common problem, that of understanding the behavior of existing ASs (e.g., humans) and the development of new ones, in the biosciences and engineering worlds.

This opportunity can take advantage of the exponential growth in the enabling computational infrastructure and technology platforms being developed outside of the autonomy community. Specifically, we call out the "architectural patterns" afforded by multitier hardware and multilayer software architectures, where we use the terms *tier* to refer to physical hardware segmentation of some of the AS functions and *layer* to refer to the logical software segmentation of AS functions (Fowler 2002). An example of such a pattern is illustrated in figure 1.8, composed of four hardware/software tiers/layers:

- *Human-Machine Interface (HMI).* Advances in HMI designs will better enable effective human-systems integration and close human-AS teaming in a manner that engenders communications, task sharing, and trust. This will allow us to communicate effectively and team with these systems, doing tasks that neither man nor machine excel at singularly and that are best served by a team effort, for example, "cyborg chess" or, as it was originally introduced, "consultation chess" (Michie 1972).

- *Autonomous System Architecture.* This is primarily a software layer designed to provide the modularity and functionality of a selected AS architecture, as we just described. Ideally, it is a reusable domain-independent plug-and-play architecture that can be used across different domains with expandable/contractible functionality. The software community typically refers to this as an *application service* or *business logic layer*. In terms of a hardware tier, this service would likely be hosted on one or more embedded computers associated with a host platform, for example.

- *Computational Methods/Algorithms.* This is primarily a software layer—although special purpose processors could be used here as additional hardware tiers[22]—providing multiple common computational approaches to

---

22. Such as the GPUs mentioned earlier that provide the parallelism needed by ANN algorithms (Janakiram 2017) or neuromorphic chips that mimic the way neurons are connected in the brain (Gomes 2017).

implementing a given function in the AS architecture, for several functions. The software community typically refers to this as a *business services layer*, or *low-level business layer*, supporting one or more higher-level functions. Included here would be not just big-data statistics/learning and probabilistic modeling/reasoning approaches (e.g., Bayesian reasoning) but also symbolic AI (e.g., rule-based systems) and subsymbolic AI methods (e.g., neural networks)—and, naturally, a variety of ML approaches to improve AS performance over time. We describe these in much greater detail in chapter 4.

- *Hardware/software platforms.* At the bottom of figure 1.8 we show hardware/software tiers and layers providing any needed software services (e.g., operating systems), computational power, and memory to instantiate the overall AS architecture, its layers and services, and sensors and effectors needed to support situated agency (see chapter 2), multi-AS operations, and human-AS teaming, augmented by high-speed and ubiquitous communications links (across agents and back to "the cloud" serving as a shared cognitive node), as well as individual "smart" nodes made possible by new software architectures, cheap computational resources, and memory (basically the Internet of Things, or IoT [Vermesen and Friess 2013]).

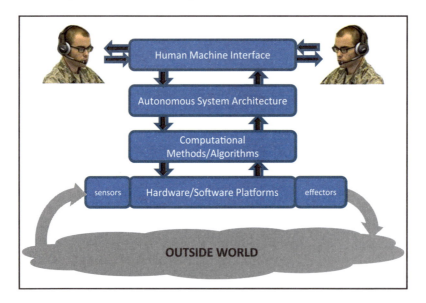

Figure 1.8. Example architectural pattern for AS development

Several benefits accrue with the definition of a well-designed architectural pattern for AS development, particularly because of the independent growth in capability in all of these areas—especially in the last two: computational methods and hardware/software platforms. To cite one example, which summarizes work by four groups at the University of Toronto, Microsoft, Google, and IBM in the application of ANNs[23] to automatic speech recognition, the researchers note, "Two decades ago, researchers achieved some success using artificial neural networks with a single layer of nonlinear hidden units. . . . At that time, however, neither the hardware nor the learning algorithms were adequate for training neural networks with many hidden layers on large amounts of data, and the performance benefits of using neural networks with a single hidden layer were not sufficiently large to seriously challenge [conventional approaches at the time]" (Hinton et al. 2012).

But significant progress over the last several decades has been made in speech recognition precision (Hinton et al. 2012) with the introduction of the following:

- DNNs incorporating many layers (tens or hundreds of layers) (Schmidhuber 2014).

- Improvements in the ML algorithms used to "train" the networks, especially in DNNs, where the large number of layers makes credit assignment and parameter selection difficult (Schmidhuber 2014).

- The availability of large datasets (big data) for training, made available via the internet and cloud-based storage—over the past 30 years, the cost per GB of hard disk data storage has halved every 14 months (Accenture 2016).

- Improvements in the underlying hardware used to host the algorithms, for example by Nvidia GPUs (Nvidia 2018) and other specialized chips (Metz 2018; Schneider 2017) for accelerated parallel processing that is particularly well suited to many ANN computational approaches.

All of these improvements—software algorithms, training data, and hardware memory and processors—have resulted in an explosion in performance of these systems, as evidenced in consumer products introduced recently, for example, Amazon's Alexa (Pierce 2018) and Google's Assistant (Bohn 2018). And they have expanded the scope of applications, from simple passive pattern recognition (including visually based object recognition) to active game play situations, like Atari (Mnih et al. 2013) and Go (Silver et al. 2016), both demonstrated by Google's Deep Mind (2018) group. It is anticipated that

---

23. Which we describe at greater length in chapters 4 and 5.

these tiers/layers will continue to evolve rapidly with the continued growth in cloud computing and data storage; improvements in communications bandwidth; the push toward fog computing for special purpose applications, for example, mobile population surveillance/identification aids deployed by Chinese police foot patrols, integrating Google Glass-Like technology, belt-worn databases, and facial recognition technology (Chin 2018); and the inevitable rise in commercial IoT applications, where MarketsandMarkets (2017) predicts a 27-percent compound annual growth rate in IoT volume, from $171B in 2017 to $561B by 2022.

To summarize, we believe that not only is there a convergence of R&D across disparate communities happening now (as illustrated in fig. 1.7) but also an acceleration of the enabling technologies and platforms occurring in a broader technology and commercial arena, supporting the layered architectures and platform-centric business models needed for effective development of ASs (as illustrated in fig. 1.8).

There is, however, one final challenge that needs to be overcome if ASs are to be developed rapidly and employed effectively by the Air Force: the need for *innovation* in our development of these systems. We believe that these information-focused systems are qualitatively different than the systems that we have developed and acquired in the past and that new methods and approaches are called for if we are to be successful in bringing ASs into the inventory. Specifically, we need to overcome limitations that we now have in three areas:[24]

- First, we need to overcome our desire to minimize development risk for new systems and our focus on incremental improvements in legacy systems, whether those efforts focus on truly legacy systems (e.g., reengining the nearly 70-year-old B-52) or are legacy in concept (e.g., recapitalizing the Joint Surveillance Target Attack Radar System [JSTARS] concept with new hardware [Deptula 2016]). A first step in overcoming this creeping incrementalism and lack of vision is to embrace **challenge problems** that would, at first glance, appear unapproachable with current systems and manning concepts. This is particularly appropriate for the introduction of ASs, given the long history of the AI community in posing and then solving what originally appeared to be intractable problems in computer vision, music synthesis, chess playing, and so forth. But it does require discarding the constraints imposed by existing systems and envisioning an entirely new way of approaching operational problem sets.

---

24. And these are by no means sufficient, since there are other issues in acquisition, logistics, training, and so forth that must also be addressed. However, they are necessary from a design-and-development standpoint.

- Second, we need to rethink our processes for exploring new technologies and applying them to the development of new materiel solutions. We need to move out from a waterfall approach in which requirements are formulated, technology solutions evaluated, systems designed, systems developed, and systems tested against requirements—all in a rigid sequential approach, often "late to market" with the original requirements made obsolete by a changing world situation. Rather, we need a much more agile set of *development processes*, with iterative wargaming, simulation, prototyping and evaluation—all constantly driving the evolution of new systems to meet changing operational needs. Learning at all steps in the process is at the center of this type of process innovation and is also at the heart of many of the newest AI techniques.[25] This is an opportunity to bring the attributes of the thing being developed (e.g., an AS) to the process used for its development.

- Finally, we need to move out of our technical discipline-focused stovepipes and take an honest, cross-disciplinary approach to the development of this new class of systems. The current AFRL structure of discipline-segregated directorates encourages this kind of stovepiping and discourages the kind of cross-disciplinary *organizational structure* and systems engineering viewpoint needed for successful early prototyping and evaluation of systems concepts. Fortunately, the *development* and *application* of autonomous systems demands a cross-disciplinary approach. The former demands it because of the need for a wide range of AS capabilities, in sensing, cognition and computation, human-system interfaces, communications, robotics, and so forth, and the latter affords it because many systems can benefit from the application of autonomy, whether they operate in the air, space, or cyberspace, as we have pointed out earlier.

Because this report is focused on the technical issues associated with AS development, we will save our discussion and recommendations regarding the above three areas until the recommendations in chapter 6. However, it seems clear to us that successful development of ASs not only demands a change in the way we do business along these three dimensions (challenge problems, development processes, and organizational structures) but also affords many new out-of-the-box opportunities for developing new systems that incorporate autonomy.

---

25. As demonstrated by AlphaGo Zero (Silver et al. 2017), in which the AI game player became its own teacher, not needing supervision by a human to perfect its play.

1.5 Future Vision

While we dominate the air, space, and cyber domains today, our adversaries have invested heavily in technologies to deny us the superiority we have come to rely on. To counter this, we must integrate our advantages across the domains in new and dramatically effective ways.

—Gen David L. Goldfein, Chief of Staff of the Air Force

At the beginning of this chapter we described a range of benefits that could accrue with the successful development and deployment of ASs. This includes impact on future *materiel solutions* across different physical platforms, in the air (e.g., Loyal Wingman), in space (e.g., self-defending satellites), and in cyberspace (e.g., autonomous rapid response defenders), as illustrated in figure 1.3. This also includes potential changes in *operational concepts*, as described in the AFFOC (USAF AFFOC 2015), which foresees considerable human-machine teaming with autonomous and semi-autonomous systems, primarily physical platforms (although, see below). The AFFOC makes little mention of changes in *organizational structures* that might result in the introduction of ASs or of potential *cultural changes*, except for an exhortation to embrace innovation and individual initiative, but that could well be accomplished in the absence of the introduction of ASs in the inventory. Finally, it is appropriate to point out that—while we are enthusiastic about the potential benefits of ASs in terms of providing improvements in situational awareness, decision quality, and speed of response, as well as supporting the removal of operators from the "dull, dirty, dangerous" mission tasks (*The Economist* 2011)—we, as an enterprise, have yet to conduct a quantitative assessment of the potential impact on overall mission success of the introduction of ASs, let alone develop comparative return-on-investment figures across different technology options.[26]

Despite this, we believe that the development and introduction of ASs will change not only how our physical (air and space) and virtual (cyber) platforms behave and operate in a larger organizational environment but also our operational concepts and organizational structures in ways not considered or foreseen in the AFFOC.[27] And much of this change will come not from turn-

---

26. Again, the AFFOC provides a few illustrative vignettes but nothing of a quantitative analytic nature (AFFOC 2015).

27. For a more forward-looking vignette, see appendix H, which includes both "at-rest" and "in-motion" AS assets that progress from an ISR mission to an area-defense mission and, finally, to a humanitarian recovery mission. The vignette displays key AS task, peer, and cognitive flexibilities discussed earlier.

ing our manned platforms (in air, space, and cyber) into unmanned ones but from introducing greater autonomy and its fundamental enabler, AI, into the information-intensive functions of warfighting, including multidomain ISR, situational understanding, operational planning and targeting, battle management, and mission assessment. The AFFOC envisions "human battle managers in control of large numbers of self-coordinating vehicles or programs [for cyber payloads]" (USAF AFFOC 2015). This is probably upside down; for a far outlook of 20 years, we should be considering ASs as central participants in the battle-management function and not merely platform-operating peripherals.[28]

The basic question then becomes: "How do we get there?" One way is to pursue the conventional physical/cyber platform-focused approach: build autonomy into our legacy manually controlled platforms (or their next-generation variants), along the lines vignetted in the AFFOC (USAF AFFOC 2015)—but build on an integrated hardware/software environment (such as that illustrated earlier in fig. 1.8) for commonality across systems, easing development time and enabling cross talk across systems, operators, missions, and domains. At the same time, pursue a platform-based business model, where *platform* in this context refers not to the physical/cyber platforms the USAF employs in its operations but rather a more generic platform that provides the services expected of the USAF as a component of the joint force. In the commercial sector, these kinds of platforms have become an industry best practice that concentrate on connecting consumers to vendors in high-value exchanges (Parker and Van Alystyne 2016; Morvan et al. 2017). The high-value exchange can be, for example, an exchange of goods for money, starkly different from the more traditional information technology view we described in the previous section since the platform business model is not necessarily focused on the specifics of a product. For example, Amazon connects vendors selling merchandise to people who want to purchase merchandise. What is it that the platform provides? In the Amazon case, it provides a way for consumers to find, review, research, purchase, and ship products of interest—all while providing buyer and seller protection, offering shipping options, collecting sales tax, and supporting several other buyer and seller experiences. Amazon also offers web services that connect organizations to big-data solutions through their scalable, reliable, big-data platform. Other commercial examples include Uber, which connects people needing transport services to people

---

28. Although in the AFFOC's favor is a callout for AS-based, front-end ISR collection and processing: "Fully integrated information systems that allow aggregation of data from a variety of classified and unclassified sources, sensors, and repositories, and a degree of autonomous processing, exploitation, and dissemination" (USAF AFFOC 2015).

capable of providing transport services in the needed timeframe. Facebook has revolutionized people-to-people connections through its social media platform. TaskRabbit connects homeowners to safe and reliable local help. On the government side, the National Security Agency (NSA) and the Defense Information Systems Agency (DISA) similarly have the Big Data Platform (BDP) that enables efficient capability development and deployment for cyber operations, which has been made available to the entire department. These are all modern platform business models that are enabled by information technology and the knowledge associated with it.

Knowledge is what an AS uses to create the meaning from its observations. An AS is a system that creates the knowledge necessary to remain flexible in its relationships with humans and machines, the tasks it undertakes, and how it solves those tasks. This concept of knowledge creation is echoed in Domingos's recent (2015) book, *The Master Algorithm*, and provides foundational understanding of knowledge representation, management, and learning. Tool-based solutions have provided considerable value over the years, but the solutions only work for a limited set of problems, since they do not scale. Knowledge provides the right mechanism to transform the traditional tools-based approach, which solves a small number of problems, to a knowledge-platform approach that is applicable to a far greater set of problems.

An Air Force Knowledge Platform (KP) should monopolize the connection of observation vendors (e.g., ISR assets, both human- and AS-based) with knowledge creation vendors (e.g., battle managers) and warfighting effects vendors (e.g., strike assets) to deliver multidomain effects. Collaborative ASs would be able to dynamically and opportunistically team and separate as necessary and as dictated by the unfolding battlespace events. Each AS would be able to communicate with other ASs to achieve some particular desired multidomain effect, through a dynamic and variable vocabulary.[29] ASs can then be focused on creating knowledge and appropriately applying that knowledge—across the AS population—to maximize their contributions to a range of air, space, and cyber operations at the strategic, operational, and tactical levels of warfare.

In essence, the KP provides the ecosystem necessary to create capabilities, and those capabilities can be used to create combat effects. We will discuss the specifics of this KP in greater detail in section 6.6. However, here we wish to merely point out its potential for transforming the USAF from an industrial-age organization focused on its hardware platforms to an information-age,

---

29. This move from a specialized mission-specific representation of knowledge to a more flexible representation and creation of knowledge is cited as what facilitated animal species in nature expanding their purview (Newell 1990).

service-oriented enterprise, enabled by AS-mediated knowledge creation and application. Figure 1.9a illustrates the Air Force's current platform-centric view of itself. Four categories of platforms—by which we mean physical entities, specifically vehicles and operations centers—are shown:

- On the left-hand side are the *sensor platforms*, including unmanned platforms like Predator, Reaper, and GlobalHawk, and *manned platforms* like Airborne Warning and Control System (AWACS), RivetJoint, and Compass Call.

- On the right-hand side are the *shooter platforms*, including close-air-support assets, bombers, fighters, and EW platforms.

- On the bottom are *support platforms*, including transport and tanker platforms.

- In the middle are *C2 platforms*, including airborne platforms like AWACS and JSTARS, and *groundborne* or *fixed platforms* like the AOCs.

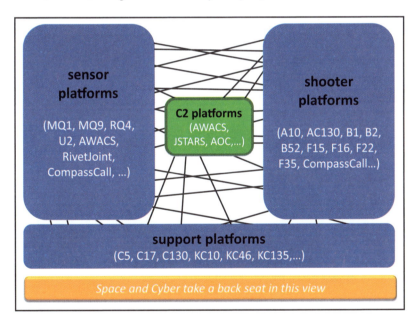

Figure 1.9a. Today's platform-centric view of the Air Force

Finally, although not called out explicitly but illustrated by the black lines, a jumble of communications networks connects some platforms to some other platforms (but none that connect all to all), with different frequencies, bandwidths, and networking protocols (Dougherty and Saunders 2003). And because this platform-centric viewpoint is focused on air platforms, space and cyber assets take a back seat.

But imagine if the sensor, shooter, and support platforms got off center stage, and we *grew* the C2 assets in the middle, as shown in figure 1.9b. We now have three categories of platforms:

- On the left-hand side, we have the *sensor/collector platforms* (labelled as such to emphasize the importance of space and cyber assets), collecting all *domains* available from the world cloud shown at the bottom.

- On the right-hand side, we have *shooter/supporter platforms* (combining the shooters and supporters from earlier because of their similar *action* roles), acting across all domains via their direct impact on the world cloud.

- In the middle, we have the critical *information-processing* and *knowledge-management functions*—functions that comprise multidomain C2—working across all domains and connecting sensor/collector platforms with shooter/supporter platforms to carry out the mission.

- Connecting them, we have *C2 guidance* driving the shooters/supporters via the *feedforward path* and the sensors/collectors via the (inner loop) *feedback path*.

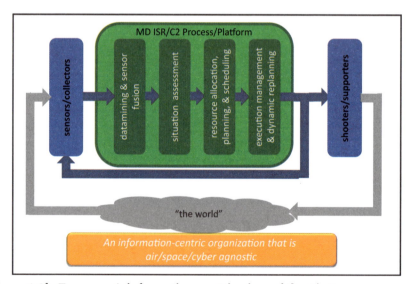

Figure 1.9b. Tomorrow's information-centric view of the Air Force

We now have an information-centric view of the Air Force system—one that is air/space/cyber agnostic and implicitly assumes *all* domains contribute to *all* components.[30] This view also forces us to focus on and deal with the critical information-processing and knowledge-management issues associated with a service organization—one built on a platform-based business model, where *platform* in this context is distinctly different from the physical platforms we focused on in figure 1.9a (and which we focus on today), where *platform* in this context is distinctly different from the physical platforms we focused on in figure 1.9a (and which we focus on today).

To move from a platform-centric, industrial-age organization to a networked, information-age enterprise requires us to focus on the following:

- *Knowledge movement*: collection, communications, and dissemination

- *Knowledge creation*: information fusion and exploitation and dynamic SA

- *Decision management*: resource allocation, planning, and scheduling

- *Execution management*: acting and monitoring

While doing this, we focus on the goals of maximizing information utility, reducing friction, accelerating processes, and winning faster and more decisively. We can leverage the best of commercial advances in information processing relying on Moore's law (Moore 1965), networking and Metcalfe's law (Hendler and Golbeck 1997), and platform-based business models (Morvan et al. 2017). However, it does require a broad architectural view centered on a knowledge platform, the introduction of greater automation and ASs throughout the organization, an infrastructure that supports ubiquitous machine-to-machine communications, and improved human-systems integration. Our vision is:

An agile, information-centric enterprise making timely decisions executed via friction-free access to exquisitely effective peripherals.

1.6 Outline

This report is organized into six chapters and nine appendices.

Chapter 1 has attempted to provide general background for the study, in

---

30. It also assumes much better connectivity than we now have, a separate technology problem that we must address if we want truly multidomain operations that are (vehicle) platform agnostic.

terms of overall motivation for the introduction of ASs into the Air Force inventory and the potential benefits that could accrue operationally. We began with a general discussion of past attempts to define the term *autonomy* and summarized the major findings of several past studies. We then focused on the operational challenges and opportunities associated with developing and introducing these systems, converging on three dimensions of desired AS behaviors, namely *properties for proficiency*, *tenets of trust*, and *principles of flexibility*. As outlined here and described in greater detail in chapters 2 and 3, these serve to define ASs in a behavioral fashion in lieu of more formal definitions proposed elsewhere. We then outlined key development challenges that must be addressed to develop these systems but also pointed out the opportunities for significant advancements in this area, including a convergence of research communities, described in greater detail in chapter 4, and the potential of one or more unifying functional frameworks coupled with a broad set of enabling technologies, discussed at greater length in chapter 5. Finally, we closed this chapter with a vision of how ASs, coupled with a global architecture incorporating a knowledge platform, could transform today's Air Force from a platform-centric, industrial-age organization to tomorrow's networked, information-age enterprise.

Chapter 2 presents four key *properties for proficiency* in ASs. While the list may not be exhaustive, we believe these to be a necessary set for ASs to realize their full potential in future defense systems. These properties are: *situated agency*, embedding the autonomous system within the environment and giving an AS access to that environment; *adaptive cognition*, providing an AS several different ways of problem solving; *multiagent emergence*, enabling AS-to-AS interactions and emergence of multi-AS behaviors; and *experiential learning*, supporting the evolution of AS behaviors based on past experience.

Chapter 3 describes four key *tenets of trust* in ASs. Again, this may not be an exhaustive list, but it is one we believe necessary for ASs to be trusted participants with humans in critical mission operations (and for ASs to be trusted by other ASs), based on our knowledge of human-human interaction and our experience in operating automated systems. These tenets are *cognitive congruence and transparency*, to support human understanding of AS behaviors; *situation awareness* of both the external environment and the internal state of the AS, to support knowledgeable decision making; *good human-system integration*, to support effective human-system interaction under adverse conditions; and *human-system teaming and training*, to support effective team operations under a range of missions and threats.

Chapter 4 provides greater background on the multiple communities that have contributed to, and continue to contribute to, the science and technology underlying our understanding of perception, cognition, learning, and action, both animal- and machine-based. The chapter describes how six distinct communities may be converging onto a common understanding of how to either develop new ASs, or understand the behaviors of existing ones, via frameworks, architectures, and even computational models and simulations. The six communities are the robotics and cybernetics communities, the cognitive psychology and neurosciences communities, and the "hard" AI and "soft" AI communities. We see all of these communities as key to the development of *proficient*, *trustworthy*, and *flexible* ASs, in the sense that we discussed earlier.

Chapter 5 discusses frameworks for the development of ASs, focusing on the category of cognitive architectures and their computational instantiations. One potential framework is described in detail, which is functionally structured rather than unstructured, makes no commitment on symbolic versus subsymbolic processing, and incorporates learning. It is deliberately engineering-focused with a strong *dataflow* orientation that has its basis in a cybernetics view of the world. In addition to identifying the broad range of functionality we believe is needed for proficient, trustworthy, and flexible AS behavior, the chapter discusses several promising technologies for achieving these different functionalities. The chapter closes with a brief enumeration of functions *not* explicitly represented in the framework, which may also be critical to AS development success and usage, to serve as motivation for additional research in this area.

Chapter 6 closes with five categories of high-level recommendations covering behavioral objectives that specify generalized design requirements for autonomous systems; cognitive architectures and enabling technologies for AS design and implementation; challenge problems for the R&D community to engage in to push forward the state of the art; development processes that incorporate more rapid prototyping, experimentation, and iterative development; and cross-disciplinary organizational structures that cut across traditional stovepipes. Also presented is the concept of a *Knowledge Platform*, which helps to integrate these different recommendations into a broader and more unified effort across technologies, processes, and organizational structures.

Chapter 2

Properties for Proficiency

In this chapter we outline four key *properties for proficiency* in ASs. Though the list may not be exhaustive, we believe these to be necessary for ASs to realize their full potential in future defense systems. As introduced earlier in section 1.3, these properties are:

- *Situated Agency.* Embedding the AS within the environment, with component abilities to sense or measure the environment, assess the situation, reason about it, make decisions to reach a goal, and then act on the environment, to form a closed loop of "seeing/thinking/doing," iteratively and interactively.

- *Adaptive Cognition.* A capability to use several different modes of "thinking" about the problem (for example, assessing, reasoning, and decision making), from low-level rules to high-level reasoning and planning, depending on the difficulty of the problem, with sufficient flexibility for dealing with unexpected situations.

- *Multiagent Emergence.* An ability to interact with other ASs via communications and distributed function allocations (e.g., sensing, assessing, decision making, etc.), either directly or through a C2 network, in a manner that can give rise to emergent behavior of the group, in a fashion not necessarily contemplated in the original AS design.

- *Experiential Learning.* A capability to "learn" new behaviors, over time and experience, by modifying internal structures of the AS or parameters within those structures, based on an ability to self-assess performance via one or more performance metrics (e.g., task optimality, error robustness, etc.), and an ability to optimize that performance via appropriate structural/parametric adjustments over time.

We describe these at greater length in the following four sections of this chapter.

2.1 Situated Agency

2.1.1 Agency and Autonomy

We start this section with the goal of trying to separate the notion of *agency* from the concept of *autonomy,* since the two are often intertwined in terms like *autonomous agents* and *agent autonomy* or a relationship is implied via definitions associated with *intelligent agents*. For example, Franklin and Graesser (1996) make note of several (then current) definitions, specifically:

- *Autonomous agents* are computational systems that inhabit some complex dynamic environment, sense and act autonomously in this environment, and by doing so realize a set of goals or tasks for which they are designed (Maes 1995).

- *Intelligent agents* continuously perform three functions: perception of dynamic conditions in the environment; action to affect conditions in the environment; and reasoning to interpret perceptions, solve problems, draw inferences, and determine actions (Hayes-Roth 1995).

- An *agent* is anything that can be viewed as perceiving its environment through sensors and acting upon that environment through effectors (Russell and Norvig 1995).

- An *autonomous agent* is a system situated within and a part of an environment that senses that environment and acts on it, over time, in pursuit of its own agenda and so as to effect what it senses in the future (Franklin and Graesser 1996).

From these consistent but not identical definitions, we can infer that **agency** has to do with interacting with the environment, in a way most clearly stated by the definitions formulated by Maes (1995) and Hayes-Roth (1995): sensing the environment, reasoning about the environment and some desired goals to be achieved in that environment, and acting upon that environment in a way that tries to achieve those goals. Because of the importance of the environment in this definition, we call this **situated agency**. Note that all four definitions above are consistent with this view of agency.

As a side note, we are specifically *not* endorsing the viewpoint that agency has to do with acting in a manner to serve the goals or desires of another agent, as, for example, the way a real estate agent might act to serve the house-hunting goals of a client (agent). This then precludes the consideration of definitions such as the following:[1]

Intelligent agents are software entities that carry out some set of operations on behalf of a user or another program, with some degree of independence or autonomy, and in so doing, employ some knowledge or representation of the user's goals or desires. (Smith et al. 1994)

This acting to satisfy the goals/desires of another adds another degree of

---

1. Aside from the fact that the definition limits agency to software instantiations.

confounding of concepts and will be addressed shortly when we talk about human-system teaming. For now, we assume that agency can occur with the existence of a single agent.

Autonomy or autonomous behavior arises in the goal-seeking behavior implicit in the Maes (1995) definition of *autonomous agent* given above, in the ability to "solve problems" in the Hayes-Roth definition of *intelligent agent,* and in the phrase "in pursuit of its own agenda" employed by Franklin and Graesser (1996). The Russell and Norvig (1995) definition of *agent* avoids the confounding of autonomy and could properly apply to a thermostat—an agent we would not consider autonomous. Essentially, then, autonomous agency implies everything we just said about agency but where the goals are effectively "owned" by the agent, unlike the situation, say, with the thermostat, where the goal is set by the human operating that thermostat.

2.1.2 Agency in Multiple Situations

The notion of *situated agency* has a long history, but one of the first serious considerations of engineering such systems was formulated by Wiener in 1948 in his treatise on cybernetics (Wiener 1948), in which he noted,

> It has long been clear to me that the modern ultra-rapid computing machine was in principle an ideal central nervous system to an apparatus for automatic control; and that its input and output need not be in the form of number or diagrams, but might very well be, respectively, the readings of artificial sense-organs, such as photoelectric cells or thermometers, and the performance of motors or solenoids. With the aid of strain-gauges or similar agencies to read the performance of these motor organs and to report, to "feedback," to the central control system as an artificial kinaesthetic sense, **we are already in a position to construct artificial machines of almost any degree of elaborateness of performance**. (emphasis added)

To graphically represent this concept of agency, consider the following sequence of figures, in which a single AS, our situated agent, is interacting with its environment. In accord with Wiener's formulation expressed in the preceding block quote, we have, in figure 2.1a, a single AS in the middle of the diagram, with sensors that are driven by some aspects of the environment (light, heat, etc.) and provide the system with "inputs." This system's effectors serve to act on the environment in some fashion (force, laser beams, etc.) and provide the system with a means of generating "outputs" to the environment. Inside the gray cloud could be other ASs and human entities, both red and blue, as well as other entities not represented in this diagram (e.g., neutrals, weather, etc.). This, in a nutshell, is situated agency as we have defined it earlier.

At the top of the diagram is also shown a human "teammate" plus a human-computer interface (HCI) connecting the two. The human teammate may be a peer (true teammate) or a supervisor (e.g., a flight lead) or a subordinate (e.g., a member of a flight package led by an autonomous package lead). Note that the human teammate is also connected to the outside world via separate channels, implying possible (and likely) disparate sources of sensory information and means of effectuation. This simple extension then lets us consider issues of human-system teaming (later in section 3.4), which we know will be central to any development and use of these ASs.

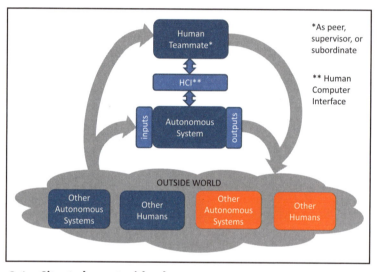

Figure 2.1a. Situated agent with a human teammate

Clearly, this graphical representation can be expanded in multiple ways to accommodate multiple ASs and/or human teammates. One way is simply as shown, with multiple entities in the cloud. Another way is to add blocks representing additional ASs and/or human teammates to the upper half of the figure, as shown in figure 2.1b.

In either case, we are recognizing that we can have any combination of humans working in concert with one or more agents:

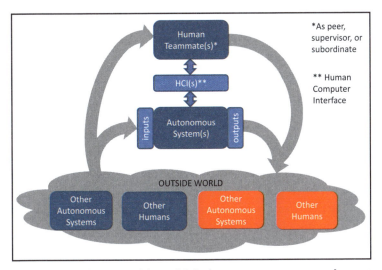

Figure 2.1b. Situated agent with multiple human teammates and agents

- 1-to-1 (1 human working with 1 agent), such as one human driver working with one situated agent, overseeing, for example, driving safety in traffic, through a variety of sensors and collision avoidance algorithms and braking/steering effectors.[2]

- 1-to-N (1 human working with N agents), such as a "swarm controller" flying a formation of agents at a high level, for example, flying the "center of gravity" of the swarm, rather than all the individual elements of the swarm, leaving the agents to move together and deconflict on their own (Zigoris 2003).

- M-to-1 (M humans working with 1 agent), such as is currently demanded of remotely piloted aircraft (RPA) operations, involving on the order of 170 humans to maintain flight operations of a single agent platform (Predator or Reaper) over extended periods of time (Zacharias and Maybury 2010).

- M-to-N (M humans working with N agents), a teaming situation that is now beginning to be explored (Chen and Barnes 2014).

---

2. The Ground Collision Avoidance System (GCAS) developed by the AFRL is an extended example of this, applied to pilot rather than driver safety (Norris 2017).

As the number of agents N becomes large, the dynamics of the inter-agent interactions start to dominate the behavior of the agent "group" (relative to the internals of the individual agents). We discuss this at greater length in section 2.3. In addition, as both M and N become large, we are entering a human-machine teaming situation where it may become difficult to ascertain who is controlling whom.

One other extension of the basic *situated agency* concept of figure 2.1a can accommodate the current focus on enhancing human performance and/or mitigating human limitations under stressors associated with the level of tasking, the time on task, the environment, and so forth. The Sense-Assess-Augment (SAA) framework described by Parasuraman and Galster (2013) and Galster and Johnson (2013) essentially involves the following:

- *Sensing* of the human's physical, physiological, and psychological state
- *Assessing* the overall state in terms of the performance objectives given the human under specific task objectives
- *Augmenting* the human's capabilities based on the assessed state, via a variety of means, including changes in the HCI, decision aiding, or even biochemical enhancers

Clearly, if the augmentation is appropriate, the effect of stressors should be reduced so the resultant sensing and assessment of the human's state should reflect this, necessitating less augmentation over time. The closed-loop nature of the process is clear.

This is illustrated in figure 2.1c, in which the SAA function is assigned to our situated agent. The inner loop arrows indicate how the situated agent might sense behavioral or physiological attributes of the human (given proper instrumentation/sensors emplaced/embedded with the human), assess the task-relevant states of the human (given appropriate situation awareness algorithms and human physiological/behavioral models), and *augment* the human in some fashion that maintains or improves human performance in the given task (via direct biochemical stimuli or indirect means through the HCI).[3] As in the earlier figures, both the human and the agent may be engaged in closed-loop interactions with the environment and with each other.

From the above discussion, *situated agency* can take on a number of roles and interactions with other agents—both human and machine.

---

3. A third option exists for augmentation if the human is operating a system to which the situated agent has access via direct interaction with the system itself, such as occurs in driving with antilock braking, stability augmentation systems, and so forth.

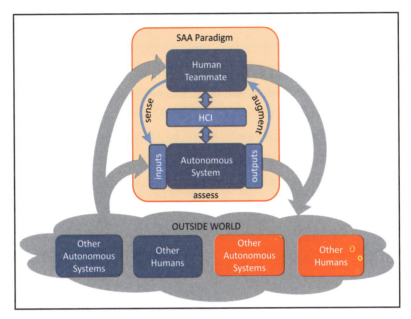

Figure 2.1c. Situated agent implementing the SAA paradigm

2.2 Adaptive Cognition

In the previous chapter, we noted the importance of cognitive flexibility in AS behavior, in how an AS should be able to change how it carries out a task, in its short-term response to a changing situation, and in its long-term experience and learning. Here, we attempt to more fully develop this idea with a brief discussion of what is meant by *adaptability* in terms of different cognitive approaches to problem solving, models of learning how to problem solve, and, finally, how metacognition and consciousness may contribute to adaptability. Note that we will be intertwining theories of how we think humans solve cognitive problems (*cognitive science*) and how we might build ASs to exhibit similar adaptive behaviors (*artificial intelligence*), since both areas are closely intertwined, as discussed later in chapter 4.

To preface this discussion, consider Minsky's insight as to why humans are so cognitively resourceful, with our shorthand tagging in brackets (Minsky 2006):

- [knowledge representation] *We have multiple descriptions of things—and can quickly switch among them.*

- [experience/memory/introspection] *We make memory-records of what we've done—so that later we can reflect on them.*

- [problem solving] *Whenever one of our Ways to Think fails, we can switch to another.*

- [goal seeking/management] *We split hard problems into smaller parts and keep track of them with our context stacks.*

- [conscious/unconscious motivation] *We manage to control our minds with all sorts of bribes, incentives, and threats.*

- [learning/metalearning] *We have many different ways to learn and can also learn new ways to learn.*

We will touch on a few of these components of adaptive cognition in this section.

Consider the *YouTube* video published by Boston Dynamics on 12 February 2018 in which the company's SpotMini dog-like robot exhibits incredible dexterity by opening a closed door using its 3-D vision system and its single attached arm to turn the handle and hold the door open for another SpotMini lacking such an arm (Boston Dynamics 2018). On the surface, the vignette plays out as if these two robots were engaging in adaptive cognition, cognitive and peer-to-peer flexibility, and team problem solving, in the short space of the half-minute this takes to play out. If these were truly autonomous robots and not controlled at some high level by off-screen human operators, one might interpret the vignette as follows:

- The first (armless) SpotMini1 enters the room and heads to the (closed) door. It stops in front of the door and stares at the door handle for a bit. As humans watching this assuming autonomous agency at work here, we might assume the following: SpotMini1 wants to go out the door.[4] Spot-Mini1 sees the door is closed but realizes it cannot get out because it has no arm/hand to turn the door handle.

- SpotMini1 then turns to another part of the room, and soon SpotMini2, sporting an arm/hand manipulator, enters the scene. From our human

---

4. In this context, *wanting* is equivalent to "having the goal of."

perspective, it appears that what has happened is that SpotMini1 reasoned that it could get out of the room by calling on SpotMini2 to open the door, effectively solving the problem in a team fashion, adding resources it did not have on its own, and changing its peer-to-peer relationship on the fly.

- SpotMini1 backs away from the door, giving access to SpotMini2, which then turns the door handle, opens the door, and, using its arm, holds the door open for SpotMini1 to go through. SpotMini2 then follows, again holding the door open for itself, letting the door close gently, and, thus, ending the vignette with both robots having left the room.[5]

We applaud the dexterity of these robots—but we should not read too much into their problem-solving abilities in this vignette, given the behind-the-scenes control by human operators that was almost certainly exercised.

However, had these robots been operating on their own—with no human supervision—and, using the working definition of the term presented in the previous chapter, acting *autonomously*, one would certainly ascribe some sort of adaptive cognition to them. By this, we mean goal-seeking behavior, which adapts as the conditions change, either in the short term in response to a changing situation or over the long term with experience and learning. In this vignette, SpotMini1, having a goal of leaving the room through the doorway, faced with a closed door and being unable to manipulate the handle, engaged in a behavior that adapted to the situation—calling upon its colleague Spot-Mini2, which had the resources to help.

2.2.1 Multiple Cognitive Approaches

Consider now *multiple cognitive approaches* to solving this type of problem.[6] One of the earliest efforts in this area was the General Problem Solver (GPS) paradigm (and program) put forward by Newell and colleagues (1959) and later extended by Laird and Rosenbloom with their Soar cognitive model (Laird et al. 1993; Laird 2012): a "production rule" framework in which goals are achieved by the successive application of IF-THEN statements (production rules) to transform a current state or situation into some desired future state or goal. In our vignette, a (complex) rule might be IF (you are in the room) AND (the door is open) THEN (exit room through the doorway). One aspect of these systems is how they handle an impasse—that is, when a goal

---

5. Engendering online cries of "robotic Armageddon," with robots escaping the labs where they're being created and endangering humanity.

6. We will not attempt to describe the potential approaches here, since these and others are noted in section 4, and the focus here is about selection and switching among approaches.

cannot be readily accomplished by the successive application of productions. What they do, loosely speaking, is create a subgoal that will support achievement of the original goal. In our vignette, we can use this approach to reinterpret what we saw in our vignette as follows:

- SpotMini1's goal is to leave the room through the doorway, which is blocked by the closed door. Knowing that one cannot walk through a closed door (an impasse), a subgoal of opening the door is generated.

- However, another impasse is reached because there are no resources in SpotMini1's production rule set to open that door directly. So another subgoal is generated: enlist the help of SpotMini2 by establishing a new peer-to-peer relationship by "teaming up."

- Which then triggers another set of productions and subgoals executed by SpotMini2: to open the door, let SpotMini1 out, and follow.

This kind of adaptive cognition formalized in the context of production rule systems has its limits, however: if the rules do not encompass potential resources or actions that may be available to the AS to accomplish its goal (that is, the AS is effectively unaware of them), then an impasse may be reached even when the AS may be physically (if not cognitively) capable of reaching its goal. For example, to extend the vignette, suppose that SpotMini2's battery died just as it was about to open the door, thus doubly blocking access for SpotMini1's egress from the room. If, at this point, SpotMini1 could *change the context of the goal-seeking behavior* it is engaging in, a simple solution might be found. For instance, if instead of fixating on its initial goal of going through the doorway, it instead *generalized the goal* to be that of exiting the room by any manner (in effect creating a supergoal rather than a subgoal), it might then explore other options—for instance, seeing if other doors might be open. If successful in achieving this supergoal, it could leave its nonfunctioning teammate behind and leave the room. This type of adaptive cognition does not require any change of basic problem-solving technique, but it does necessitate a reframing of the problem to change the context—something that humans, let alone goal-seeking AI programs, have a hard time doing.[7]

GPS and Soar are two examples of a cognitive model built on "productions"

---

7. For humans, this is exemplified by the cottage industry in the business world focused on reframing problems. (See, for example, Wedell-Wedellsborg 2017.) In regard to the goal-seeking AI programs, this is one of the main criticisms of early expert system implementations of production rule systems: the human designer needs to keep adding rules to account for "off nominal" situations that were not considered during the initial design, like battery failures of a teammate. (See, for example, Duda and Shortliffe 1983; Bell 1985.)

to get us from one problem state to another, based on a few simple but unifying concepts and mechanisms (and many "rules"). At the other end of the spectrum lies Minsky's *Society of Mind* (SOM) theory of cognition (Minsky 1986, 1991, 2006), in which the mind is viewed as an interacting *society of agents*, each agent performing some simple and specialized mental function; their interactions and resulting "societal" behavior—that is, communicating, cooperating, interfering, and so forth are the bases for cognition, in Minsky's view (Minsky 1986, 1991, 2006). This almost *ad hoc* assembly of different agents is justified by Minsky in the following passage:[8]

> *Mental activities are not the sorts of unitary or "elementary" phenomenon that can be described by a few mathematical operations on logical axioms. Instead, the functions performed by the brain are the products of the work of thousands of different, specialized sub-systems, the intricate product of hundreds of millions of years of biological evolution. We cannot hope to understand such an organization by emulating the techniques of those particle physicists who search for the simplest possible unifying conceptions. Constructing a mind is simply a different kind of problem—of how to synthesize organizational systems that can support a large enough diversity of different schemes, yet enable them to work together to exploit one another's abilities.* (Minsky 1991)

The span of SOM behaviors is large, encompassing not just rational goal-based reasoning (like the door-opening problem above) but also a wide range of human mental activities, including perception, language, learning, emotion, consciousness, and goal setting. Moreover, Minsky proposes, *for each of these mental activities there are likely many different ways of performing them*, since no one approach is likely to work in all contexts. This notion of adaptive cognition, with multiple types of agents available to perform the same mental function, across a range of mental functions, is at the heart of SOM. Minsky justifies this approach with the following hypothetical discussion among different specialties (Minsky 2006):

- *Mathematician*: It is always best to express things with logic.

- *Connectionist*: No, logic is far too inflexible to represent commonsense knowledge. Instead you ought to use Connectionist Networks.

- *Linguist*: No, because Connectionist Nets are even more rigid. They represent things in numerical ways that are hard to convert to useful abstractions. Instead, why not use everyday language—with its unrivaled expressiveness.

- *Conceptualist*: No, language is much too ambiguous. You should use Semantic

---

8. While the first sentence may appear to be aimed squarely at GPS and Soar, Minsky goes on to castigate "pure" connectionists as well—that is, those looking for neural network models of cognition untainted by other methods. His major point is made in the last sentence of this passage.

Networks instead—in which ideas get connected by definite concepts!

- *Statistician*: Those linkages are too definite and don't express the uncertainties we face, so you need to use probabilities.

- *Mathematician*: All such informal schemes are so unconstrained that they can be self-contradictory. Only logic can ensure us against these circular inconsistencies.

However, avoiding a cacophony of individual agents (or teams of agents) proclaiming a solution to a given problem becomes a central problem for SOM. Minsky proposes a variety of approaches, including sensor and suppressor agents to squash unproductive approaches based on past experience in similar situations, self-reflective agents that terminate bad internal processes (like looping), specialized experience-based agents that can remember roughly analogous situations—or cases—that can serve as the basis of a close solution in a similar situation,[9] and *mental managers* that control how we select specific knowledge instances and specific problem-solving techniques (Singh 2003). Clearly, a plethora of solution approaches that provide cognitive flexibility comes with the mental cost of solving another problem: figuring out how to manage and adjudicate across the solution options.[10] But if we want that flexibility in our ASs, we need to invest in this aspect of the problem.

Unfortunately, there has been little follow-on work to turn the SOM theory into computational models of cognition (like Soar did with GPS), so it is unclear, for example, how one might best develop an appropriate computational executive "selection" mechanism or intervention by other agents. However, IBM's Watson, developed to address the game of *Jeopardy!*, provides some insight how this might be accomplished in the confines of a "question answering" (QA) function. This may, at first blush, sound like a rather narrow application, but the team that developed it brought to the problem backgrounds in "natural language processing, information retrieval, machine learning, computational linguistics, and knowledge representation and reasoning" (Ferrucci et al. 2010). At the heart of the design is the use of "more than 100 different techniques for analyzing natural language, identifying sources, finding and generating hypotheses, finding and scoring evidence, and merging and ranking [potential answers]" (Miner et al. 2012). Final selection of an answer, from among those provided by many different experts/specialists, is based on a *selector*

---

9. As implemented in an approach called case-based reasoning (Carbonell 1983; Kolodner 1992, 1993); see also chapter 5.

10. We discuss this further in section 2.3 on multiagent emergence and its relation to consciousness in a SOM context.

developed via a machine-learning approach (Jacobs et al. 1991) and trained over a known set of answers; confidence estimates (in the answer) are also generated to support the actual game-playing strategy (e.g., to answer or not). Many Watson-based QA applications have been initiated since (Olavsrud 2014; Keim 2015), with this same multi-expert selector concept for answer adjudication.

2.2.2 Learning How to Problem Solve

However, adjudication and switching are not the only problems in managing multiple cognitive approaches to knowledge representation and problem solving. The community now recognizes that, for any given single approach, it is unlikely that we will be "hand coding" all the knowledge needed by an AS to operate in anything but a toy environment;[11] we will need to provide an AS with models for *learning how to problem solve* and improve proficiency over time and tasks, operating as a situated agent as discussed in the previous section.[12]

We discuss learning at greater length in section 2.4 below, but here we wish to point out that each knowledge representation approach calls for its own learning method. Domingos (2015) makes this quite explicit in his vision of a *master algorithm*, when he, like Minsky above, describes five different "tribes" of knowledge representors:

- *Symbolists*: symbolic representation, reasoning via deduction, and learning via induction

- *Connectionists*: artificial neural networks, inferencing via forward propagation, and learning via backpropagation

- *Evolutionaries*: genetic/evolutionary algorithms, and learning by selection for fitness

- *Bayesians*: Bayesian belief networks, inferencing by forward propagation, and learning of network structure/parameters via training

- *Analogizers*: case-based (or analogical) reasoning and learning via support vector machine (SVM) techniques

Each tribe comes with its own way of representing knowledge, making inferences, and learning across experience (or memories of experiences). So,

---

11. For example, in a "Tower of Hanoi" problem solving instance; see Butterfield (2017).

12. And, of course, learning how to solve problems is only one aspect of learning for the mind, including learning "percepts" from sensory clusters (Modayil and Kuipers 2008), building mental models that correspond with events and entities in the outside world (Johnson-Laird 1983), and so forth.

with multiple representations comes multiple learning algorithms. Domingos (2015) reasons that, because some representational approaches excel over others in certain situations (e.g., Bayesians over Symbolists when uncertainty is a key feature of the problem set), perhaps some learning approaches excel over others in certain situations. He goes further, however, in proposing that one might use a single hybridized and structured *master algorithm*, which builds on specific learning methods associated with the five reasoning approaches cited above, suitable for learning across all specialist reasoners. We will not detail the approach here, since it is a conjecture and an active area of research; however, it may provide a way forward for an efficient learning algorithm that can support multiple knowledge representation approaches, foundational to cognitive flexibility.

Finally, we should note that when we look for insight by closely examining human inferencing processes and learning across experiences,[13] we see many issues, especially when humans are faced with judgment and decision making under uncertainty. As Tversky and Kahneman discovered and explained over a nearly 50-year span of research (Tversky and Kahneman 1973, 1974; Kahneman et al. 1982; Kahneman 2011a, 2011b), humans can be notoriously illogical, biased, overly influenced by stories rather than data, and overly dependent on heuristics as a short-cut to "logical" reasoning of the sort espoused by Domingos's five *tribes* or Minsky's five *specialists*. Kahneman, in more than 400 pages, describes the many different heuristics that humans use that often drive them to "non-rational" choices.[14] How those heuristics are learned is an entirely different field of study but may be deeply influenced by human desire to make sense of the world in a most economical and believable story-like form.

> *Confidence is a feeling, one determined mostly by the coherence of the story and by the ease with which it comes to mind, even when the evidence for the story is sparse and unreliable. The bias toward coherence favors overconfidence. An individual who expresses high confidence probably has a good story, which may or may not be true.* (Kahneman 2011a)

Creating new stories—and fitting them into the tapestry of earlier ones—may be one of our fundamental learning mechanisms as humans, quite different from the ones noted above (see, for example, Herman [2013]). This may provide us with additional insight into how we model our future learning mechanisms in ASs, where storage efficiencies and recall speed could be of the essence.

---

13. In contrast to when we appeal to "first principles" of knowledge representation and inferencing.

14. Kahneman won the Nobel prize in economics in 2002 for having created the field of behavioral economics, which models actual human economic behavior more closely than idealized earlier versions (see Kahneman 2011b).

2.2.3 Metacognition and Consciousness

We close this section with a very brief foray into *metacognition and consciousness*, two attributes that may have a potential for improving adaptive cognition in ASs.

The term *metacognition* was introduced by Flavell (1979) to describe the act of "thinking about thinking," specifically one's own thinking, a concept that is probably centuries old, with informal names like *reflection, introspection*, and so forth. *Metacognition* is modeled with two components (Flavell 1979):

- Metacognitive knowledge (both declarative and procedural) is defined as "that segment of your . . . stored world knowledge that has to do with people as cognitive creatures and with their diverse cognitive tasks, goals, actions, and experiences."

- Metacognitive experiences are defined as "any conscious cognitive or affective experiences that accompany or pertain to any intellectual enterprise."

According to Flavell, metacognition can occur consciously or unconsciously, but when the experience enters consciousness, it can have an important impact on cognitive goals, cognitive actions, and metacognitive knowledge. For example, recognizing that your cognitive actions are getting you nowhere in a given task may change your goals (as we ascribed to our robots in the opening of this section) or your methods for achieving those goals. It may also be saved for future tasks of a similar nature as new metacognitive knowledge.

Computational theories and models of metacognition have been in development in the AI community since at least the 1950s. As noted in an extensive review of metacognition by Cox (2005):

> From the very early days of AI, researchers have been concerned with the issues of machine self-knowledge and introspective capabilities. Two pioneering researchers, Marvin Minsky and John McCarthy, considered these issues and put them to paper in the mid-to-late 1950's. . . . Minsky's [1968] contention was that for a machine to adequately answer questions about the world, including questions about itself in the world, it would have to have an executable model of itself. McCarthy [1968] asserted that for a machine to adequately behave intelligently it must declaratively represent its knowledge. These two positions have had far-reaching impact.

Cox goes on to review more recent efforts in the AI community, covering areas like *belief introspection* (both about the "outside world" and about an agent's internal states), *metareasoning* (that is, reasoning about the reasoning system itself), thinking versus doing via *anytime systems* (that is, trading off

more contemplation of an action/answer vs acting/answering now), and, finally, functional models of the self's mental reasoning processes, particularly
for model-based reasoning and case-based reasoning approaches (Cox 2005).
Current trends, as of the time of the publication, are also covered and show
promise, but as Cox notes:

> *Thus again I emphasize that metacognition in its many forms has limitations. As noted
> above in the general case metareasoning is intractable. But at the same time, it has the
> potential to provide a level of decision making that can make an intelligent system robust
> and tolerant of errors and of dynamic changing environments.* (Cox 2005)

We close with a few comments on consciousness, since it seems to be inex-
tricably intertwined with metacognition, self-awareness, self-reflection, and
deliberative mental activities in general. For instance, Flavell (1979) claims
that metacognitive experiences are primarily conscious experiences, and consciousness is required to understand the experience and, from a learning
point of view, benefit from that experience. Likewise, Cox (2005) points to the
extensive literature on consciousness in philosophy, cognitive science, and
the neurosciences (Metzinger and Chalmers 1995), and the beginnings of trying to understand how to provide machine self-awareness and self-reflection,
via metaphors with human consciousness.

As Evans and Stanovich (2013) point out, there is strong evidence for a
"dual-process" theory of the mind—commonly referred to as the conscious/
unconscious nature of the mind—in which clusters of attributes are associated with each process. Those attributes are detailed in table 2.1 below, which
is a slightly modified version of one presented by Evans and Stanovich (2013).
It is not our intent to discuss this duality of mind in any detail but simply to
point out that much of our mental processing occurs under the auspices of the
Type 1 intuitive/unconscious system.[15] This makes many of the "snap decisions" that occur in the (usually) successful pursuit of our daily lives but can
result in many types of errors and biases—behavior that is well-documented
by Kahneman (2011a). In contrast, much of our work in cognitive modeling
and artificial intelligence has focused, since the 1950s, on the activities of the
Type 2 rational/conscious system,[16] presumably because: (a) it is easier to do,
being founded on "rational" first-principle logic and mathematics; and (b) it
holds the promise of coming up with the "right" answers in reflective and
"conscious" decision making (e.g., winning at a series of Go games). However,

---

15. Although Evans and Stanovich are quick to point out that there may not actually exist
any systems per se but merely types of processing.

16. See, for example, the links Stanovich (2011) makes between rationality and consciousness and introspection.

it is worth noting the dichotomy between what seems to drive much of human "intelligent behavior and common sense" and what the AI community is pursuing.

Table 2.1. Clusters of attributes associated with dual-process theories of cognition

Dual-Process Systems	Type 1: Intuitive/ Unconscious	Type 2: Reflective/ Conscious
Key Features	Does not require working memory	Requires working memory
	Autonomous	Uses mental simulation
Typical Correlates	Fast	Slow
	High capacity	Low capacity
	Parallel processing	Serial processing
	Uses implicit knowledge	Uses explicit knowledge
	Biased responses	Unbiased responses
	Contextualized	Abstracted
	Automatic	Controlled
	Associative	Rule-based
	Experience-driven decision making	Algorithmically driven decision making
	Independent of cognitive ability	Dependent on cognitive ability
	Deals with simple emotions	Deals with complex emotions
Evolutionary History	Evolved early	Evolved late

Building on this apparent duality of mind, Sloman (2001) introduced a three-level model,[17] in which, starting from the bottom, a reactive "A-Brain" interacts with the external world (for example, in traditional stimulus/ response fashion). Above that, a deliberative "B-Brain" introduces consciousness of what is happening with the A-Brain activities and engages in more "thoughtful" activities (for example, solving calculus problems). Finally, above that, a reflective "C-Brain" provides the introspection needed to improve on B-Brain performance (for example, learning from B-Brain errors in inferencing). This layered view was compared with others in a review by McCarthy et al. (2002) and extended by Singh (2005) in an architecture including five

---

17. The A-, B-, and C-Brain nomenclature used here follows the terminology provided by Minsky (2006) in his description of the same model.

layers: reactive, deliberative, reflective, self-reflective, and self-conscious, with methods for "activating" and "deactivating" each layer. These early modeling efforts have spawned many others and resulted in two broad reviews of the area of "artificial consciousness" conducted over the last decade: one by Chella and Manzotti (2007) and one by Reggia (2013), covering a growing community working in this area. These efforts have served as a means to understand human consciousness (for example, Gelepithis 2014; Sloman and Chrisley 2003), and to provide machine intelligence with greater self-awareness (for example, Brown 1995; McCarthy and Chaudhri 2004; Hesslow and Jirenhed 2007; Haikonen 2013; da Silva Simoes et al. 2017; Sanz et al. 2007), as has been our focus here.

We conclude our brief comments on metacognition and consciousness by noting that, for our purposes of developing resourceful and adaptive ASs, the term *artificial consciousness* may be far too loaded, given that human consciousness covers a broad span of activities, mental states, and philosophies, including a "sense of self," "free will," and illusions of same (Dennett 1991, 2017). However, we do believe that some form of metacognition, perhaps implemented in a layered architecture that supports self-reflection, will be necessary for adaptive cognition in future cognitively resourceful ASs.[18]

To conclude this section, we believe that we need to address the question posed by Hernandez (2017): Can robots learn to improvise? We believe they need to, by embracing *cognitive flexibility* in addressing a problem; bringing to bear different cognitive approaches, learning new ones and improving on old ones; and using metacognition for self-reflective improvement. Different cognitive approaches can range from simply reframing a task's intent or goal to bringing to bear multiple, simultaneous cognitive techniques and adjudicating among multiple candidate solutions that differently account for the situation, the need to trade optimality for timeliness, and a number of other contextual factors. Learning can range from acquiring new heuristics to honing existing skills to learning entirely new techniques for problem solving. Finally, metacognition can range from simple postmortem reviews of what went wrong (or right) to elaborate self-reflection on how to improve on framing a problem, goal setting, decision making, and task execution monitoring.

---

18. We discuss some of these issues in other sections, including in section 3.2, where we note that situation awareness pertains to awareness of the external world and the internal world of the AS. In section 5, we describe several different modes of thinking about the problem, from low-level rules to high-level reasoning and planning—depending on the difficulty of the problem—and the need for flexibility in dealing with unexpected situations.

2.3 Multiagent Emergence

The ability to reduce everything to simple fundamental laws does not imply the ability to start from those laws and reconstruct the universe. . . . At each level of complexity entirely new properties appear. Psychology is not applied biology, nor is biology applied chemistry. We can now see that the whole becomes not merely more, but very different from the sum of its parts.

— Philip W. Anderson

Anderson writes about how the "laws" of science that we have identified tend toward a reductionist view and do not equate to a constructivist approach (P. Anderson 1972). These laws do not inform us about what happens as the number of entities or agents increases and they interact with each other. From an observer's perspective, complex systems and patterns will arise out of multiple and relatively simple interactions among entities in a system. The externally observed behavior of the multiple entities is an *emergent behavior* or *emergent property*.

Note that *emergent* is not resultant (Mill 1843)—composites are not mere aggregates of the simples. This captures that emergence is not simply laws of how single entities interact but concerns itself with the definition of the single entities and their interactions manifesting in an unexpected outcome.

Emergence has been a topic of research in multiple fields including psychology (P. Anderson 1972), biology (Braitenberg 1984), philosophy (Bedau and Humphreys 2008), and computing (Woolridge 2009; Brooks 1986). It is the discussions and developments in philosophy and computing that we focus on here, since they are most relevant to the topic of autonomy and interacting autonomous agents.

2.3.1 Emergence within ASs: Consciousness

The previous section discussed metacognition and consciousness from a perspective of relatively recent work on "intuitive" and "reflective" mental processes. But there exists a longer history—starting in the 1800s—in the philosophy of the mind discourse, centered on *emergentism*, which seeks compatibility with physicalism while also being nonreductive and a means for discussing the underpinnings of *qualia* (Korf 2014). The connection here is that the qualia—subjective nonphysical qualities experienced by an individual via stimuli—are the core of consciousness (Cowell 2001) and may originate in an emergent system. This places emergentism in the middle ground between these two camps on the road to explaining and, for our purposes developing, consciousness in ASs.[19]

There are three concepts that are prevalent in the discussion of emergent intelligence. They are supervenience, downward causation, and reductionism.

Broadly, *supervenience* is a relation between upper-level properties being determined by their lower-level properties. In emergentism, this appears in the argument that mental states can occur only in states that have physical properties (Humphreys 1997) and is the property that connects emergence with physicalism. Formal definitions of supervenience tend to reject reductionism, or the ability to describe the behavior of the collective whole from the sum of its parts.

Downward causation is the connection between macro causation and micro causation (Bedau 2008). For example, we may describe an agent in emergent terms—it is afraid of the light—and then note the micro causations: that because it is afraid, the way that the effectors are adjusted results in the agent going to the same location each time the light appears. The macro cause is also responsible for the micro piece. This argument combats the physical determinism argument that for every physical event y, some physical event x is causally sufficient for y (Humphreys 1997).

The terms *weak* and *strong emergence* are common in the discussion of downward causation and directly relate to the third concept of emergent intelligence: *reductionism*. *Weak emergence* is when the agent is reactive rather than proactive. In weak emergence, there is a potential to derive knowledge of the macro behavior (anticipate what the emergence behavior will be) from the system's microdynamics and external conditions. But when we observe the interactions of the microstates, we still witness unexpected phenomena (Bedau 1997; Chalmers 1996). In the case of weak emergence, a reductionism that explores the micro causations and connections to the macro causations can yield some insights (Silberstein and McGeever 1999). Strong emergence applies when the behavior observed is not deducible from the component truths (Chalmers 1996). Reductionism fully breaks down in the face of strong emergence that is irreducible (O'Connor 1994). Strong emergence is associated with higher-level cognition.

In addition to the argument that all emergence is weak and reducible, another argument against emergentism is the causal exclusion argument (Kim 2006);

---

19. We describe *qualia* as the "vocabulary of consciousness" at greater length in appendix F.1.3, where we outline a challenge problem focused on developing a conscious computing framework for advanced-and-aware autonomous agents.

a detailed discussion and responses surrounding this complex argument is found in Wong (2010).

Because consciousness is an emergent property of the brain (Bedau and Humphreys 2008), and because we know that consciousness is critical to human rationality and metacognition (as discussed in the previous section), it would seem that investments in understanding emergence could significantly benefit the development of cognitively resourceful ASs.

2.3.2 Emergence across ASs

One of the most common examples of emergence in nature is the flying-wedge behavior exhibited by flocking birds. This begins when a bird flaps its wings downward, so that as it pushes air downward, it creates a low-pressure region above one of higher pressure (due to the added air). This pressure difference produces a small updraft that a bird that is trailing and to the side can use to lift its own wing, thus saving some energy. In addition, if birds prefer seeing the horizon ahead of them, instead of another bird, then they will seek out locations that both save energy and provide an unobstructed view. When all birds have these motivations, the group self-organizes into a flying wedge.

The flying-wedge emergent behavior has been demonstrated in robotics (Mataric 1995; Balch and Arkin 1998; Fredslund and Mataric 2002; Spears et al. 2004). The manner of developing the agents tends to focus on four primary rules that the robots use to generate their next motor command. They are:

- *Homing*: navigate to a specific place

- *Dispersion*: maintain distance to avoid conflicts

- *Aggregation/Attraction*: stay together to make a group

- *Self-wandering*: avoid collisions with obstacles

From these four simple rules, swarming, flocking, and formation traversals have all been demonstrated. The resultant system is robust to changes in terms of the number of available robots, failing robots, and environmental changes in terms of obstacles (Mataric 1997). In addition, this is all achieved through the emergence of collective, without having explicitly accounted for all of the environmental dynamics when developing the individuals. These advantages have motivated many of the recent investments in unmanned swarming systems in the air, on the ground, and on the water by the military services and the Defense Advanced Research Projects Agency (DARPA), and development and experimentation continue (Pomerleau 2015).

The flocking example covers a number of the points just presented. First, the observer is a requirement to describe the emerged behavior (the flying wedge). Second, we define the component, the bird/robot, in terms of its micro causalities (distances to goal, neighbors, and obstacles), and that the interactions between the components result in the macro causality.

However, we should not be misled into assuming that the components must be homogeneous. Robot and agent development in behavior-based robotics (Brooks 1986) explores the philosophical discussion of how emergence can lead to intelligence. Braitenberg, a biologist, refers to emergence with the following: "Get used to a way of thinking in which the hardware of the realization of an idea is much less important than the idea itself" (Braitenberg 1984). In essence, the macro observed behavior supersedes the micro.

The example we will use is Vehicle #3, "Love," from Braitenberg's study entitled Vehicles: *Experiments in Synthetic Psychology*, in which he uses simple connections between light sensors and motor actuators to discuss how the micro engineering we do results in emerged macro behaviors. In the case of Love, left and right light sensors are connected to left and right motor actuators, respectively. The connection allows for excitation and inhibition. The motors receive an excitation signal when the light is far away and an inhibition signal when the light approaches an ideal distance for Love to charge its batteries. The robot will be observed to move toward the light, approach it, and then stop. If the light moves, the robot will follow it. The observed behavior is that the robot *loves* the light (fig. 2.2).

Note that when we talked about the design of the swarm bird/robot above, we leveraged Love: the robots were both attracted to and repelled by their friends. In designing swarm robots, behavior-based robots and the emergence in the individual are commonly used as a layer in the larger collective emergence (Brambilla et al. 2013). The layering of emergence that results in ever-more capable behaviors is pervasive in the philosophical literature.[20]

The first demonstration of emergence in a robot with a behavior-based system is Brooks' *subsumption architecture* (Brooks 1986), which tightly couples sensory information to action selection in a bottom-up layered fashion. These systems incorporate several sub-behaviors that are small and simple to implement. An example behavior is "avoid an obstacle": when an obstacle is sensed, the behavior component outputs an action command that ensures the robot does not move closer to it. Other behaviors include wander, explore the world,

---

20. As it is in the layered cognitive architectures we discussed in the previous section, going from a two-layered structure of unconscious/conscious to a five- or six-layered one, including deep self-reflection and metacognition.

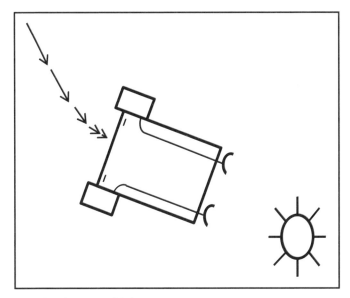

Figure 2.2. Braitenberg's vehicle "Love"

or find the charging station. Each behavior examines the current sensory information and makes an action choice. The final action choice emerges from the inhibition and subsumption structure in the layers. The inhibition and subsumption layers allow the robot to balance goal-seeking behaviors with critical safety behaviors. For example, the robot avoids (avoid obstacle) the person walking down the hall on the way to the mailroom (go to location). The iRobot Roomba uses a subsumption architecture consisting of more than 60 behaviors (Cagnon 2014).

There have been several behavior-based systems developed. They all leverage emergence, combining multiple simple behaviors to result in a complex macro behavior (Arkin 1998). Common to all of the systems are robots that are situated, embodied, and emergent. By leveraging the combination of behaviors and focusing more on the protocol interconnection and letting the "world be its own best model" (Brooks 1986), they are able to act in situations the designer may have not had the forethought to explicitly program them to address (Woolley 2009). This is in direct contrast to "model-based approaches" that are predicated on developing an internal mental model of the external world, "solving" for a goal-seeking behavior against the mental model, and then acting out the solution against the "real world"; see for example Johnson-Laird (1983) and Gentner and Stevens (1983).

We believe that further research into emergence across ASs is called for, for two major reasons. First, the assemblage of multiple simple agents or components has the potential for dealing with larger, more complex, and/or unanticipated problems than originally envisioned by the designers working on the individual agent designs. This creates an interesting situation where many simple systems working together with emergent behaviors can achieve more than one massively engineered system. Second, because it is unlikely that there will be a single "designer" in charge of an assemblage of heterogeneous friendly ASs (say, converging in an operation involving the separate services)—even without considering additional interacting humans and adversary systems—the potential for the emergence of unanticipated outcomes is multiplied far beyond what we might expect with a homogeneous set of ASs under our "control." As the DOD Office of Net Assessment (ONA) has recognized,

> *The prospect of unintended interactions among intelligent machines, and between intelligent machines and humans, is one of the largest risks associated with AI in the years ahead. This risk can be manifested in almost countless ways—which is exactly the problem.* (DOD ONA 2016)

Significant additional research will need to be conducted to address this potential explosion of possible emergent outcomes to, in some way, bound the outcome space, especially in the face of adversary actions deliberately aimed at degrading the performance of our AS teams.

2.3.3 Design for Emergence

The key to designing emergent systems is considering how the entities *interact*, not just in terms of the internals of the entity and how it interacts with the fixed environment (Mitchell 2006). Some of the characteristics are a decentralization of control, component connectivity, environmental dynamism, redundancy, coherence, and diversity. Consistent across the literature is that the development of emergent systems focuses on two primary concerns: the relationships between the entities and the shared rules and utilities that guide the resultant emergent system toward the designer's goal.

Couture adds to these two concerns a third: the number of items (Couture 2007). Mitchell considers it from a networking perspective and as four principles of the emergent systems, as follows (Mitchell 2006):

1. Global information is encoded as statistics and dynamics of patterns over the system's components

2. Randomness and probabilities are essential

3. The system carries out a fine-grained, parallel search of possibilities

4. The system exhibits a continual interplay of bottom-up and top-down processes

Mitchell also notes that complexity and emergence rarely depend on the number of entities and how intricate they are—but rather more on their inter-action protocols (Mitchell 2006). The four principles are also useful for considering the application of emergent systems to optimization problems.

Emergent agents have been developed to aid in the design of complex systems because their flexibility and scalability allow for expansion and contraction of the systemic environments, such as smart cities (Patrascu et al. 2014; Ishtiaq 2012) and aircraft (Balestrini-Robinson et al. 2009).

The most complicated design decision is in the social rules: What local utilities—those objectives/goals that an individual entity seeks to achieve—do we create to enable the emergence of an entity-wide behavior that achieves the large-scale goal that we are after? Not surprisingly, this is an open research question (Chatty et al. 2013). Said another way, "The problem is to determine the individual agents' local utilities and the strategies that they use to select actions that maximize their utilities in such a way that the overall system utility is maximized" (Kroo 2004). The identification of the utilities and rules is most often accomplished in a trial-and-error fashion, via simulation (Osmundson et al. 2008; Balestrini-Robinson et al. 2009).

One conceptual hurdle that designers must overcome, which is often quite hard, is to keep the entities themselves behaviorally simple and let the emergence happen by appropriate design of the interactions available. In teaching students about developing emergent systems, many of the assignments are about moving from sequential control toward more parallel and emergent controls (Horswill 2000; Ziegler et al. 2017). This is done by providing means to quickly experiment with multiple interaction strategies and providing tools that do provide some reducibility to aid in debugging when the global emergent property is not as intended.

Finally, we should note that not all emergent behavior needs to be "designed in" from the beginning. If we provide a capability of experiential learning at the individual agent level, then new behaviors can be acquired, which, in turn, can lead to desired systemic behaviors over time. We discuss this further in the next section on learning.

2.3.4 Evaluation of Emergence

The formalization of an emergence test is the Design, Observation, and Surprise test (Ronald et al. 2007). In this test, the designer first describes the local elementary interactions in a language L_1. The observer is aware of the design but describes global behaviors of the running system using a different language L_2. Because the design language L_1 and the observation language L_2 are distinct and the causal link between the elementary interaction in L_1 and the behaviors observed in L_2 is "non-obvious" to the observer, the observer experiences surprise because of the cognitive dissonance between the observer's mental image of the individual's design and his observations of group behaviors. The test has been used to evaluate both the flocking and Braitenberg's Love vehicle as exhibiting emergence.

A demonstration of an implementation of the test in an air combat system appears in Mittal et al. (2013). They specifically developed their environment abstraction to be able to detect emergence that is only at a level above the local interacting agents situated in their environment.

Other research topics related to emergentism and emergent systems are complexity theory, complex adaptive systems, and chaotic systems. An example relationship is that a hallmark of a chaotic system is sensitive dependence upon initial conditions whereby two initial states infinitesimally close can result in very different evolved states later in time. We can see the same divergence in multiagent systems and the performance of those agents when implemented with a behavior-based system (Islam and Murase 2005).

2.3.5 Implications of Multiagent Emergence

In addition to consciousness being an emergent system, humans themselves in social conditions—and in interactions with objects[21]—exhibit emergent behaviors. In particular, research and development is an emergent knowledge process: problem interpretations, deliberations, and actions unfold unpredictably, over time and society, in a uniquely human-driven way (Markus et al. 2002). Human society and organizations have also been shown to repeat emergent patterns throughout history (Read 2002). The question naturally arises: can we better develop artificial intelligence as an emergent property of several agents interacting, along, say, the lines espoused by Minsky (1986)? As noted by Pfeifer and Bongard,

> *Embodied artificial intelligence is closely connected to the philosophy of embodied cognition which postulates that intelligence is not a discrete, centralized property that exists within an agent, but instead is an emergent property of an inherently distributed system that possesses many loosely coupled, system-wide processes.* (Pfeifer and Bongard 2007)

In discussing emergence, it is important to temper the idea that more of the same will lead to consciousness. This is because each computational approach

---

21. Emergent behavior can be quite complex when a user adapts a product to support tasks that designers never intended for the product (Johnson 2016).

is within a supervenience level. What this means, for example, is that ever deeper deep-learning neural networks of the same layer types will not lead to consciousness or a cognitively flexible AS. This is illustrated in behavior-based robotics: when used in conjunction with other components (such as planners, sequencers, coordination mechanisms), the result is more capable AS behaviors (Gat 1998; McGinn et al. 2015).

The examples we have for robotics (Gat 1998) and the considerations of embodied intelligence (Pfeifer and Bongard 2007) underscore that we should explore how emergence occurs in multiagent systems. Emergence also plays a role in other artificial intelligence research areas. In machine learning, ensemble methods are individually trained statistical machine-learning algorithms that, when combined, perform better than the individuals (Optiz and Maclin 1999). These ensemble methods also appear in planning (Helmert et al. 2011), constraint satisfaction (Xu et al. 2011), and in IBM's Watson (Ferrucci et al. 2010), where multiple solvers are combined into what are commonly termed portfolio-based systems. The combined strength of the portfolio enables them to solve problems that are more complex than addressable by a single approach. An interconnection issue these portfolio systems must address is load balancing and tuning of parameters in terms of which algorithm gets compute resources and when. This problem is related to the utility assignment problem discussed earlier.

Another example of supervenience is the collective migration problem: how the destination-setting behavior comes from emergence of leaders in a collective. In a flock, the migratory behavior is often led by a small group (Guttal and Couzin 2010). To develop ASs with emergent properties, we must explore how to achieve strong emergence over the weak emergence that is demonstrated with simple swarming behaviors (Pais 2012).

In closing, we note that emergence can lead to an AS (or systems) that exhibits all three behavioral flexibilities. For example, peer flexibility could arise with a change in multiagent swarm leadership as conditions change (Pais 2012). Task flexibility appears in autonomous robot systems (Gat 1998). Finally, cognitive flexibility arises directly from the structure and behavior of portfolio-based GPSs (Optiz and Maclin 1999; Helmert et al. 2011; Xu et al. 2011).

2.4 Experiential Learning

Learning in humans refers to a relatively permanent change in behavior due to experience and not due to fatigue nor maturation. As we touched on briefly in section 2.2 in our discussion of adaptive cognition, the cognitive process guides acquisition and adaptation of knowledge and understanding. Learning and the development of real-world expertise involves a shift in cognitive processing from an initial blend of analytical cognition (conscious deliberation) and intuitive cognition (unconscious situational pattern recognition) to a blend that emphasizes more intuitive cognition as expertise progresses (Reyna and Lloyd 2006; Reyna et al. 2014). Thus, as expertise develops, human cognition becomes more unconscious. This shift in cognitive processing toward intuitive cognition is driven by the accumulation of situational patterns whose structure and meaning have been implicitly encoded over time as an individual engages with, and navigates, his/her domain of expertise (Klein 1997, 1998, 2008).

In nature, knowledge is created in only one of three ways. Knowledge is created via evolution where species capture knowledge and pass it to their progeny. Knowledge is also created by experience. An agent that learns captures experiences and subsequently uses that knowledge to respond appropriately to a stimulus. In nature, knowledge can also be created by culture: e.g., agents use communication to pass knowledge to each other. Each source of knowledge creation is an order of magnitude faster than the prior one (Domingos 2015). It is conjectured that in the future machines will create most of the new knowledge on Earth:

> [N]ow or in the near future most of the knowledge in the world will be extracted by machine and reside in machines. It's inevitable. An entire industry is building itself around this, and a new academic discipline is emerging. (Zajac 2014)

The encoding of knowledge in an AS enables it to generate *meaning* and *understanding*. The meaning of an object or event would be its interpretation by an agent as a sign denoting some other object or event. For example, the meaning of a traffic jam during a morning commute could be its interpretation by the traveler as a sign denoting that he or she will be late for work (Patterson and Eggleston 2017). Understanding refers to the capability and knowledge, based on meaning making, to assess a given situation to accomplish a task. In an ANN, that meaning can be taken to be the activation of the "neurons," and understanding would be how those activations are used to accomplish some task, like identifying an object. However, artificial meaning making in any kind of machine is thought to require, at a minimum, the grounding of the internal symbol system to objects and events in the outside real world (Harnad 1990).

Machines create knowledge faster even than culture, nature's fastest solution to knowledge generation (LeCun 2016). The development of a means to refine machine-generated knowledge to achieve understanding and improve the ability to accomplish complex tasks is a key remaining challenge.

ASs create knowledge through ML where the *learning* portion of ML consists of two components: *representation* and *search* (Russel and Norvig 2010). Both representation and search are system design considerations left to the designer to define. Representation has embedded in it a fixed *model* and a fixed approach to *search*. These have as consequence a fixed representation. For example, the designer could choose an ANN as the model (LeCun et al. 2015), where the designer encodes *knowledge* using that model's representation language, for example, the values of the interconnection weights, the connections themselves, and the mechanism to present the stimuli and feed those stimuli through the network. The designer uses *search* to encode the knowledge via learning the weights that enable the system to do a specific task more proficiently over time. For example, in an ANN, the designer can use Gradient Descent by way of the backpropagation training algorithm (Werbos 1974) or a stochastic search, such as Generalized Simulated Annealing (GSA) (Fletcher 2016) to update the weights in a regression or classification task. Once the model is selected and the learning has occurred (or is allowed to continue learning during use), the learned knowledge can be used to perform the task for which training was originally provided.

ML can be parsed in a fashion similar to nature's knowledge acquisition alternatives: evolutionarily, experientially, or culturally. In an *evolutionary* approach, machine knowledge can be embedded into the machine, based on what the designer knows when creating the machine. In this case, the human designer provides the machine with all the wired-in knowledge that the designer thinks those machines need to function acceptably. This off-line pre-programmed approach has dominated ML research and fielding (Winston 1992; Vapnik 1998; Russel and Norvig 2010). With an *experiential* approach, machines capture knowledge by having been programmed to learn during use. A recent spectacular success in this type of (reinforcement) learning is the Google Deep Mind AlphaGo system that defeated a human, world champion Go player (Silver et al. 2016). This on-line approach relies on machines that interact with their environment, using those interactions and experiences to change their stored knowledge. Lastly, with a *cultural* approach, machines can provide knowledge to each other or from human teammates if they are designed to accept those sources of new knowledge after fielding. Exciting research vectors attempting to reduce the impedance between machines and humans exist and could facilitate human-machine teaming (discussed further in section 3.4), to include capabilities like zero-shot and one-shot learning (Russel and Norvig 2010).

ML is one of the disciplines in the field of AI,[22] and we discuss much of the history of AI later in section 4.3. However, it is worth pointing out here that there is a rich history in the development of AI-based systems, closely driven by ML, and which can be described in a series of waves: describe, classify, and explain as illustrated in figure 2.3 (Launchbury 2016).

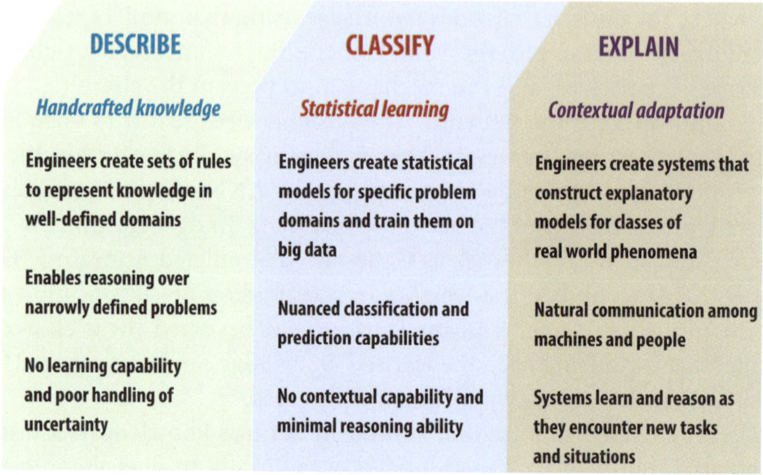

DESCRIBE	CLASSIFY	EXPLAIN
Handcrafted knowledge	*Statistical learning*	*Contextual adaptation*
Engineers create sets of rules to represent knowledge in well-defined domains	Engineers create statistical models for specific problem domains and train them on big data	Engineers create systems that construct explanatory models for classes of real world phenomena
Enables reasoning over narrowly defined problems	Nuanced classification and prediction capabilities	Natural communication among machines and people
No learning capability and poor handling of uncertainty	No contextual capability and minimal reasoning ability	Systems learn and reason as they encounter new tasks and situations

Figure 2.3. Waves of artificial intelligence. (Launchbury 2016)

The first wave of AI ("describe") was characterized by humans handcrafting the knowledge that was put into machines (Winston 1992). Expert systems epitomize this wave; an example is given in figure 2.4. Engineers create sets of rules (based on what a domain expert tells them) to represent knowledge, and these systems function acceptably in well-defined applications for precisely defined problems. Traditionally, they have been costly to build and have exhibited little flexibility to a changing environment and are thus brittle since they are poor at handling situations that were not anticipated in the original design (Winston 1992). To "learn" from new situations, the developers themselves adapt the AI system to performance shortcomings by adjusting or extending the machine's knowledge base; there is no ML, so that performance improvement is limited by the learning/adaptation effort expended by the developer.

---

22. Which can be generally described as the field of computer science concerned with trying to get computers to "behave" like humans, in terms of perception, comprehension, decision making, and action.

Figure 2.4. Expert Systems TurboTax®, an example of the first wave of AI. (Winston 1992)

The second wave of AI ("classify") is taking place now and is dominated by statistical, analogical, evolutionary, and neural-inspired ML approaches (Vapnik 1998; Fogel 2002; Russell and Norvig 2010; Keller et al. 2016). For example, statistical approaches have brought great value in recommender systems in extraction of patterns of life in real-world data (Liao et al. 2015). Analogical approaches include SVMs and have been popular because of their ability to learn using a small number of examples (Vapnik 1998); early spam filters and recommender systems are typical applications. Evolutionary approaches leverage knowledge of how nature adapts to changing environmental pressures evolving life to fill niches of opportunity; they have brought value to many areas including new drug development (Ecemis et al. 2008). Neural-inspired approaches, particularly deep-learning neural networks, have provided the most recent breakthroughs and are currently dominating second wave AI (LeCun et al. 2015); see figure 2.5.

Systems in this second wave of AI are at the core of most of the recent innovations in speech processing, image processing, speech translation, and the sensory processing at the core of autonomous vehicles (LeCun et al. 2015). As impressive as the second wave solutions are on average, individually they are often unreliable or unacceptable, since the knowledge generated by these systems is not readily aligned with human knowledge on related tasks. This is an inherent flaw in designing systems that generate knowledge driven by an objective function (such as minimizing the mean squared error), versus understanding a task from the perspective of the human. For example, image classifiers can be easily fooled into misclassifying an object with the addition of appropriate classes of noise, whereas a human seeing the same noisy image is not fooled (Amodei et al. 2016; Tran 2016; Savage 2016).

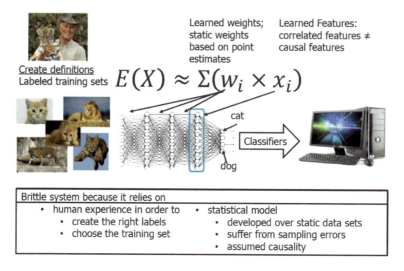

Figure 2.5. Example of second wave of AI: deep learning. (Tran 2016)

Systems in this second wave of AI are at the core of most of the recent innovations in speech processing, image processing, speech translation, and the sensory processing at the core of autonomous vehicles (LeCun et al. 2015). As impressive as the second wave solutions are on average, individually they are often unreliable or unacceptable, since the knowledge generated by these systems is not readily aligned with human knowledge on related tasks. This is an inherent flaw in designing systems that generate knowledge driven by an objective function (such as minimizing the mean squared error), versus understanding a task from the perspective of the human. For example, image classifiers can be easily fooled into misclassifying an object with the addition of appropriate classes of noise, whereas a human seeing the same noisy image is not fooled (Amodei et al. 2016; Tran 2016; Savage 2016).

Many in the AI community believe that these (and other) shortcomings will be overcome with a third wave of AI ("explain") characterized by machines that not only learn over time but that are also more compatible as teammates with humans (Launchbury 2016). They will be able to learn and reason as they encounter new tasks, including tasks with few examples, in situations they were not programmed to address. They will also be able to generate knowledge that includes explanatory models (i.e., be able to explain, to a human, the basis of their reasoning) and more natural forms of human-machine communications for real-world applications; see, for example, figure 2.6. We discuss these issues further in section 3.3 and 3.4.

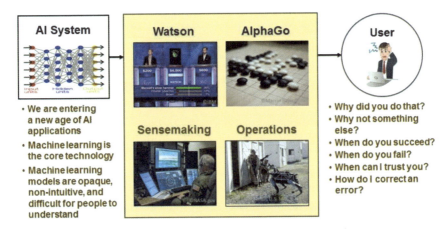

- The current generation of AI systems offers tremendous benefits, but their effectiveness will be limited by the machine's inability to explain its decisions and actions to users.
- Explainable AI will be essential if users are to understand, appropriately trust, and effectively manage this incoming generation of artificially intelligent partners.

Figure 2.6. The need for explainable AI. (Gunning 2016)

The general principles that will guide the third wave have yet to be identified. Some believe it may be a hybrid of the first two, that is, the more traditional semantically based and logically driven "descriptive" wave and the current statistical and neural-inspired "classification" wave (Launchbury 2016). To develop such systems, the issue of meaning making within a common frame and context, by both human and machine, must be solved.

The discussion to this point has focused on single ASs or agents. There is a school of thought that another "wave" could best be based on multiagent systems (MAS) (Balestrini-Robinson et al. 2009) that are coupled together and that interact for the completion of far more complicated tasks (as discussed earlier in section 2.3). The idea is that if appropriately constructed, such systems can overcome the limitations of an individual agent's limited knowledge base and/or capacity for reasoning. MAS also offers some protection from failure because there is more than one agent working to complete a task or set of tasks.

One of the complicating issues that arises, however, is in MAS learning (Panait and Luke 2005). For MAS success, each agent has to account for the agency of the other agents as they form competitive-cooperative teams—self-driving cars that not only have to deal with the relatively fixed road environment but also with the agency of the other vehicles in the environment, both self-driving and manually driven; personal AI assistants that have to interact with the world on behalf of their humans; or swarms of UAVs that must coordinate and negotiate and learn—to be effective as a team. As discussed earlier, the cultural approach to learning, that is, transmitting learned knowledge from one agent to another, suggests that learning in a MAS environment could be major step toward advancing the development of AI. This assumes, of course, not only compatible agent-to-agent communications channels but also common representational frameworks for expressing and transmitting learned knowledge. Another issue concerns the limited "field of view" of any one agent in a MAS: each agent only observes a portion of the environment. The observations may overlap or may be mutually exclusive, and as such, the learning problem is partially observable. The implication is that conventional learning approaches that assume stationarity are no longer valid; the MAS is unable to integrate the partially learned models of the individual agents into a cohesive whole. New approaches to learning that address the stability of representation and the globalization of individually learned knowledge are required if we are to identify the principles that will guide the theory and practice of MAS learning.

It is our belief that the discovery of the principles and practices of the third wave of AI and MAS learning is likely to occur through the convergence of several communities: robotics and cybernetics, hard AI and soft AI, and cognitive psychology and neurosciences. The convergence of these communities may result in a common understanding of how to either develop new ASs or understand the behaviors of existing ones, via frameworks, architectures, computational models, and simulations. This is discussed at greater length in chapter 4.

Chapter 3

Tenets of Trust

Chapter 2 has just described the necessary *properties for proficiency* needed in autonomous systems, but, based on our considerable experience with simpler automated systems, simple proficiency will not be sufficient for system acceptance and wide-scale usage within the DOD without human user trust in these systems (Kirlik 1993; Muir 1994; Dzindolet et al. 2001; Lee and See 2004). Accordingly, this chapter examines *tenets of trust* that will contribute to user acceptance of, and reliance on, future ASs.

Many commercial AS applications take place in relatively benign environments where the system accomplishes well-understood, safe, and repetitive tasks, as illustrated in figure 3.1a. Not only does this set a relatively low bar for performance and therefore a relatively low expected failure rate in the system's mission/tasking, but also, even if the system does fail, the consequences of failure are low (e.g., a pallet may fall off an autonomous forklift). A low failure rate and a low cost of failure naturally lead to a low expected utility (loss) function. In contrast, many DOD missions and tasks (such as illustrated in fig. 3.1b) occur in dynamic, uncertain, complex, and contested environments—setting a relatively high bar for performance of both humans and ASs—and with potential life-and-death consequences. In other words, the expected utility (loss) function is high, and a decision by an AS could lead to high-regret actions. As a consequence, trust will become central to any future employment of an AS by the DOD.

Figure 3.1. (a) Benign commercial environment; (b) adversarial defense environment

We can gain insight into key trust issues in dealing with ASs by recognizing that trust is a fundamental social psychology concept and is a critical factor in a number of areas outside of autonomy, including interpersonal relationships, economic exchanges (among firms, customers, management, and staff), organizational productivity, cross-disciplinary and cross-cultural collaboration, and electronically mediated transactions (Lee and See 2004). Moorman et al. (1993) note that the importance of trust will grow with environmental uncertainty, task flexibility, and team structures as organizations move away from hierarchical structures.

There are many different definitions of trust, but one that is widely cited is

*willingness of a party to be vulnerable to the action of another party based on the expectation that the other will perform a particular action important to the trustor, **irrespective of the ability to monitor or control** that party.* (emphasis added) (Mayer et al. 1995)

The emphasized portion of the quote is particularly relevant given our focus on ASs.

Hoff and Bashir (2015) suggest three different types of trust, namely: dispositional trust, situational trust, and learned trust. Hergeth (2016) showed that trust in automation can sometimes be assessed via eye movements and gaze behavior (e.g., frequency of monitoring the automation). Interestingly, in a brain imaging study, Adolphs (2002) obtained results that suggested judgments about trustworthiness may involve both deliberate and emotional evaluations, which are processed in different brain regions. This dual-processing account of trust judgments seems very similar to the analytical versus intuitive cognition distinction found in the decision-making literature and discussed earlier in section 2.2.

There is also a significant body of research in trust determinants, based on human-human interactions, which have been applied to assessing human-system trust; see, for example, the extensive review by Madhaven and Wiegmann (2007). Lee and See (2004) conclude, for example, that human trust of a system is enhanced when the system appears to have the following positive attributes:

- *Competence*: the system is *competent* in its domain and is used by its human partner as intended

- *Dependability*: the system *operates reliably* (i.e., no surprises) and has a good past history of performance

- *Integrity*: the system is not *failed* or *compromised*

- *Predictability*: the operator has a good *mental model* of system behavior (e.g., state transitions)

- *Timeliness*: the system can provide an *anytime response*, where longer times yield better answers

- *Uncertainty reduction*: the system works to *reduce uncertainty*, not add to it

Clearly, these are all closely related to system proficiency—and the properties discussed in the previous chapter—so that the more proficient a system appears to the human, the greater the level of trust accorded it.

However, there are additional barriers associated with human trust of systems, several of which were identified in the 2016 DSB report on ASs (DSB 2016). These include:

- *Lack of analogical "thinking" by the AS.* When the AS approaches and/or solves a problem in a fashion that is not at all like a human would attack the problem, trust can become an issue because of human concern that the approach may be faulty or unvalidated.

- *Low transparency and traceability in the AS solution.* Lacking an ability to "explain" itself, in terms of assumptions held, data under consideration, reasoning methods used, etc., it is difficult for the AS to justify its solution set and, thus, engender human trust.

- *Lack of self-awareness or environmental awareness by the system.* In the former, this might include AS health and component failure modes, while in the latter, this might include environmental stressors or adversary attacks. Either may unknowingly affect performance and proficiency and overstate the confidence in an AS-based solution made outside of its nominal "operating envelope."

- *Low mutual understanding of common goals.* When a human and AS are working together on a common task, a lack of understanding of the common goals, task constraints, roles, etc., can lead to a lack of trust on the part of the human in terms of the system's anticipated proficiency over the course of task execution.

- *Non-natural communications interfaces.* The lack of conventional bidirectional multichannel communications between human and system (e.g., verbal/semantic, verbal/tonal, facial expressions, body language, etc.) not only reduces communications data rates but also reduces the opportunity to convey nuances associated with operations by well-practiced and trusting human-only teams.

- *Lack of applicable training and exercises.* Lack of common training and practice together reduces the opportunities for the human to better understand the system's capabilities and limitations, as well as how it goes about "problem solving" and, thus, opportunities for understanding a system's "trust envelope"—that is, where it can be trusted and where it cannot.

Expanding on the last item, figure 3.2 illustrates a very simple model of a partitioned trust-reliability space, based on the work by Kelly et al (2003). Along the x-axis is a measure of system proficiency, in this case reliability; along the y-axis is the human's trust in that system. The upper-right quadrant shows appropriately high trust by the human of a high-reliability system; conversely, the lower-left quadrant shows appropriately low trust in a low-reliability system. The two other quadrants show cases of inappropriate trust by the human: the upper left designating a situation of overtrusting a system and the lower right a situation of undertrusting a system. Both of these situations are to be avoided, since the former can lead to catastrophic errors and the latter to underuse of the system (Parasuraman and Riley 1997).

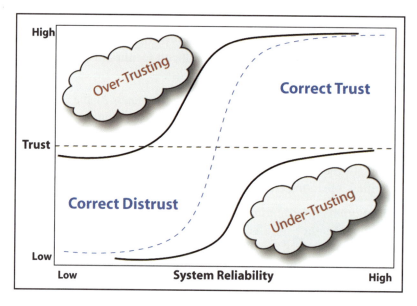

Figure 3.2. Trust-reliability space and quadrants of appropriate and inappropriate trust

Inappropriate trust in systems has historically led to many operational errors and some outright catastrophes that could have been avoided (Mosier et al. 1998). For example, two different DC-10 mishaps were a result of overtrust. In one, a DC-10 landed at Kennedy Airport, touching down about halfway down the runway and about 50 knots over target speed. A faulty auto-throttle was probably responsible. The flight crew, who apparently were not monitoring the airspeed, never detected the overspeed condition (Ciavarelli 1997). In a more serious incident, a DC-10 crashed into Mount Erebus in Antarctica. The accident was primarily due to incorrect navigation data that was inserted into a ground-based computer and then loaded into the onboard aircraft navigation system by the flight crew. The inertial navigation system, erroneously programmed, flew dutifully into the mountain (Ciavarelli 1997). Other examples are articulated by Weyer (2006) and are representative of the problems faced with the introduction of automation in the commercial cockpit in the 1990s. More recently, the automotive industry experienced its first death due to overtrust and misuse of an automotive autopilot system in a Tesla Model S (Yadron and Tynan 2016). The driver was reportedly watching a movie and relying solely on the autopilot system to safely deliver him to his final destination. The driver was killed in an otherwise avoidable accident when the system failed to identify a tractor trailer blocking its path; the National Transportation Safety Board (NTSB) blamed the truck driver for failing to yield the right of way but also blamed the Tesla driver for overtrusting the system, citing

> the car driver's inattention due to overreliance on vehicle automation, which resulted in the car driver's lack of reaction to the presence of the truck. Contributing to the car driver's overreliance on the vehicle automation was its operational design, which permitted his prolonged disengagement from the driving task and his use of the automation in ways inconsistent with guidance and warnings from the manufacturer. (NTSB 2017)

Casner et al. (2016) provide additional examples of "driver error" and overtrust induced by different levels of automation/autonomy in self-driving automobiles. The automotive industry is now learning what the commercial aviation industry has already learned. We can expect this trend to accelerate as more semi-autonomous and autonomous systems are fielded, in both the commercial and military arenas.

Certainly, the current focus is on autonomous automobiles. Many questions are being raised, including whether the onboard ASs are sufficiently proficient to make the "right" decisions all the time[1] and what the "value system" behind those decisions is, especially when it must select a course of action when all outcomes are less than ideal (Kirkpatrick 2015)? And how do humans "take over" if they need to, especially given the issues we have had with aircraft automation failures, misapprehensions of the situation, and failures of control transfer? Some of these issues are already being addressed via extensive simulation testing: Waymo simulates, on a daily basis, up to 25,000 virtual cars driving 8 million miles (Cerf 2018); and by "fleet learning," as each Tesla "learns" something new about the environment, all other Teslas acquire the same knowledge through big-data processing and networking

---

1. And if they do not, who is legally to blame? *See* Greenblatt (2016).

(Frommer 2016)—and certainly by extensive accident investigations conducted by NTSB and others, as we noted earlier.

We expect to see similar trends as weapon systems incorporate more autonomous capabilities. Right now, in the infancy of development of these systems, articles are being published with controversial headlines. The Association of Computing Machinery (ACM), in its flagship publication, *Communications of the ACM*, had these five article titles in recent years:

- "Potential and Peril" (Underwood 2017)
- "Toward a Ban on Lethal Autonomous Weapons" (Wallach 2017)
- "Can We Trust Autonomous Weapons?" (Kirkpatrick 2016)
- "The Case for Banning Killer Robots" (Goose and Arkin 2015)
- "The Dangers of Military Robots" (Arquilla 2015)

And, from one of the fathers of AI, Stewart Russell, in *Scientific American*:

- "Should We Fear Supersmart Robots?" (Russell 2016)

To deal with these concerns, many have begun proposing a variety of guidelines, for both design and operation. Horne (2016) recognizes the centrality of trust and proposes that we develop systems with the following three characteristics: ability, integrity, and benevolence. *Ability* refers to competence and proficiency, aspects of which we discussed in the previous chapter. *Integrity* refers to the values held by the system as well as the transparency of that system (i.e., a trusted AS should have no hidden values[2]). As noted by Dietterich and Horvitz (2015), integrity of ASs can be compromised by design or programming errors in the software or by the unanticipated impact of cyberattacks by adversaries. Conventional software design approaches of adding on security *after* the functional design is completed may fail us completely in situations where part of the design is completed by training the software (as in deep ANN learning algorithms) on patterns or situations it needs to recognize:

> Malicious inputs specially crafted by an adversary can "poison" a machine learning algorithm during its training period, or dupe it after it has been trained. (Klarreich 2016)

This area of "adversarial machine learning" has only recently gained the attention it deserves, especially given the "black box" nature of deep and highly trained ANNs (McDaniel et al. 2016; Papernot et al. 2016). Kott et al.

---

2. Unlike the HAL 9000 sentient computer in Arthur C. Clarke's *2001: Space Odyssey*.

(2015) point out that the integrity of these systems can also be compromised via conventional cyberattacks on AS members of the C2 network.

Finally, benevolence refers to a belief that the AS wants to do no harm to the human teammate, a concept that combines intent and values. Dietterich and Horvitz (2015) note that AS formation of intent, or understanding of its human teammate's intent, is key to "intelligent behavior," an aspect of general autonomous behavior we noted in section 1.2 earlier. And not fully understanding a machine's intent can be troublesome, which Russell (2016) made clear in an article quoting Wiener:

> If we use, to achieve our purposes, a mechanical agency with whose operation we cannot efficiently interfere . . . we had better be quite sure that the purpose put into the machine is the purpose which we really desire. (Russell 2016)

The values held by an AS are also beginning to receive attention by the research community, with Kaplan (2017) calling for "programmatic notions of basic ethics to guide actions in unanticipated circumstances, Etzioni and Etzioni (2016) exploring ways of designing systems to obey our laws and values using "AI guardians," and Scharre (2016) reporting on the operational risk of employing these weapon systems and the potential of exacerbating a given situation. More recently, Kuipers (2018) points to the need for programming in social norms, including morality and ethics, perhaps directly via techniques like case-based reasoning. Finally, the DOD Directive 3000.09 (2012) provides broad guidance, which "establishes DoD policy and assigns responsibilities for the development and use of autonomous and semi-autonomous functions in weapons systems." All three components of ability, integrity, and benevolence contribute to the trust of these systems.

Lastly, several recent autonomy studies have also made suggestions on how to ensure appropriate trust in these systems (e.g., DSB 2012, 2016). For those studies reviewed in appendix A, we have summarized, in table A.2, desirable AS behavioral objectives for effective and trusting human-system teaming; these focus primarily on SA (of the environment, of any teammates, and of self), good human-systems integration design practices, effective communications across teammates, and peer flexibility.

Trust is critical to effective human-system teaming, and it will have a significant impact on the development, proliferation, and use of future ASs. Human-system trust is not a resolved problem. Trust is not absolute; it can change over time, is different for each human, and will depend on the proficiency and transparency of a given AS. Trust is also situationally dependent, and it should be taken into consideration when balancing the tradeoffs of the cost and benefit of designing for trust. It is our goal that the Air Force not relearn the importance of trust in dealing with automation and its impact to fielding new AS capabilities. So what can be done to enable trust? It is our position that, in designing ASs, trust considerations need to be addressed from the beginning of the design process, concentrating on four areas, which we label *tenets of trust*:

- *Cognitive congruence and transparency.* At a high level, the AS operates congruently with the way humans parse the problem, so that the system approaches and resolves a problem in a manner analogous to the way a proficient human does. Whether or not this is achievable, there should be some means for transparency or traceability in the system's solution, so that the human can understand the rationale for a given system decision or action.

- *Situation awareness.* Employing sensory and reasoning mechanisms in a manner that supports situation awareness of both the system's internal health and component status and of the system's external environment, including the ambient situation, friendly teammates, adversarial actors, etc. Using this awareness for anticipating proficiency increments/decrements within a nominal system's "operating envelope" to support confidence estimates of future decisions and actions.

- *Human-systems integration.* Ensuring good human-systems interaction design to provide natural (to the human) interfaces that support high-bandwidth communications if needed, subtleties in qualifications of those communications, and ranges of queries/interactions to support not only tactical task performance but also more operational issues dealing with goal management and role allocation (in teams).

- *Human-system teaming and training.* Adapting or morphing human-system team training programs and curricula to account for the special capabilities (and associated limitations) of humans teaming with ASs. Conducting extensive training so that the team members can develop mutual mental models of each other, for nominal and compromised behavior, across a range of missions, threats, environments, and users.

We now address each of these areas in the remainder of this chapter.

3.1 Cognitive Congruence and Transparency

In this section, we discuss two important factors contributing to trust in human-system teaming, namely cognitive congruence and cognitive transparency. *Cognitive congruence* refers to the degree to which the AS and human possess correspondence in their underlying cognitive representations and processes. It is an extension of the notion of cognitive (or psychological) consistency *within* an individual and the problems that occur when that individual has to deal with cognitive dissonance (Festinger 1957). A lack of cognitive congruence can influence joint meaning making (i.e., sense-making) and subsequent trust by the human of the AS's understanding of the situation, and acting on it, in a fashion similar to what the human would do in a similar situation. *Cognitive transparency* refers to the degree to which the reasoning and actions taken by the AS are intelligible and obvious to the human, even if there is little cognitive congruence. Transparency provides the human with a means of following and validating the AS's assessment and reasoning "audit trail" and will also contribute to trust, even if the AS approach to the problem differs from the human approach. We begin with the topic of congruence, followed by transparency.

3.1.1 Cognitive Congruence

The degree of congruence between the AS and human cognitive representations can affect joint meaning making. This is important because, in many applications, both the AS and human may need to come to a common understanding of the situation, that is, make the same meaning about the same objects, events, or situations presented them, for the team to be successful. Brooks, founder of iRobot, presents a number of compelling instances of potential human-machine misunderstandings likely to occur between self-driving cars and human drivers/pedestrians when confronted with the same traffic situation, because of a lack of congruence between human and machine cognitive representations:

> If a semiautonomous car is not playing by the unwritten rules, bystanders will probably blame the person using the car. But they won't have that choice if the car is fully autonomous. So in that case, they will blame the car. (Brooks 2017)

Understanding meaning making is therefore critical if we are to provide for correspondence.

Many studies have shown that human cognition strives to make meaning of objects, events, and situations in the world (Klein 1998; Patterson 2012). Meaning making can be conceptualized as *sign interpretation*, which is called semiosis (Hoopes 1991; Peirce 1960). The meaning of an object, event, or situation lies in its interpretation by an individual as a *sign* denoting some other (determining) object, event, or situation. In a sense, meaning making via sign interpretation involves relationships (Bains 2006), namely, the relation between the sign and its denoted object or event. Humans are very adept at meaning making via sign interpretation (Patterson and Eggleston 2017).

Meaning making in the real world can involve linguistic sign systems. In such a system, for example, human language can be grounded to an AS's internal representation of the physical world, and to perception and action, by the use of collaborative techniques (Chai et al. 2016). However, meaning making in the real world also involves nonlinguistic sign systems. Collectively, environmental sights, sounds, smells, and so forth can serve as situational patterns (signs) whose recognition can denote some outcome. For example, the meaning of a traffic jam during a morning commute would be—in its interpretation as a sign—denoting that the person will be late for work.

Meaning making can also be conceptualized as *frames* (Minsky 1975). Frames are remembered data structures representing stereotypical situations adapted to a given instance of reality. Higher levels of a frame represent context (things that are always true about a given situation), and lower levels represent terminals instantiated by specific data. Concepts such as scripts and plans are frame-based (Schank and Abelson 1977). During meaning making, frames may help define relevant data, while data may drive changes to existing frames (Klein et al. 2006).

Sign interpretation and frames can be seen to represent complementary approaches to meaning making. Assigning meaning to an object, event, or situation by an individual in terms of a *sign* denoting some outcome would depend on context or frame. Alternatively, the context or frame can help determine which objects, events, or situations are interpreted as signs and what those signs denote. In the previous example, the meaning of a traffic jam *in the context or frame of a morning commute* would be its interpretation as a sign denoting lateness for work. If the frame were different (e.g., en route to dinner or a movie), the sign would still denote lateness, but it would be lateness for a different event, which would be contingent upon the given frame.

For the AS to work effectively with the human, the actions of the AS and human need to be generated within a *common frame* (i.e., recognize the same context at any given moment). This is what we mean by congruent cognitive representations between the AS and human. A common frame would enable several ways for the AS and human to interact:

- The AS and human could interpret the same objects, events, and situations as the same signs denoting the same outcome.

- The AS and human could interpret different but correlated objects, events, or situations as the same sign denoting the same outcome.

- The AS and human could interpret different (and unrelated) objects, events, or situations as different signs (still in the same frame) denoting different outcomes, a conflicted situation that would then have to be resolved.

In general, the AS and human may need to make the same meaning of common or related objects, events, or situations; to do so, the reasoning and actions of the AS and human must be generated within the same frame or context—they must have cognitive congruence. The challenge here is to define at the appropriate level of abstraction, and with the appropriate elements, exactly what is frame or context and how it could be common to both AS and human.

3.1.2 Cognitive Transparency

Cognitive transparency can also affect trust in the use of automated systems and in working with ASs. Trust can be enabled by designing the AS so that its reasoning—from perception to goal generation to action selection—is accomplished by understanding how the system works (transparency) and the ability to trace any decision it makes (traceability). When transparency does not exist, humans can not only fail to understand the AS's reasoning behind an assessment or decision, but they also may attribute capabilities to the AS that it does not have.[3] This happened early on with Weizenbaum's ELIZA program in the 1960s that simulated a psychoanalyst, using a very shallow natural language processing program based on simple pattern matching and substitution, but that fooled many a "patient" interacting with the program (Weizenbaum 1966). This has currently raised issues in the community about developing anthropomorphic robots that appear and behave like humans: they may facilitate human-machine interaction (Duffy 2003), but they may also hinder transparency and understanding of their limitations (Zlotowski et al. 2015). Walsh (2016) has gone so far as to propose a "Turing Red Flag Law":

> An autonomous system should be designed so that it is unlikely to be mistaken for anything besides an autonomous system, and should identify itself at the start of any interaction with another agent. (Walsh 2016)

Were this guidance to be followed by designers of future systems, it remains to be seen how this will balance out against the human tendency to anthropomorphize animals and nonliving agents, in general.

Which brings us back to transparency. Current research supports the idea that the more humans accurately understand the AS's decision-making process, the more humans will trust them (Wang et al. 2015; Lyons et al. 2016a;

---

3. Which brings us squarely into the "overtrusting" region of the trust space sketched in the previous section.

Knight 2017b). This view is also supported in recent studies related to the Auto Ground Collision Avoidance System (Auto-GCAS) used in the Air Force F-16 (Lyons et al. 2016b). Transparency has been effectively implemented by, for example, hand-crafted explanations as described by Dzindolet et al. (2003). Support for traceability also occurs for ethical reasons, as described by Microsoft CEO Satya Nadella in his *10 Rules for Ethical AI* (Reese 2016). DARPA also places emphasis in this area through its E*x*plainable *A*rtificial *I*ntelligence (XAI) program, a program that concentrates on explainable models, interfaces, and the psychology of explanation (Gunning 2016).

Some authors have argued that designing for transparency and traceability for improved human-autonomy trust could result in nonoptimal results. Norvig makes this observation as it relates to the success of probabilistic models trained with statistical methods for language modeling (Norvig 2016). Similarly, LeCun recently said that for the most part it does not matter: "How important is the interpretability of your taxi driver?"[4] (LeCun 2016). LeCun's statement suggests that interpretability of the machine may not be important, at least for certain applications. However, there are many applications where decision explanation, understanding, and an ability to correct the decision-making process—and not optimality—are the primary motivating factors.

In the early adoption of any AS, trust will be a key issue to get people to use the system. Building systems with transparency will likely improve early adoption. Care must be taken so that systems do not appear to be too human-like (too much transparency), otherwise users may overestimate the system's capabilities. As research in this area grows, there will be instances where significant improvements in trust will occur that also result in optimal solutions, making them even more valuable to the warfighter.

3.2 Situation Awareness

Situation awareness is a term that has been the subject of considerable research and attention over the past 30 years (Endsley 2015c). It is a topic that has contributed to several domains, including the aviation community. But what is it? In its basic form, SA is being aware of your surroundings and understanding the impact of those surroundings. Often images of awkward (and often deadly) situations are tied to poor SA. For example, birdwatching in a prime location only to *not* realize the bird you are watching for is watching you (fig. 3.3a). Other, more serious, examples include a cat hunting a bird,

---

4. Until, of course, he takes you to the wrong local airport when you're in a rush to catch your flight.

but the bird is a bald eagle (fig. 3.3b). The former is a possible missed opportunity while the latter likely a fatal error.

Figure 3.3. (a) A case where a lack of situation awareness has comical consequences. (b) A case where a lack of situation awareness has potential fatal consequences.

SA has roots in psychology, but there is a difference of opinion within the psychology community centered on two dominant frameworks: the perceptual-cycle framework (PCF) and the Human Information Processing Framework (HIPF).

The PCF is based on the work in Neisser (1976), which also has roots in applied perception (Gibson 1947) and fundamental areas of perception (Gibson 1966). Neisser's PCF model contains three elements: schema, exploration, and object. The interaction of those is represented in figure 3.4. The *schema* is the internal human process that directs the exploration of the human with the environment. *Exploration* is the use of the senses (e.g., sight, sound, touch, etc.) to gather representative samples of objects the human interacts with. *Object* is the thing that is explored, and what the senses gather about that object is used to update the schema. Significantly, PCF places emphasis on environment interaction in a closed-loop control perspective. It does provide less detail within the schema, which represents internal human processes. However, the work by Neisser (1976) does state that there is no attempt to deny the existence of an internal cognitive process and that it would be highly structured.

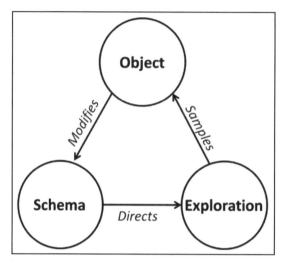

Figure 3.4. Example perceptual-cycle model. (Redrawn from [Neisser 1976].)

The HIPF was popularized by Neisser's cognitive psychology work (Neisser 1967), which is inspired by the use of the computer as an analogy to how humans process information. HIPF-based approaches use symbolic representations for human information processing and then manipulate that representation (Lachman et al. 1979); an example model is presented in figure 3.5.[5] In this model construct, the symbolic components one would manipulate are perception, attention, working memory, and long-term memory.

The most popular HIPF-based SA model is Endsley's 1995 model, shown in figure 3.6, which forms the basis for the material in *Autonomous Horizons*, volume I (Endsley 2015c). This model provides a theory of SA for a single human in a dynamic environment, such as an air traffic controller or a pilot. In this model, SA is defined as (Endsley 1995a):

> *the perception of the elements in the environment within a volume of time and space, the comprehension of their meaning, and the projection of their status in the near future.*

Here, one should consider that *volume of time* and space could be logical and is not necessarily restricted to being physical. It is also important to note that Endsley defines SA relative to dynamic decision making, as is shown in the figure, via the closed loop activity of performing actions on the environment, and then perceiving the consequences of those actions, as illustrated in

---

5. Note that this is a special case of the situated agent model construct of section 2.1.

figure 3.5 and as described earlier in section 3.1 on situated agency (Endsley 2015c).[6]

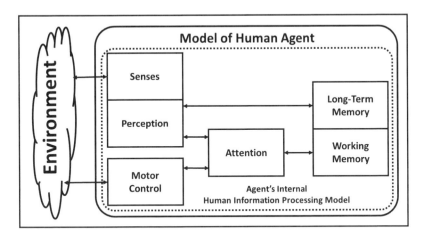

Figure 3.5. Example Human Information Processing Model that demonstrates the interaction of a human with the environment. (Lachman et al. 1979).

An important takeaway is that SA is relative to a specific task and takes into account goals and objectives. Also shown in figure 3.6 are a number of external "moderators" that affect SA, including system capabilities and automation, operator stress and workload, interface design, and so forth, as well as several internal processing functions, including information processing, long-term memory, skill levels, and more.

As noted by Endsley (Endsley 2015c), many systems have not been designed to provide operators with adequate SA regarding a system's internal status. This has led to a poor understanding—on the part of the human—of a system's intent, current behavior, and potential future actions and, as noted at the beginning of this chapter, lead to disastrous consequences. This issue is

---

6. A 2015 *Journal of Cognitive Engineering and Decision Making* special issue is dedicated to the Endsley SA model. The diversity of material in support of and in disagreement with the model demonstrates the existence of the disagreement between the PCF and HIPF approaches. This special issue is bookended by two papers by Endsley. The opening paper (Endsley 2015b) reviews and responds to criticisms of the SA model, while the closing paper (Endsley 2015a) provides closing remarks and final comments.

compounded by increased system complexity,[7] poor interface design, and a lack of human-machine training.

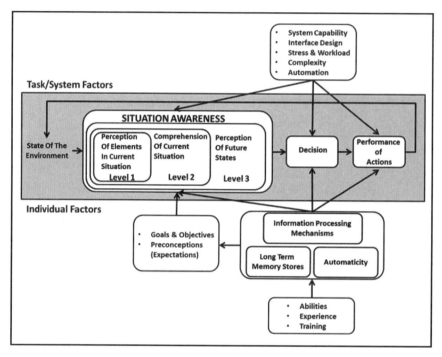

Figure 3.6. Model of situation awareness in dynamic decision making. (Endsley 1995a)

Significant research (Endsley 1987, 1995b) has been conducted on what factors drive "good" SA (that is, awareness that corresponds with the actual situation and the likely fashion in which it will unfold over time). For example, traditional approaches to automation can lead to "out-of-the-loop" errors where the human has low situational awareness of the mission (Endsley 2015c), because of the human's lack of vigilance and complacency at his/her mission or task. Changes to information feedback also have an impact on out-of-the-loop problems, which highlights the importance of balancing tightly coupled versus loosely coupled human-autonomy interaction. Finally, there sometimes is a need to embed the human "in the loop" so he/she can take an

---

7. Especially as AI systems become more complex and widespread.

active vice passive role. When the human is out of the loop, he/she tends to be slow to detect, diagnose, and fix problems.[8]

We hypothesize that any AS, like any independent human agent, will also need good SA to: (a) proficiently perform complex tasks in dynamic uncertain environments in an autonomous fashion and (b) be perceived as trustworthy by any human agents engaged in the same task. We illustrate this notion in figure 3.7, which shows, in the upper left quadrant, a human and AS teaming to accomplish the same mission while operating in a common environment. In the lower right quadrant, we show that, for the human to be successful, he/she needs to have adequate SA of the environment, the mission, his/her own status, and, critically, awareness of the status of system he/she is teaming with. Likewise, the AS needs adequate awareness of the environment, the mission, its own status, and, ideally, awareness of the status of the human it is teaming with. In effect, SA needs are mirrored between human and AS. In an ideal situation, we would have human and system SA of the environment and mission the same: the human's SA of self the same as the system's SA of the human; and the human's SA of the system the same as the system's SA of self. This is exceedingly unlikely in all but the most trivial of situations. The key design issue, of course, is how much "SA correspondence" is needed for both proficient human-system team performance and human-system trust. Certainly, the system must have some level of self-awareness (e.g., understanding of its health or ability to complete a task) and it must have some awareness of the environment the human and the system are operating in. Otherwise, the human will not trust the system and likely not use it.

In considering SA of the self, whether human or machine, we naturally touch on metacognition, that is, thinking about one's own thinking, and consciousness, two topics briefly touched on earlier in section 2.2. As noted there, we have some qualitative understanding of what these terms mean from a human standpoint, based on our experience in engaging in metacognition and in being conscious agents. What these terms mean in terms of developing and operating ASs is, however, quite another matter. But we realize some sort of self-assessment function is needed for metacognitive behavior, and some sort of consciousness function (called "artificial consciousness" in the literature, which may be a misnomer) is needed for self-awareness.[9] We

---

8. As we recounted earlier with the introduction of automation in the commercial aviation cockpit and will no doubt see with the introduction of more "self-driving" cars.

9. SA, of both the environment and the self, has been traditionally defined as a Type 2 mental process (recall discussion in section 2.2) involving conscious awareness (Endsley 1995b). Its measurement frequently relies on an individual's working memory and verbalization (e.g., Durso et al. 2007; Endsley 1995a). Accordingly, conceptualizations and measurements of

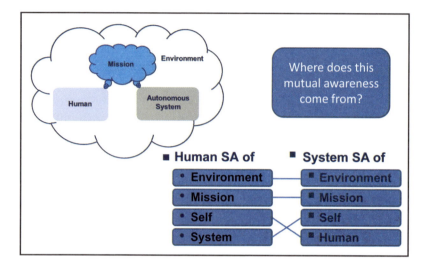

Figure 3.7. SA is critical to human-autonomy teaming. (Endsley 2015c)

have described, albeit briefly, in sections 2.2 and 2.3 how artificial consciousness might arise, via multilayer architectures and "emergentism," respectively.
Here we describe a third, promulgated by Tononi over the last decade (Tononi
et al. 2016; Oizumi et al. 2014). He calls it Integrated Information Theory
(IIT), which has five essential properties (Koch and Tononi 2017):

- *Every experience exists intrinsically (for the subject of that experience, not for an external observer).*

- *Each experience is structured (it is composed of parts and the relations among them).*

- *It is integrated (it cannot be subdivided into independent components).*

- *It is definite (it has borders including some content and excluding others).*

- *It is specific (every experience is the way it is, and thereby different from trillions of possible others).*

In essence, IIT starts with the characteristics of consciousness and then *derives*
mathematically "the requirements that must be satisfied by any physical substrate [system] for it to support consciousness" (Koch and Tononi 2017). IIT
turns on its head the conventional approach of building an AS that is sufficiently complex and behaviorally adept and hoping that consciousness—or

---

SA may not be reflecting Type 1 intuitive cognition and, thus, may be capturing only a portion
of an individual's SA; this is an open area of research.

some partial form of it—emerges; rather IIT starts with the consciousness phenomenon and specifies—in principle—the system that needs to be built that will provide that consciousness. The operative phrase here is *in principle* since IIT is still a theory, but its continuing development may provide us with an alternative to "building and hoping on the outcome" of some measure of consciousness and self-situation awareness in future ASs.[10]

Returning to the issue of *mutual human-system* SA, figure 3.8 provides a more detailed view, in which the three stages of SA are illustrated (perception, comprehension, and projection) and some of the components for each stage are indicated (Endsley 2015c). The essential message is that most of the components for both human and AS are the same, except for the reciprocal components indicated by the highlighted boxes: the human's need for SA on the status of the system, like the system's need for status of the human; and the human need for understanding the impact his or her tasks will have on those of the AS, and conversely the AS need regarding its own task activity impact on the human.

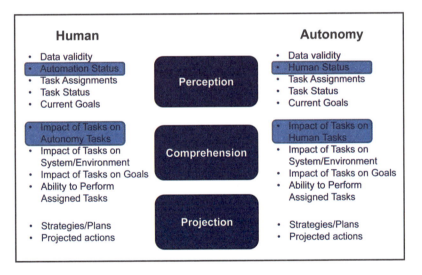

Figure 3.8. SA stages for both humans and AS and their components. (Recreated from [Endsley 2015c].)

---

10. In fact, IIT predicts that "conventional digital computers running software will experience nothing like the movie we see and hear inside our heads" (Koch and Tononi 2017).

The main takeaway here, of course, is that if we expect the system to be as proficient and trustworthy as another human teammate, we need to provide it with the same span and depth of information needed by the human to maintain adequate levels of SA. But there's a secondary message here as well: if either agent (human or machine) needs to have some understanding of the other agent, it necessarily must have some representation or *model* of the other agent. But if the other agent must likewise have a model of its teammate, then the first agent needs a *model of the model*. And so on.[11] This infinite regression of modeling needs for SA related to a teammate is illustrated in figure 3.9, in which Alice has a model of the rabbit having a model of Alice having a model of the rabbit, and so forth; any practical implementation will likely stop at a single depth of models.

Figure 3.9. Infinite regression of two teammates' internal models of one another

Shared SA, as described in (Endsley 2015c), which also appears as Team SA (TSA) in (Endsley 1995a),

> *is fundamental to supporting coordinated actions across multiple parties who are involved in achieving the same goal and who have inter-related functions such as those that occur with flexible autonomy.* (Endsley 2015c)

This is an important topic in human-autonomy teams. Some approaches to TSA consider it as the intersection of each team member's individual SA (see, for example, Endsley 1995a, fig. 2). A key to establishing and maintaining TSA in human-human teams is communication between the team members (Endsley 1995a; Salas 1995). Demir and colleagues demonstrated this experi-

---

11. And the problem quickly grows as the number of teammates multiplies, because of the combinatorics.

mentally with human-autonomy teams and concludes that information push (anticipating a teammate's needs) is more important than information pull (teammates requesting information) in human-human teams (Demir et al. 2017). Interestingly, they also found that human-autonomy teams had less pushing and pulling than all-human teams, clearly pointing to a lack of adequate communication. Work by Salas (1995) also considers a team situation assessment process that incorporates information processing functions, team processes, preexisting knowledge, predispositions, task interdependence, and team characteristics.

Others working in team SA suggest there is a need to integrate individual SA into cooperative activity by way of mutual awareness (Shu 2005). Although the work of Demir et al. (2016) can form the foundations for understanding human-autonomy TSA, there exists a need for additional research emphasis in human-autonomy teams in both the experimental and theoretical domains so that adequate models can be created and used by an AS to improve TSA. Other areas to be explored concern the method and frequency of interaction between the human and the AS: when initiating communications, the AS must know when the interaction is too much or too little, so as not to overburden the human teammate while still supplying him or her with adequate information to maintain adequate levels of SA.

The discussion of these key issues of human-systems integration and teaming is continued in the next two sections.

3.3 Human-Systems Integration

3.3.1 General Considerations

HSI is "a robust process by which to design and develop systems that effectively and affordably integrate human capabilities and limitations" (DHPI 2008). HSI covers nine specific domains: personnel, training, human factors engineering (HFE), manpower, occupational health, environment, safety, habitability, and human survivability.[12] Of these, the first three are particularly relevant to engendering human trust in AS operation. For example, the personnel's dispositional trust and other individual differences and skills can influence the extent to which the human relies on autonomy when operating a system (recall fig. 3.2). Training is important to ensure the human under-

---

12. By incorporating the study of multiple domains, HSI helps assess the complex relationship between humans and system components (automated and otherwise) being operated, supervised, and maintained. See also (Booher 2003).

stands the capabilities and limitations of autonomous components, to better maintain an appropriate level of trust, as system capabilities vary over situations; we discuss training at greater length in section 3.4. Finally, HFE is especially critical in that it is the mechanism for *how* the other tenets of trust are enabled. Informing the human's trust with cognitive congruence/system transparency (section 3.1) and SA (section 3.2) is only possible by providing the appropriate *display* of information to the human operator/teammate and enabling the human to make appropriate *control* inputs into the system, both of which are key design objectives in HFE. The remainder of this section will focus on HFE and its impact on system trust.

Much of HFE is centered on the design of the user interface (UI), the set of the displays and controls with which a human user interacts with a device or a system. For simpler mechanical interfaces associated with a single platform, the term human-machine interface is often used (Nelson-Miller 2016). For more complex systems where the interface is more computationally driven, the term HCI is often used (Farooq and Grudin 2016); we will use this latter term because of the complexity and computational intensity of current and projected ASs.

The usability of a system's HCI is paramount to engendering an appropriate level of human trust. Moreover, the human's reliance on an AS can be influenced by interface design. For instance, the human may have an appropriate suspicion of the AS's decision accuracy and want to intervene, but cumbersome interfaces can impede the human's efficiency, resulting in increased workload or the human electing not to inform or redirect the AS (resulting in potential safety implications). The degree to which HCIs are intuitive also has implications with respect to other HSI domains. Interfaces that are well designed with respect to HFE principles will reduce manpower, personnel, and training requirements (Dray 1995). Moreover, intuitive usable interfaces can lower error rates and improve efficiency that, in turn, can reduce task execution time (Hardman 2008).

However, HFE also reflects "under the hood aspects" of a system's design, such as the type of "cognitive aiding" it might be providing (see below), its reliability, its cognitive congruence (as discussed in section 3.1), and other less visible design aspects of a system. The HFE of both the surface aspects and the deeper constructs of an AS can therefore drive usability, comprehensibility, and ultimately trust in AS operations. But HFE considerations must also account for the human's capabilities and limitations as well. Consider the following example: A system's autonomous component presents its mathematical calculations on a display in the workstation with the intention of making the AS more transparent (as recommended in section 3.1) and enhancing the human's SA of the AS (as recommended in section 3.2). But this can instead have negative impacts. First, the additional information may add clutter to displays, making the human's retrieval of critical information more difficult and workload intensive. Also, the computational information may impose new demands on personnel selection and training for the human to be able to understand how the calculations relate to the AS's decision making and system state. Clearly, the HFE aspects need to be taken into consideration in AS design.

Most basically, good HFE design should help support ease of interaction between humans and ASs. For example, better "etiquette" often equates to better performance, causing a more seamless interaction (Parasuraman et al. 2004; Dorneich et al. 2012). This occurs, for example, when an AS avoids interrupting its human teammate during a high workload situation or cues the human that it is about to interrupt—activities that, surprisingly, can improve performance *independent of the actual reliability* of the system. To an extent, anthropomorphism can also improve human-AS interaction, since people often trust agents endowed with more human-like features (Waytz et al. 2014). However, as we discussed in section 3.1, anthropomorphism can also induce overtrust, as demonstrated early on by Weizenbaum's 1960s-era program ELIZA, which emulated a psychotherapist with a shallow program that merely parroted and transformed the human patient's statements (Weizenbaum 1966) and more recently by a security piggybacking agent successfully disguising itself as a food delivery robot (Booth et al. 2017).

HSI has been conceptualized over the years with the use of taxonomies. One of the more successful taxonomies of HSI interaction, based on suggestions by Sheridan and Verplank (1978) and Riley (1989), comes from Parasuraman et al. (2000); but see also Kaber and Endsley (2003). In Parasuraman et al.'s taxonomy, types and levels of automation both influence the degree of automation. Types of automation represent various aids that can assist certain cognitive functions, a scheme based on a multistage framework of human information processing. Four types of automation were proposed, with each linked to a different stage of information processing: information acquisition (filtering of external information, helping selective attention), information analysis (integrating information, assisting perception and working memory), decision and action selection (deciding action, based on the information analysis stage), and action implementation (implementing action, based on the decision stage). Within each automation type, the level of automation can range from low (full manual) to high (full automation). An extension of this approach has been proposed by Miller and Parasuraman (2003) to bring in task dependence (in addition to dependence on information processing stage dependence) and a more direct ability to "delegate" tasks to automation systems that can support those tasks.[13]

We discussed metacognition and consciousness earlier in section 2.2, in the context of a "dual process" theory of mind described by Evans and Stanovich (2013): a Type 1 intuitive/unconscious system and a Type 2 rational/conscious system.[14] Recently, Patterson (2017) considered the implications of this dual process model on Parasuraman's automation taxonomy (Parasuraman 2000), noting that Type 1 intuitive cognition may be closely related to unconscious situational pattern recognition, whereas Type 2 rational or analytic cognition may be focused more on deliberation associated with decision making and action implementation. As a result, Parasuraman's taxonomy may be too simplistic, since intuitive cognition needs to be included in the scheme. When one does so, it becomes apparent that the characteristics of intuitive cognition match up well with automation aids focusing on information acquisition and information analysis. Here, intuitive cognition could be encouraged whenever the AS fosters a quick grasp of the meaningful gist of information based on experience or perceptual cues, without placing demands on the human's working memory or requiring precise analysis. This type of cognition may lend itself well to automation developed at lower levels of the taxonomy, where the computer can organize and integrate inputs using meaningful perceptual cues, especially in light of the recent advances in machine-based pattern recognition. Conversely, the characteristics of analytical cognition seem to fit in well with decision-making and action-implementation automation types. Here, analytical cognition should be encouraged whenever the AS requires the human to read, remember information via working memory, or engage in rule-based reasoning, hypothetical thinking, deliberation, or precise analysis. This type of cognition may lend itself well to automation developed at higher levels of the taxonomy where the AS could offer various decision alternatives or the AS could execute an action under varying constraint and interact with the human using text or other symbols or rules. Figure 3.10 summarizes how Parasuraman's taxonomy might be modified by these considerations.

---

13. In contrast, a simpler approach has been taken by the Society of Automotive Engineers (SAE), in which a six-level framework is used to define automobile automation (https://www .sae.org/misc/pdfs/automated_driving.pdf), where zero denotes full manual, five denotes fully autonomous, and levels in between progressively aid the driver in driving. This is a start at clarifying what the industry is offering the consumer but has a way to go in terms of the subtler aspects of aiding, automation, and autonomy (Shladover 2016).

14. Although, as we pointed out earlier, Evans and Stanovich (2013) are quick to point out that there may not actually exist any systems per se, but merely rather types of processing.

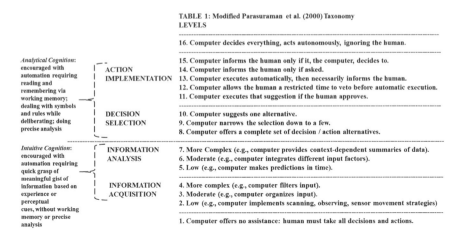

TABLE 1: Modified Parasuraman et al. (2000) Taxonomy

LEVELS		
		16. Computer decides everything, acts autonomously, ignoring the human.
Analytical Cognition: encouraged with automation requiring reading and remembering via working memory; dealing with symbols and rules while deliberating; doing precise analysis	ACTION IMPLEMENTATION	15. Computer informs the human only if it, the computer, decides to. 14. Computer informs the human only if asked. 13. Computer executes automatically, then necessarily informs the human. 12. Computer allows the human a restricted time to veto before automatic execution. 11. Computer executes that suggestion if the human approves.
	DECISION SELECTION	10. Computer suggests one alternative. 9. Computer narrows the selection down to a few. 8. Computer offers a complete set of decision / action alternatives.
Intuitive Cognition: encouraged with automation requiring quick grasp of meaningful gist of information based on experience or perceptual cues, without working memory or precise analysis	INFORMATION ANALYSIS	7. More Complex (e.g., computer provides context-dependent summaries of data). 6. Moderate (e.g., computer integrates different input factors). 5. Low (e.g., computer makes predictions in time).
	INFORMATION ACQUISITION	4. More complex (e.g., computer filters input). 3. Moderate (e.g., computer organizes input). 2. Low (e.g., computer implements scanning, observing, sensor movement strategies)
		1. Computer offers no assistance: human must take all decisions and actions.

Figure 3.10. Modified automation taxonomy. (Parasuraman et al. 2000, including intuitive cognition. Figure taken from Patterson et al. 2017.)

There are a number of implications with the introduction of the concept of intuitive cognition to the design of human-ASs. For example, consider mental workload, which has been linked to working memory (Parasuraman et al. 2008). Because working memory seems to play little or no role in intuitive cognition (Patterson and Eggleston 2017), performance of tasks that rely mostly on intuitive cognition should contribute less to workload, and such performance should remain relatively unaffected if workload increases due to other factors. However, because working memory plays a large role in analytical cognition, performance of tasks that rely mostly on analytical cognition should contribute significantly to workload, and such performance should become degraded if workload increases.

3.3.2 HFE Design Approaches for Human-AS Teaming Interfaces

For human-AS teaming to be a success, an HFE design process needs to be executed to ensure that any associated HCIs provide coordination and collaboration support functions. Woods and Hollnagel (2006) have described these as:

- *Observability* into the assumptions, objectives, processing, roles, and actions of all team members to serve as a basis of a shared mental model

- *Directability* via mechanisms by which the human can interact with the autonomous components to set limits, provide updates, set priorities, and guide processing

- Ability for teammates to *attract each other's attention or processing* to critical information and required actions

- *Shifting perspectives* whereby each team member's involvement in joint problem solving ensures multiple courses of actions and a variety of viewpoints are considered for system operations

These support functions are aligned with characteristics of successful multiagent teams working together, sharing responsibility for ensuring successful system operation and capitalizing on their joint knowledge and skills so that system operation is improved with teamwork and better able to react to novel situations. HCI design needs to support the development of mutual trust between team members (humans and ASs) by providing mechanisms such that each can suggest, prioritize, remind, critique, and caution each other during task performance, with the result being optimized human-system team operation. This includes ensuring that the communications do not negatively interrupt task completion (e.g., not burden the human with a suggestion that is not critical to system operation when workload is high or the human is tending to a critical process).

To design HCIs that better support human-AS collaboration and coordination, it is first important to understand the respective capabilities and limitations of each team member. For the skills and backgrounds of humans, both experienced subject-matter-experts (SME) as well as those not as experienced but slated for the application environment need to be examined. SMEs can identify challenging environmental situations and plausible off-nominal, unplanned events to inform HCI design requirements, in addition to training and evaluation protocols. In contrast, understanding candidate human members will help recognize contemporary skills that should be considered in HCI design (e.g., cell phone touch-based gestures for display manipulation). To support this examination, there are a variety of useful user and task analysis instruments (Hackos and Redish 1998), including cognitive task analysis techniques to identify cognitively demanding tasks in the application environment and the cognitive skills required to perform tasks (Klein 1989; Militello 1997; Nehme 2006). These approaches also identify features and events in the application environment pertinent to developing a cognitive domain ontology that represents the knowledge that an AS can use to categorize situations, develop hypotheses, and plan and recommend alternative courses of action (Atahary 2015). For instance, for a mission planning application involving multiple unmanned vehicle types, a task analysis can identify that the AS needs to be able to determine which vehicle is most likely to find a specified target, versus which would utilize the least fuel reaching the search area or reach the search area in the least amount of time (Hansen 2016).

The affordances and constraints of each AS component also need to be understood. There are a variety of taxonomies that have been used to describe the features, compositions, and computational approaches of intelligent agents, in addition to their ability to interact with external sources. However, most of these taxonomies reflect the total system in the context of its application. Instead, a taxonomy designed to guide human-AS HCI design can help evaluate the relevant tradeoffs in terms of the AS's agility (ability to respond in an effective manner to new inputs within a short time), directability (ability for an external party [e.g., the human] to influence the AS's operation), observability (level to which the exact state of the AS can be determined), and transparency (ability to provide information about why and/or how the AS is in its current state) (Hooper 2015). These descriptors augment the AS principles described earlier in section 1.3: task flexibility, peer flexibility, and cognitive flexibility. Armed with these characterizations of the AS, the designer is better equipped to determine how HCIs should be designed. Additionally, the ability of the AS to report the degree to which it is functioning within its competency boundaries (i.e., providing a self-assessment of its confidence in any recommendation or action) (Hutchins 2015; Guerlain 1999; McGuirl 2016) would help calibrate the human's trust of the AS.

Typically, the next step in developing a new HCI is to employ a user-centered (Norman 1986) design process where the end-user influences how the design takes shape. However, the Air Force's emphasis on human-AS teaming for future decision making and system operations (USAF AFFOC 2015) complicates the notion of a "user." Instead, an alternative *use-centered design* is recommended that focuses on the application's mission or goal rather than distinct user properties (Flach 1995). With this perspective, attention can be better devoted toward the integration of the human and AS team members in joint problem solving and task completion, rather than concentrating on the distinctions between human and AS team members. Application-focused use cases, scenario vignettes, and task analyses can help guide HCI designs that provide the needed mechanisms for human-AS teams to work together and adapt to meet dynamic requirements of future complex Air Force missions.

For identifying detailed HCI specifications, there are excellent sources that can provide guidance (Wickens 2004; Federal Aviation Administration, n.d.). The principles of Ecological Interface Design should also be applied to better support direct perception (Vicente 1990) and direct manipulation (Schneiderman 1983) concepts intended to lessen the human team member's cognitive demands (Kilgore 2014). This approach also helps maximize stimulus-response compatibility and increases the predictability of the system's response. To help ensure the usability of designed HCIs, the heuristics suggested by Nielsen and Mack (1994) should be consulted. Six of the heuristics address human errors when interacting with HCIs: steps to prevent errors (e.g., macros for frequent tasks, use of consistent platform conventions, and help documentation) and recover from errors (undo functionality and detailed error messages). Two deal with information presentation: employing a minimalist design by ensuring only necessary information is presented (with supplementary information easily retrievable), and offering option choices and examples to minimize the human's memory load.

The final two heuristics are particularly relevant to human-AS teaming interfaces: feedback and communications. Feedback of system status is critical. The example provided by Nielsen and Mack (1994) is to present a progress indicator to communicate the degree to which a task is completed. For engendering a human's trust in an AS teammate, the HCI design needs to also provide feedback on the AS's purpose, process, and performance with respect to task completion (Lee 2004; Hoff 2015). Feedback is also needed on the current progress of respective human and AS agents in jointly performed tasks, including who is doing which tasks or which of multiple steps of the same task. Finally, the "real-world agreement" or communications heuristic is particularly challenging to address for systems centered on human-AS teaming (Nielsen and Mack 1994). The heuristic recommends that the system's communication should be in the user's language rather than computer-oriented terms and that the information communicated should be consistent and not mislead the user's mental model. This requirement is difficult to address for future systems, given the increasing capabilities of ASs along with the limited progress toward achieving natural language processing/dialog. To date, dialog systems (both speech- and chat-based) for human-AS teaming are rigid and limited in the activities, team processes, and tasks they support. Current approaches fail to enable the fluid and timely coordination interactions required for dynamic task management whereby the human-AS team can adapt on a moment-by-moment basis to changes in the application environment. Thus, besides addressing the language used, significant articulation work is needed before the human and AS team members can perform task management activities "aimed at functionally decomposing a task, negotiating goals, identifying dependencies, and divvying up who will do what and when" (Rothwell 2017).

The Air Force's Future Operating Concept (USAF AFFOC 2015) centers on future systems supporting mixed-initiative interaction (subsuming adaptive and adaptable autonomy [Calhoun 2016]) such that the complementary strengths of both human and autonomous team members can be harnessed.

To this end, AFRL has engaged in a long-term program focused on human-AS teaming in the C2 of multiple UAVs. More recently, a system was designed to implement a "playbook" delegation type architecture (Parasuraman 2005; Miller and Par 2007) extended to enable more seamless transition between control levels (from manual to fully autonomous). At one extreme the human can manually control a specific unmanned vehicle (UXV).[15] At another level, the operator makes numerous inputs to specify all the details of a "play" that defines the tasking that one or more UXVs autonomously perform once the play is initiated. At the other extreme, the human can quickly task UXVs by specifying only two essential details (play type and location) and then a "C2" AS determines all the other tasking details. For example, if the human operator calls a play to achieve air surveillance on a building, the C2 AS recommends which UXV to use (based on sensor payload, estimated time en route, fuel use, environmental conditions, etc.), a cooperative control algorithm—another limited-capability AS—provides the shortest route to get to the building (taking into account no-fly zones, etc.), and the C2 AS monitors the play's ongoing status (e.g., alerting if the vehicle will not arrive at the building on time) (Draper 2017). The HCIs also support the human communicating any other play detail, and this additional information informs the AS on how to optimize the recommended plan for the play. For example, the human may have information that the AS does not, like the target size and current visibility. With these HCIs, the human operator can, at any time, tailor the role of the AS depending on the task, vehicle, mission event, or the human's trust in the AS (or a unique combination of these dimensions). More detailed descriptions of the HCIs and the rationale for their design are available in appendix D (Calhoun 2017).

There are still many *challenges* in developing effective HCIs as systems grow in complexity and exhibit more autonomous behaviors. There exists, for example, a clear need for advances in display design to effectively fuse, for the operator, information coming from a variety of sources (e.g., multiple ASs), without simply multiplying the number of display interfaces; likewise, there's a need to limit the number of controls or interaction modalities when the number of AS entities increases (Calhoun et al. 2016). This becomes increasingly important as the roles of multi-UXV missions continue to develop and mature. Likewise, as data increases with these multirole missions, there exists a need for better collection, analysis, and display back to the operator, with more integrated displays (and controls). We discuss these and other AS-related HCI design issues further in section 5.3.

---

15. X is used to denote Air, Ground, Sea, or Undersea vehicle (A, G, S, U).

Regarding trust calibration, one critical element to realize effective human-AS teaming is bidirectional communication that approximates the characteristics of human teams. Additionally, the human must have efficient access to any information desired, and the AS needs to be programmed to reason and respond to the human's inputs based on the current context of the task or application environment. Explicit cues for supporting a shared mental model with respect to the current system state and task allocation can be gleaned by the ongoing human-AS dialog, the displayed information, and the human's interactions with the HCIs. To engender joint problem solving, "what-if" querying needs to be supported (Calhoun 2016). In this manner, the human-AS team can pool resources to explore predicted consequences of system state changes or events in the application environment. This capability can also be used as a tool for the human-AS team when negotiating the next actions for system operation.

In terms of real-time operations, the HCIs need to support ongoing communications of intent, i.e., mechanisms by which the human and the AS can frequently share their respective assumptions and bases for proposed tasks including recommendations on how to accomplish system goals. The means by which an expressed intent/goal is translated into actionable tasks depends on the established role and/or authority of each human and AS team member. The terms "on the loop" versus "in the loop" have been used to differentiate systems where the human supervises the AS's operation, versus being directly involved in task completion. For the vision of human-AS teaming, this dichotomy is less useful. Rather, the HCIs should support a wide spectrum of control levels such that, depending on specific task demands, either the human determines how much to be directly involved in task completion or the human-AS team negotiates how the tasks are shared. Thus, the HCIs need to provide functionality by which the human can specify exactly what level of AS support is desired (per task, event, resource, etc.), from absolutely none to having the AS completely responsible.

To capitalize on joint human and AS capabilities, many tasks will involve both human and AS team members. The extent to which the AS is relied upon can be influenced by the human's corresponding trust in the AS, although there are other mediating variables (e.g., workload, task criticality, control modality, etc.). Actually, the more the human is involved in task completion, the higher the human's engagement and awareness of the degree to which the AS's recommendation/action is accurate and appropriate (Endsley 2015c). A human's inappropriate reliance on the AS has been described as automation complacency. However, there are a host of "unintended and unanticipated" ways in which the AS can impact the human's role in system operation (Parasuraman and Manzey 2010). A more detailed examination indicates that the "brittleness" of ASs can influence a human's attention, perceptual, memory, and/or problem solving processes (Smith 2017). For instance, the AS can induce a perceptual framing effect when presenting recommended courses of action to the operator. The bias is that the human is likely to only consider the AS's honed list of candidate options and fail to consider other options despite the known difficulties implementing an AS that is resilient to real-world uncertainties. Further research is needed to explore how to minimize the potential for hypothesis/plan fixation by the human, such as determining the ideal content and timing of the AS's support provided via the human-AS interfaces (Smith 2017).

The work station's HCIs, including the dialog functionality, need to support a flexible, mixed-initiative approach (Bradshaw 2008) by minimizing the cognitive demands and workload involved in establishing who is doing what and when. Ideally, for an effective human-AS team, some division of tasks is prespecified (each member starts with assigned tasks, e.g., certain task types, environmental events, resources, and/or geographical areas). During system operation the HCIs should provide the means by which the operator and the AS can suggest task reassignments, recommend changes in how tasks are completed, as well as propose entirely new tasks/actions. To help inform task sharing/coordination as well as calibrate trust levels, each team member's state, actions, and rationale need to be available/communicated (via displays or dialog). One example is the AS warning the human operator that its recommendation is based on stale data. Any ongoing feedback serves as the means by which trust between team members is established and reinforced and can be verified. In fact, without effective human-AS teaming interfaces, informed by HFE, the human's trust in the AS will be degraded, increasing the likelihood that subsequent support from the AS will be discounted and the human's workload will increase.

3.4 Human-System Teaming and Training

We now discuss human-system teaming and training and the importance of accounting for special issues associated with teaming between humans and ASs. We begin by extending the discussion of the previous section on human-system integration, focusing on design for effective human-AS teaming. We then explore general issues associated with teams and team training and discuss how they may be extended to human-system team training, in particular. We conclude with recommendations for research into how system designs and training protocols might be best evaluated.

3.4.1 Design for Effective Human-AS Teaming

In addition to the basic HFE procedures described in the previous section, there are also design considerations unique to human-AS teaming. Fortunately, a body of relevant research as well as thoughtful proposed guidelines inform the design and evaluation process. Here, four of note are briefly introduced. The first was published over a decade ago: Klein and coauthors identified 10 challenges for making autonomy a teammate (Klein et al. 2004). We list these below, reordered, aggregated, and with pointers to relevant sections of this report where these are discussed further:

1. Forming and maintaining the basic contract (between teammates; see this section, 3.4)
    a. Engagement in goal negotiation
    b. Controlling the costs of coordinated activity
2. Forming and maintaining adequate models of others' intentions and actions (3.2)
    a. Effective signaling of pertinent aspects of status and intentions
    b. Observing and interpreting signals of status and intentions
3. Maintaining predictability without hobbling adaptivity (2.3)
    a. Maintaining adequate directability
4. Autonomy and planning technologies that are incremental and collaborative (5.3)
5. Attention management (2.2)

A publication by O'Hara and Higgins (2010) includes a tabular summary of general principles for supporting teamwork with machine agents. More recently, design principles for human-AS teamwork have been discussed with an applications focus by including lessons learned from a robotics challenge sponsored by DARPA (Johnson et al. 2014). A publication (Joe et al. 2014) examines in more depth the differences between envisioned human-AS teams and teams where all members are human. New insights are provided as to whether or not design principles will translate well to *both* human and AS team members. Finally, a recent study by the DOD ONA points to the success of "centaur" chess, in which human-AS teaming is most effective when tasking is broken down hierarchically (DOD ONA 2016):

Human strategic guidance combined with the tactical acuity of computer was overwhelming.

If this generalizes, it may provide additional task-specific design guidance for human-AS teaming.

In addition to these studies, the studies reviewed in appendix A have also made recommendations for ensuring effective human-system teaming, which we have summarized in table A.2. As noted earlier in this chapter, these recommendations focus primarily on SA (of the environment, of any teammates, and of self), good HSI design practices, effective communications across teammates, and peer flexibility.

We have also noted, in section 3.3.2 above, four key design issues that enable effective human-AS teaming:

- Effective bidirectional communication that approximates the characteristics of human teams, to support team-member trust

- Some means of communicating task intent, ongoing assumptions for task accomplishment, and general approaches to task completion

- Task sharing agreements on how to share the workload, according to task criticality, trust, workload, and so forth

- Incorporation of a flexible mixed-initiative approach that can suggest task reassignments and recommend changes in how tasks are completed as well as propose entirely new tasks/actions

Workload and cognitive demands under team operation can also be influenced by the modalities utilized for human-AS interfaces. For instance, research has shown that that use of auditory and tactile displays may help avoid overloading the human's visual channel. Likewise, there are multiple input modalities (speech, touch, gesture) that leverage the human's natural communication capabilities to make inputs to the system (Oviatt 1999; Calhoun and Draper 2006). Before applying these to human-AS interfaces, however, candidates need to be evaluated to ensure that they will support quick and accurate inputs. The human's trust in system operation can be undermined if a speech system's recognition of an utterance is poor or a touchscreen's detection of an input is imprecise. Moreover, speech-based control is most ideal when the human's hands and/or eyes are busy or there is limited screen real estate to exercise control (Cohen and Oviatt 1995). Touch input is more useful with smaller reach distances or when inputs are not frequent (Vizer and Sears 2015). In a recent evaluation, mouse and keyboard inputs were found more efficient than both speech and touch input for a seated human making frequent inputs on a monitor (Calhoun et al. 2017a, 2017b). These results suggest that the emphasis should be on determining which input modalities are optimal for specific types of tasks/application environments, rather than spending resources to implement multiple modalities across all tasks. There are also combinations of input modalities that merit further examination. One example is the integration of sketch and speech inputs on a computer-generated map, whereby verbal commands can be efficiently associated with spatial locations indicated with gesture inputs (Chun et al. 2006; Taylor et al. 2015). Any candidate control modality, though, should also be evaluated with an HSI perspective by considering the entire system environment (e.g., auditory alerts may be less useful for task environments with frequent auditory communications).

There are still many *challenges to be addressed* for effective human-AS teaming. A fundamental issue focuses on how tasks are shared and roles allocated and how that is communicated/displayed across multiple human and AS team members. Dynamic hand-off or swapping of subtasks across team members can help balance task loading and has the potential to improve system operation. However, it can also complicate the ability to maintain shared awareness across team members. Besides the coordination involved in role allocation and task assignments, interfaces that support human-autonomy collaboration in other joint problem-solving activities in the work domain are required (Bruni et al. 2007). The implications of co-located team members versus those working in a more distributed system also need to be addressed.

Another challenge already mentioned is the need to improve natural language processing and dialog capabilities, especially with unanticipated dialogue-mediated tasks (Rothwell and Shalin 2017). For example, design instances should be employed where the existing human-AS teaming strategy is perturbed by introducing a problem or manipulating information available to each partner. This will drive requirements for team members to engage in give-and-take communications to determine what task steps should be changed (Rothwell and Shalin 2017). It would be useful to also leverage other domains in developing effective dialog systems such as Clark's theory of Common Ground (1996) and Grice's Conversational Maxims (1975). Incorporation of natural multimodal inputs in dialog may also help communicate intent, such as when a human designates an area on a map with sketch- and gesture-based input concurrent with a human speech utterance "loiter here" (Chun et al. 2006; Taylor et al. 2015) or an AS's reply with speech-generated status information specific to an AS-highlighted map region.

Another challenge reflects the limited focus of most current HCI designs on the next immediate step to be accomplished in task completion. Further research is needed on how interfaces can best support communication of temporal parameters, for example, specifying when each human and AS team member ideally should complete each substep of a task, if not immediately. This could be accomplished, for example, by having temporal details communicated by adjusting symbology on a simple timeline interface, where past, current, and projected time-related information is shown, or via the use of other novel temporal interfaces (Cummings and Brzezinski 2010; Cook et al. 2015). Because there may be instances when the communication link between the human and AS is interrupted or degraded, such timelines could also support querying between human and AS team members to regain a shared SA of the system's past operation.

Exploring human-AS HCI design for more complex applications will likely complicate information portrayal necessary for maintaining shared SA. Typically, a central display serves as a common operating picture showing the status of all resources as well as the current tasks under way (for example, the role of a map display in multi-unmanned vehicle management). Such a display helps ground system-relevant communications between the human and AS and support queries to obtain more detailed information on the status of resources as well as task progress and system status. But such displays can become rapidly cluttered and difficult to understand when the number of entities increases; individual entity status is displayed, for example, using a single glyph to depict entity type, current tasking, fuel state, payload state, and overall status in supporting the mission (Calhoun and Draper 2006); or other mission functions are "layered on," such as showing future mission plans or conducting "what-if" queries between team members. Continued research and development will be required to address the trade-off between an HCI's information density and its rapid comprehensibility.

3.4.2 Training for Effective Human-AS Team Performance

As human-AS teams are brought into operations, it will be necessary to adapt or morph training programs and curriculums to account for the capabilities and limitations unique to the human-AS team. Extensive human-AS team training will be required so that the team members can develop mutual mental models of each other, for nominal and compromised behavior, across a range of missions, adversaries, and environments. This training will be necessary for the team to understand common team objectives, individual roles, and how they co-depend. Effective human-AS teaming will be accomplished by enabling mutual understanding of common and complementary roles and goals, to support ease of interaction between humans and ASs. Evidence appears to suggest that training can help calibrate human-machine trust, as it provides operators with an understanding of the limits of the AS capabilities (Lyons et al. 2016a).

Team performance (among humans) has been a subject of research at least since the 1950s, as tasks became more complex, teams became more commonplace, and errors became more consequential. Salas et al. (2008) summarized some of the main research findings over that time span, beginning with some basic definitions, including

- *Teams* are social entities composed of members with high task interdependency and shared and valued common goals (Dyer 1984).

- *Taskwork* is defined as the components of a team member's performance that do not require interdependent interaction with other team members.

- *Teamwork* is defined as the interdependent components of performance required to effectively coordinate the performance of multiple individuals.

- *Team performance* [arises] as team members engage in managing their individual and team-level taskwork and teamwork processes (Kozlowski and Klein 2000).

A number of research themes are identified by Salas et al. (2008), including the importance of "shared cognition" as a driver of performance, where components of shared cognition include shared mental models, common or "team" SA, and understandable communications, components that have been discussed in earlier sections. One finding of particular importance, however, is simply that "team training promotes teamwork and enhances team performance," a fact not lost on high-performing teams in, for example, professional sports or small-unit military teams. While team training may support practice and improvement in an individual team member's taskwork, the main focus is (or should be) a focus on developing the component competencies associated with teamwork—that is, with coordination across members of the team.

Stevens and Campion (1994) propose a knowledge, skill, and ability (KSA) framework for defining the necessary competencies covering both self-management (individual-focused) and interpersonal (team-focused) KSAs, composed of, for the latter category, conflict-resolution KSAs, collaborative problem-solving KSAs, and communication KSAs. A more recent review of the literature by Salas et al. (2005) identifies the following needed team competencies:

- *Team leadership*, or the ability to direct, motivate, and coordinate other team members and provide them with the necessary KSAs

- *Mutual performance monitoring*, or the ability to assess teammate performance and develop a common understanding of the environment

- *Backup behavior*, or the ability to assess others' needs and provide support and workload balance as needed

- *Adaptability*, or the ability to adjust team strategies or reallocate intrateam resources in response to environmental demands

- *Team orientation*, or the belief in the importance of the team's goals over that of the individual's

With these team-focused component competency definitions, it then becomes a question of how to select or train for them. A meta-analysis of team interventions conducted by McEwan et al. (2017) identifies four classes of interventions aimed at improving team performance:

- *Didactic education in a classroom setting* of "good" team behaviors, such as clear communications or supportive workload balancing.

- *Team training in an interactive workshop format*, with discussions of common goals, specific issues, and so on.

- *Simulation-based training*, involving all members of the team, with experiential enactment of specific scenarios and situations.

- *After-action reviews of actual team performance* and individual behaviors experienced in the "real" team operating environment.

Based on a review of 51 studies that covered 72 unique interventions, McEwan et al. (2017) concluded that while classroom education of good team behaviors was ineffective, the latter three interventions could be used, to varying degrees, to good effect to improve team performance.

If we now consider these results in terms of their applicability to training for effective human-AS team performance, we immediately recognize that, without a well-developed set of natural language processing skills on the part of the AS team members, neither the *workshop format* nor the *after-action reviews* can be used for team training. This leaves us with simulation-based training as our only feasible means for human-AS team training, at least for the near future.

Two distinct simulation-focused communities have contributions to make in this area. The first, simulation-based training in the military, has a long history, going back to the original Link Trainer developed in 1929 to teach pilots basic instrument flying skills (American Society of Mechanical Engineers [ASME] 2000). Over time, these simulators grew more sophisticated, and with the advent of SIMNET in the 1980s (Miller and Thorpe 1995) and the support of the Defense Modeling and Simulation Office in the 1990s,[16] they became networked and much more interactive (National Research Council 2006). This allowed for the development of today's complex real-time simulations of thousands of entities (both actual and computer generated, or "constructive") interacting over large swaths of territory. The focus of these Live-Virtual-Constructive (LVC) simulations[17] has been on training and, specifically, training human operators of conventional systems like aircraft and decision makers in INTEL and operations. The other simulation community, one with a much shorter history, is the AI/AS community, which can be broken down into two subgroups. The AI community, particularly the two-person game-focused community (e.g., Chess, Go) has made good use of learning by having its AI play against itself, in effect, simulating its opponent with a clone of itself.[18] In this way, thousands or millions of games can be played in a short period of time without having to endure the slow response times of a human opponent. Likewise, the AS community, particularly those developing self-driving cars, simulate a driving environment filled with other simulated pedestrians and cars and embed a simulation of the AS to learn and/or be "tested" (see, for example, Cerf 2018). Again, by running these "constructive" simulations (that is, ones that are void of "slow thinking" humans) at many times real-time, the AS can experience millions of virtual hours of driving time, something that would take years of actual road time.

If we can bring these two communities together effectively, we will be able to continue the successes we have already had in the traditional military simulation-based training community. One approach is to add simulated and trained ASs into existing LVC simulations; conceptually, this is merely an extension of current efforts with "constructive" agents, where the next generation of agents would be autonomous. Another approach would be to embed simulated humans into existing AS simulations, as those simulations and their agents get developed; again, this is not a great leap from what is being done on the commercial side, when self-driving car simulations embed simulations of thoughtless pedestrians, children playing in the street, and so forth. The former approach would help train humans to deal with introduced ASs at a pace that matches the human's learning rate. The latter would help train the ASs to team

---

16. Now the DOD Modeling and Simulation Coordination Office (DMSCO), https://www.msco.mil/.

17. **Live**: A simulation involving real people operating real systems. **Virtual**: A simulation involving real people operating simulated systems. **Constructive**: A simulation involving simulated people operating simulated systems. http://www.acqnotes.com/Attachments/DoD%20M&S%20Glossary%201%20Oct%2011.pdf.

18. For the game of Go, for example, see Silver et al. (2016, 2017).

with humans at rates that would be much faster than real-time, assuming adequate modeling fidelity of human actions in these situations.[19]

3.4.3 Evaluation of Human-AS Team Performance

All HCI designs and training paradigms should undergo evaluation, ideally using human-in-the-loop testing methods with the actual autonomous technologies integrated in the system. If the latter are not yet mature, it may be possible that the anticipated autonomous capabilities can be simulated or represented with "Wizard of Oz" techniques (Riek 2012). The experimental setting should also include instances in which the AS's reliability is imperfect, as would be the case when the data utilized by the AS is inaccurate/stale or the AS is operating outside its competency boundaries. Likewise, unexpected events should be included as well as off-nominal operations to challenge the human-AS team. Experimental manipulations that enable the examination of how the human and AS handle disagreements or lapses in their shared mental model would also be informative.

Research is also needed to determine how best to evaluate the performance and effectiveness of human-AS teams (Bradshaw et al. 2004; Billman 2000). Examining task performance that reflects the joint contributions of human and AS team members may not provide information at the level of detail needed to indicate, for example, how the HCI can be improved. Another challenge to consider is that of how the data will highlight the relative contributions of each AS member participating in the mission, relative to the human team's involvement. Besides task performance and joint problem-solving ability, definitive methods for assessing the human's trust are also needed. However, the dynamic nature of trust complicates its measurement, and research is still ongoing to better understand the myriad variables that can influence how an appropriate level of trust can be gained and maintained (Hoff 2015). Consideration also needs to be given regarding how the HCIs can be better designed/employed, and/or the training paradigms modified, to support better SA through better mutual mental models and the long-term development of trust during human-AS team operation.

---

19. Modeling of human behavior in military environments is discussed further in Pew and Mavor (1998) and Zacharias et al. (2008).

Chapter 4

Convergence of Communities

We believe the time is right to begin to develop a common framework for describing, developing, and assessing autonomous systems, motivated by the research activities of a number of seemingly disparate communities. In our view, these communities appear to be converging toward very broad common frameworks that describe the problem of autonomous behavior from radically different viewpoints—but that may provide the basis for architectures and functional components that can enable the more rapid development of conceptually well-founded future autonomous systems. To motivate more discussion across these communities, we present here very brief and broad-brushstroke histories of these communities, their achievements, and their implications for framework development.

Figure 4.1 (which repeats fig. 1.7 introduced earlier in section 1.4) attempts to illustrate, at a very high level, how six distinct communities may be converging onto a common understanding of human behavior *and* autonomous system behavior via the development of common computational models of cognition. This, in turn, can provide the basis for developing new autonomous systems and understanding the behaviors of existing ones, via frameworks, architectures, computational models, simulations, and fully instantiated systems. The six communities are:

- At the top, the robotics and cybernetics communities, which have driven a better understanding of machine-based autonomy and human-system integration

- At the bottom, the cognitive psychology and neurosciences communities, which bring us closer to understanding human cognition via concepts, cognitive architectures, models, and simulations

- And in the middle, the AI communities, both "hard" and "soft," which continue to provide us with nontraditional computational approaches to perceptual/cognitive problems that only recently were thought to be unaddressable via machines

In the sections to follow, we provide a brief history of each of these communities, along with observations of what we believe to be future trends toward one or more common frameworks.

We believe that the payoff in striving toward a common autonomous system framework is threefold:

- From a theoretical point of view, it would be particularly appealing to attempt to unify a wide range of research efforts now under way, exploring what constitutes autonomous behaviors, as well as an equally diverse number of efforts attempting to build and test such systems. The unification of some of these efforts under a common "architectural umbrella" would drive both the research and the engineering, and perhaps support faster development along both fronts, with lasting tenets that might eventually serve as the basis of a "science of autonomy."

- From a practical point of view, common architectures can provide insight into "best practices" in terms of explaining observed behaviors and/or supporting the development of new autonomous systems across different domains and for different applications. Common architectures can also make possible the "reuse" of solutions in one domain that have already been developed in another. And, at the very least, they can encourage, through the introduction of common constructs and nomenclature, greater communication between different groups pursuing what, on the surface, may appear to be significantly disjointed aspects of the same or a closely related problem.

- And in terms of the opportunity, we believe that the time is ripe for attempting to develop one or more unifying architectural views, because of the long-term and now rapidly accelerating "convergence" across multiple R&D communities focusing on many different aspects of what may very well be a common problem—that of understanding the behavior of existing autonomous systems (e.g., humans) and the development of new ones, in both the biosciences and the engineering world.

We recognize the pitfalls that can accrue in adopting an architectural viewpoint before the basic science is well understood or the fundamental engineering guidelines are well developed. On the other hand, attempting to develop one or more architectures may serve to drive the research and support the engineering in a more focused fashion, providing for a more general approach to the understanding and development of autonomous systems, in contrast to the "one-off" approach characterizing many of the efforts ongoing in the (different) communities today.

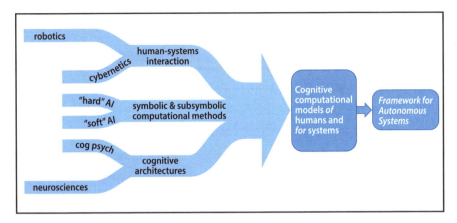

Figure 4.1. Multiple streams of research and development leading to a common framework for autonomous systems

4.1 Robotics and Cybernetics

The **robotics stream**, shown at the top of figure 4.1, has a long history, going back, in the defense world alone, to the late 1800s with remote control torpedoes (wired and wireless), followed in the early 1900s with remote control airplanes (Yuste 2004; Torres 1903) and rockets. This focus on telerobotics (basically, remote control via a human operator in the loop) gave rise to more autonomously operating robots, starting in the 1930s with "open-loop" (preprogrammed) industrial robots performing repetitive industrial tasks in a well-controlled environment. The flexibility of setup and manipulation accuracies improved with the introduction of numerically programmed industrial robots in the 1950s, but the real advances in behavior repertoires came with the gradual introduction of sensors, including imaging, on (or near) the robot. Most of these industrial robots have been stationary and nonlocomoting but have most recently taken on locomotion in certain situations (e.g., Kiva's mobile robots in Amazon warehouse distribution centers [Kim 2016]).

Robotics focused on locomotion got started in the 1940s with Machina Speculatrix, which searched for light and power sources in its environment (Walter 1950). Other simple robots followed, but most were limited by computational power. More autonomous locomoting robots came on the scene in the 1970s, enabled by better sensors and on-board computers that were sufficiently low size weight and power (SWAP). Some focused on using symbolic processing as their "thinking" part (see section 4.3 below), but Brooks in 1990 finally put that dependence to rest with effective mobile robots that did not rely on symbolic processing; rather they depended on the classic "see/think/do" loops (illustrated earlier in chapter 2), where the actual world kept its own model of itself and the robot sensed what it needed in order to locomote and interact with it (Brooks 1990). With this "actionist" approach, we are now seeing an explosion of robotic locomotion applications, particularly in the commercial sector, focusing on ground vehicles (personal, taxi, trucking, etc.). The DOD sector is pushing this forward as well, via such activities as DARPA's Robotics Challenge (Guizzo and Ackerman 2015), but at a considerably slower pace, in the ground, air, sea and undersea, and space domains.

As these systems become more sophisticated in their sensor, computational, and locomotion capabilities and exhibit a greater range of behavioral repertoires, there exists a greater demand for assisting the human operator, teammate, or passenger to better understand the intent of the system (i.e., goals), its understanding of its environment (situation awareness), and its anticipated future actions (plan). This has led to a greater emphasis on human-systems interaction research and development to provide the necessary interfaces and control modes for interacting with these systems (Murphy 2002).

The **cybernetics stream**, also shown at the top of figure 4.1, started in the 1940s, when Wiener defined *cybernetics* "as the scientific study of control and communication in the animal and the machine" or what we would now call "human-system interaction" (1948). It was heavily based on a systems theory framework, incorporating signal processing, closed-loop feedback, and estimation/control theory and treated each element of the "total system" (human or machine) as a functional block with inputs/behaviors/outputs with communications lines between blocks supporting the flow on information throughout the system. From the 1950s through the 1970s, manual control research and modeling dominated the cybernetics branch, focusing on "continuous signal" man-machine problems like bike riding, aircraft piloting, the development of realistic flight simulators, and so forth (Ashby 1956). The focus was largely on a single operator operating with a single system, with continuous control as the dominant mode of operation (Diamantides 1958).

As systems became more complex starting in the 1970s and as more automated/digitized systems came on board (e.g., flight management systems in commercial cockpits, nuclear power plant management systems, etc.), the research focus shifted toward "supervisory control" where loop closures were done by automation and the human operator focused on how the automation was controlling the basic system (Sheridan 1976). In effect, the focus went from the human "in the loop" to "on the loop." In addition, because of increasing system complexity with time and the need for a team of operators, efforts have been increasingly devoted to understanding team interactions and multi-operator supervision of modern systems (Cooke et al. 2000, 2006).

In both the early phase of manual control and the later and current phase of multi-operator supervisory control, a major focus of research has been understanding effective human-system integration,[1] from the basic human factors of effective controls and displays design, to the "shaping" of systems dynamics (e.g., an aircraft's handling qualities), to the development of an understanding of "joint cognitive systems" (Woods and Hollnagel 2006) as the human-system tasks become increasingly cognitive and less manually focused.

There has also been a parallel progression in the development of computational models used to describe the interactions of humans with systems. Early work in the manual control area in the 1950s onward saw the rise of the classical "crossover model," a frequency domain approach to modeling human-system dynamics in relatively simple systems (McRuer et al. 1965); more complex systems were later modeled starting in the 1960s using the "optimal control model," a modern control theory approach to dealing with more complex dynamic systems (National Research Council 1990). More recent efforts in supervisory control modeling that started in the 1980s focused on developing "cognitive models" of the operator(s) whose structures were heavily influenced by systems theory and engineering (National Research Council 1990; Sun et al. 2006). Reviews are provided by the National Research Council (Pew and Mavor 1998; Zacharias et al. 2008). In short, the cybernetics stream gave rise to the development of a series of computational models of human "cognition," when humans are faced with extremely well-defined tasks and situations (e.g., aircraft flight control, supervisory detection of automation anomalies, etc.).

4.2 Cognitive Psychology and the Neurosciences

The **cognitive psychology** stream, shown at the bottom of figure 4.1, focuses on the study of mental processes such as attention, working and long-term memory, perception, problem solving, the development and use of language, metacognition, creativity, and so forth. The definitive text of its time, *Cognitive Psychology* by Neisser, defines cognitive psychology as fundamentally a mental information-processing function occurring in the brain:

The term "cognition" refers to all processes by which the sensory input is transformed, reduced, elaborated, stored, recovered, and used. It is concerned with these processes, even when they operate in the absence of relevant stimulation, as in images and hallucinations. (Neisser 1967)

---

1. As discussed earlier in section 3.3.

Starting in the 1950s and 1960s Broadbent and colleagues (Broadbent 1958; Treisman 1964) began measuring human attention and developing theories of attentional processes: how they worked across multiple sensory modalities (e.g., vision, audition, tactile, etc.), how attention can be driven by external cues (e.g., alarms) or by internally directed deliberation, and how theories of attention can be expanded beyond sensory processing and applied to multi-tasking. Attentional models served to predict human performance under a wide variety of attention-focused tasks, such as display vigilance tasks (e.g., that of a sonar operator).

Models of perception were also being developed at that time, with theories to account for how one or more sensory channels of information are "fused" (in today's terminology) to give rise to a "percept" of an object existing or an event occurring in the world outside of the observer (Russell and Norvig 2010). Concurrent development of models of memory broke down memory into short-term (now labeled working) memory and long-term memory. In recent times, memory systems have been subdivided into declarative memory (composed of episodic and semantic memory, both involving conscious recollection), procedural memory (unconscious memory of invariant, relational knowledge that supports skill development and behavioral dispositions and is acquired and tuned through experience), and several other memory systems (Squire 2004, 2009). In combination, perception and memory models set the basis for understanding procedure-guided actions in the face of event-driven sensation, in effect providing a path parallel to that afforded by a cybernetics viewpoint of human-system interaction. A good summary of many of these efforts was provided by Newell's *Unified Theories of Cognition* (1994).

More recently, there has been a focus on *metacognition*, that is, thinking about thinking, as we discussed earlier in sections 2.2 and 2.3; for example, identifying one's own shortcomings about problem solving and developing "workarounds" to compensate. This type of executive cognition, thinking about "how" to go about solving a problem—more than just diving in and attempting to solve it directly—can lead to significantly improved performance (Swanson 1990); it also may have significant implications for the development of synthetic cognitive systems, for example, in the identification of a system's limits in a given situation and the bringing forth of alternate cognitive strategies to deal with the situation.

Starting in the 1980s and continuing into the present, significant efforts have been devoted to converting many conceptual cognitive models into *computational* cognitive models whose goal is to explain, replicate, and "predict" human behavior under different circumstances and tasking. Card et al. (1986) described their Model Human Processor (MHP), which draws a direct analogy between human processing (with its limited perceptual, cognitive, and motor capabilities) and computer processing (with its finite central processing unit [CPU] speeds and memory limits) and, via appropriate parameterization, provided a means of modeling response times and error rates for a given set of tasks. Around the same time, significant effort was devoted to developing the Adaptive Control of Thought-Rational (ACT-R) cognitive architecture (Anderson et al. 2004). A perceptual module serves to connect outside world events to perceptions, a cognitive module uses procedural and declarative memory to generate desired actions, and a motor module serves to translate those desired actions into actual effects on the outside world (e.g., verbalizations, motor control actions, etc.). Central to its operation is a production rule system (see section 4.3 below) operating on symbolic (i.e., linguistic) knowledge. A very large research community has expanded its application domains over the years, covering basic human performance in attention, memory, natural language, as well as more complex tasks like driving cars and flying UAVs. The Soar cognitive architecture (Laird et al. 1993; Laird 2012), based on Newell's unified theories of cognition work, is also fundamentally a production rule system, with "problem solving" afforded by effectively discovering the path from a "current state" to a desired "goal state." Soar has recently been extended to incorporate nonsymbolic approaches to information processing (Laird and Mohan 2018).

As discussed earlier in section 2.2, human cognition can be considered in terms of two cognitive systems[2]: analytical cognition (conscious deliberation) and intuitive cognition (unconscious situational pattern recognition), the latter of which is likely subserved by procedural memory and developed, in part, via implicit (unconscious) learning of statistical regularities in the environment (Patterson and Eggleston 2017). It turns out that the simulation of implicit learning and intuitive cognition cannot be implemented in a straightforward way in the ACT-R framework, which may need to be expanded to include abstract representations of statistical regularities (Kennedy and Patterson 2012). A recent model of human cognition involving both the analytical and intuitive components comes from Helie and Sun (2010).

A wide variety of other cognitive architectures, models, and simulations have been developed over the years since the 1980s. Extensive surveys are provided by Pew and Mavor (1998), Sun (2003, 2004, 2006, 2007), and Zacharias et al. (2008).

---

2. Clearly an oversimplification, but a useful one for many studies, for example, human decision-making (Tversky and Kahneman 1973, 1974).

The **neurosciences** stream, shown at the bottom of figure 4.1, focuses on the functioning of the nervous system via a number of approaches, including analysis of the neuroanatomy though observation, correlation of deficits in cognition associated with postmortem brain damage, and through the use of modern imaging techniques such as functional magnetic resonance imaging (fMRI) (Kandel, 1992).

In the late 1800s, with the development of a staining procedure by Golgi, Ramon y Cajal determined that the functional unit of the brain is the neuron (illustrated in fig. 4.2),[3] which became known as the neuron doctrine (Kandel 2013). This doctrine was supported by the experiments by Galvani (Galvani 1791), who conducted electrical stimulation of muscles, and by Helmholtz, who later demonstrated that neurons were electrically excitable and that their activity affected the electrical states of neighboring neurons (Kandel 2013). Much later, the introduction of electron microscopic imaging revealed the structure of synapses both on dendrites and on axons and provided a means for understanding how neurons communicate and a basis for a possible model for learning via synaptic efficacy. Neurons communicating via both electrical and chemical means along with experiments demonstrating that the activity of one cell not only impacts the activity of connected cells but also facilitates a more efficacious impact on those subsequent neurons' activity (Hebb 1949).

In the last half of the nineteenth century, there was a rapid sequence of discoveries that dramatically advanced the neurosciences and laid the basis for the concept of functional modules of the brain. The first advance was in 1861 when Broca, a French neurologist, theorized—based upon his work with brain-damaged patients—that functions were located in specific regions of the brain (Broca 1865). He demonstrated that the left frontal lobe was responsible for speech in several patients who understood speech but could not talk. And the fact that some of these patients could sing clearly demonstrated the speech motor apparatus was fine. Postmortem examination of the brain revealed a lesion in the posterior region of the frontal lobe, later named Broca's Area. The next step was taken in 1876 by Wernicke, who described a stroke victim who could speak, but the speech made no sense; in effect, he could execute speech but not in a comprehensible fashion (Wernicke 1908). These works added to the concept of functional localization in the brain, since the Wernicke lesions were in a different location than the Broca lesions, as illustrated in figure 4.3.

---

3. Golgi and Ramon y Cajal later shared the 1906 Nobel Prize for their descriptions and categorizations of neurons. The neuron doctrine is the principle that neurons are the building blocks of the nervous system.

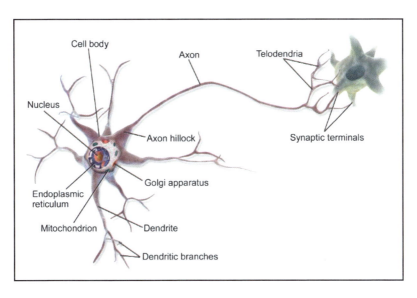

Figure 4.2. Neuron. (From medicalsciencenavigator.com)

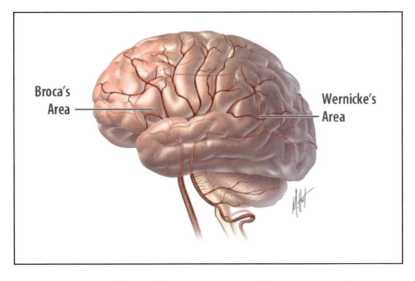

Figure 4.3. Broca's and Wernicke's Areas. (From http://www.strokecenter.org/)

Around 1952 Hodgkin and Huxley conducted experiments at the cellular level (using the neurons of the giant squid) and developed a nonlinear differential equation model of how action potentials in neurons are initiated and propagated (the Hodgkin-Huxley model) (Hodgkin 1952). Barlow later experimented with the horseshoe crab using optical fibers in direct contact with the cornea, delivering light to a single retinal receptor (Barlow 1977). His work revealed the use of lateral inhibition and excitation in neural networks, setting the foundations for postulating the existence of complex interactions among large numbers of neurons and providing a conceptual basis for functional specialization and localization identified in the earlier cited studies.

In the 1950s neurosurgeon Penfield—sometimes called a "neural cartographer"—revolutionized brain surgery and made major discoveries about cognition, memory, and sensation (Penfield 1950). Applying only local anesthesia, he would probe exposed brain tissue, and, guided by the response of the patients, he would decide what parts of the brain could be removed in an attempt to treat epilepsy. During these surgeries, he demonstrated that electrically stimulating human patients produced *specific* memories. For example, a particular stimulation would cause the patient to call out the word "grandma," leading Lettvin to propose around 1969 in his MIT course titled "Biological Foundations for Perception and Knowledge" the idea of a *grandmother cell*, a specific location where all the information about a person's grandmother was located (Barlow 1995; Gross 2002). Support for this notion has since been provided by the activity of a subset of medial temporal lobe (MTL) neurons that fire selectively to "strikingly different" facial images of a specific individual and even their names (Quiroga 2005). It is not much of a leap to speculate that the MTL serves as a repository for people that one "remembers" and recognizes when cued. The nervous system does not represent an object by the activity of a single cell but rather does so by an ensemble, distributed code. This distributed code can involve many neurons and requires complex connectivity.

The notion of *receptive fields* helps bridge the gap between what we know about low-level individual neurons and what appear to be higher level, functionally specific *regions* of the brain. Basically, sensory neurons respond (i.e., change firing rates) to changes of the associated stimulus in the neuron's receptive field, so that, for example, a retinal neuron will change firing rates to changes in levels of brightness of a visual stimulus, if that stimulus is in the receptive field of that cell (Alonso and Chen 2008). Hubel and Wiesel (1962) extended this concept by noting that more complex visual receptive fields of cells at one level could be formed by taking in the outputs of cells at a lower level, with this "bottom-up" signal flow structure replicated all the way down to the lowest level sensory cells—in this case, retinal cells. Hubel and Wiesel also determined that neurons collectively were connected in a columnar fashion (roughly normal to the surface of the visual cortex) with each column tuned to a particular type visual stimulus, as illustrated in figure 4.4. Their discovery of the orientation selectivity of cortical columns led to their Nobel Prize award in 1981.

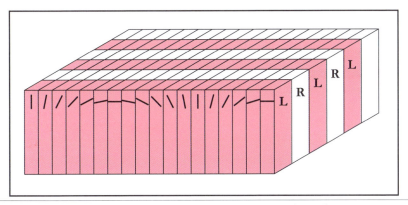

Figure 4.4. Columnar organization of visual receptive fields[4]

There have also been attempts at defining the neural basis for higher functions like consciousness. Most famously Crick and Koch (1990) did groundbreaking work on an approach based on synchronized oscillations of neuron firing. Although there has been considerable speculation about what consciousness is and how it is supported by the physics of the neural substrate (e.g. the mind-brain dichotomy), there are no currently accepted models that provide a definitive explanation of mechanisms of consciousness—or even of its function. This may prove to be the ultimate barrier to developing truly autonomous agents modeled after what we understand to be the basic neural circuits and capabilities of the mammalian brain: *a highly complex and capable agent that is unconscious may never satisfy our trust in its capabilities to understand and act under a wide range of unforeseen situations and assigned tasks.*

As we have attempted to illustrate, cognitive science is not a new field.[5] Many have sought to extract information about the relationship between the brain

4. Kate Fehlhaber, "Hubel and Wiesel and the Neural Basis of Visual Perception." Knowing Neurons, 29 October 2014, https://knowingneurons.com/2014/10/29/hubel-and-wiesel-the-neural-basis-of-visual-perception/.

5. For a review of modern neural science principles, the textbook by Kandel (2013) is regularly revised and up to date. For an introduction to the nervous system, see Hilyard (1993); information about how neurons communicate, Catterall (2012); early perceptual processing, Hilyard et al. (1998); object recognition, Haxby (1991); spatial cognition, Whalley (2013); language, McGettigan and Scott (2012); memory, Montalbano et al. (2013); attention, Posner (2012); executive function, Dong and Potenza (2015); and emotion and social cognition, Schupp et al. (2006).

and the mind by observation and experiments. The concept of functional modules had its original basis in observations that combined neuroanatomy and coarse functionality deduced from examples of brain damage. And, up until recently, the idea that different regions of the central nervous system are specialized for different functions has been the accepted cornerstone of modern brain science. However, an increasing number of neural scientists believe that cognitive functions cannot be understood when considered as separate entities (e.g., memory relies both on attention and perception). Much current work is thus focused on attempting to define neural assemblies that coordinate activity from disparate regions of the cortex (Fuster et al. 2000).

Nonetheless, we believe that the neurosciences can provide us with plausible structures and perhaps even an architecture for understanding cognition, one that can serve as a basis for developing synthetic analogs for agents that can behave in an "autonomous" fashion, working with human teammates across a limited set of tasks and situations. The vision is that these neurally inspired cognitive architectures will converge to the behaviorally inspired architectures developed by the cognitive psychology community discussed above and provide a sufficiently robust and detailed framework for follow-on autonomous agent development.

4.3 Symbolic Logic and Subsymbolic Logic

The **"hard" AI stream** shown in the middle of figure 4.1 is so named because of its strong foundations in symbolic logic and linguistics and the clarity ("hardness") with which inferences or deductions are made on the basis of declarative knowledge (i.e., facts, data) and procedural knowledge (logic and processes) for dealing with that declarative knowledge.[6] The tag also serves to differentiate this branch of AI from the "soft" branch (see below).

One could claim beginnings going back to Euclid's geometrics proofs and Aristotle's formalized logic, but hard AI got its initial beginnings as a "challenge problem" when, in 1950, Turing introduced the dialog-based *Turing Test* (Turing 1950) as a way of asking if machines can think, which effectively reduced the question to one of assessing input/output language-based behaviors—that is, "understanding" a human query and responding appropriately. This naturally attracted researchers with strong logic and linguistics backgrounds as computational capabilities grew and as the underlying hardware and soft-

---

6. For a more in-depth history of the development of hard AI, see "History of Artificial Intelligence" entry in Wikipedia, accessed 9 January 2019, https://en.wikipedia.org/wiki/History _of_artificial_intelligence.

ware demanded just these backgrounds for the development of the underlying logic circuits, operating systems, and programming language compilers. In 1956, AI got its formal start at the Dartmouth Summer Research Project on Artificial Intelligence (McCarthy et al. 1955), with discussions covering computer-based intelligence, natural language processing, neural networks, and other computer-based approaches to "intelligence." In 1959, Newell and Simon introduced *General Problem Solver* as a "universal problem solver" program based on symbolic processing and logic (Newell et al. 1959). It could solve toy problems (or "microworlds" as they were labeled) like the Tower of Hanoi[7] but few "real-world" problems that we would consider of interest today.

However, this groundbreaking work set the stage for much of the AI research conducted in the 1960s and 1970s, which was symbolically/linguistically focused, with formal algorithms used to propagate declarative knowledge and generate conclusions (Fikes and Nillson 1971; Schank and Tesler 1969). Significant optimism followed early successes, but the first "AI winter" (so named because of dramatic reductions in research funding starting around 1975) soon followed after a number of research failures due to combinatorial explosions associated with formal logic approaches, a lack of being able to imbue machines with "common sense" knowledge and reasoning (McCorduck 2004), and limited computational/software resources to execute programs, among other factors. There were also a number of failures at building robots that depended on symbolic processing for functions such as visual perception, route planning, and locomotion, all to be done in real-time (McCorduck 2004; Moravec 1988). This led to a "branching off" of robotics researchers using different paradigms for perception and control, epitomized by Brooks' pioneering work in the 1980s and onward (see earlier section 4.1).

Hard AI found success, however, in the 1970s and 1980s with the development and application of "expert systems" (ES), which explicitly separated declarative and procedural knowledge via the incorporation of a *production rule engine* and a separate *knowledge base* of given and declared (produced) knowledge (Hayes-Roth et al. 1983). Successes were initially demonstrated by Feigenbaum and colleagues in the 1970s with the AI programs DENDRAL and MYCIN for analysis of spectrophotometer readings and diagnosis of blood diseases, respectively (Lindsay et al. 1980; Buchanan and Shortliffe 1984). Later in the 1980s, specialized Lisp-based machines were built for improved programming and processing, and industry invested millions in highly specialized ESs for their operations (Dyer, n.d.). These efforts also

---

7. For more on the background of this puzzle, see the "Tower of Hanoi" entry in Wikipedia, accessed 22 December 2018, https://en.wikipedia.org/wiki/Tower_of_Hanoi.

drove the need for "knowledge engineers" to painstakingly populate an ES with the domain-specific declarative facts and production rules for each application, based on interviews with domain experts, incremental development, and re-interviews, as the knowledge engineer discovers increasingly deeper knowledge constructs/models as the process unfolds (McCorduck 2004; Crevier 1993).

A second "AI winter" followed in the early 1990s, with a significant drop in research funding and industry interest in ESs. There were several factors contributing to this, including:

- The cost of an extensive knowledge engineering effort to deal with "all" possible situations occurring in real-world applications and the need to update the ES as system changes were introduced

- Issues associated with "brittle" production rules that effectively "broke" when provided with combinations of declarative knowledge rule antecedents that were not anticipated in the initial design[8]

- Limitations in dealing with uncertainty and probabilities that characterize dealing with the real world, that is, uncertainties in perception and uncertainties of action outcomes[9]

- A lack of "common sense" knowledge about the world, which restricts generalization by any domain-specific ES

- And, perhaps most critically, the inability of ESs to encapsulate knowledge that was simply not amenable to semantic encoding, like visual or auditory perception, manual control, locomotion, and the like, all critical functions to the robotics and HCI communities

With respect to the "common sense" issue, it is worth noting that a heroic attempt at encapsulating common sense knowledge in a hard AI system[10] has been going on since 1984 under the Cyc project (Lenat 1998; Curtis et al. 2005). The goal is to encode, via knowledge engineering, enough declarative facts in Cyc's knowledge base to give the system the same level of common sense about the world as is held by a human. Currently, the system is estimated to hold over a million common sense assertions. Unfortunately, little is known about the details of the knowledge base, the "engine," or the overall

---

8. The tag of "hard" AI is also associated with the brittleness of ESs, a subset of same.

9. Even though certainty factors introduced by MYCIN (Buchanan et al. 1984), Dempster-Shafer theory (Sentz and Ferson 2002), and fuzzy expert systems (see below) were introduced to overcome these limitations.

10. In this case, one running an engine based on predicate calculus.

performance of the system, and it is the subject of considerable criticism and controversy (see, for example, Davis and Marcus 2015; Domingos 2015).

Although these issues reduced the level of research and broader application of ESs (and traditional hard AI systems in general), a wide variety of other hard AI-based or related technologies grew, starting around the 1990s, including case-based reasoning (Aamodt and Plaza 1994), intelligent tutoring systems (Nwana 1990; Freedman et al. 2000; Nkambou et al. 2010), forward/backward chaining systems (Feigenbaum 1988), constraint-based programming (Mayoh et al. 2013), planning and scheduling systems (Ghallab et al. 2004; Allen and Hendler 1990), and the like, even though the AI tag was (and is) often missing because of past "bubbles" and "winters." In effect, significant hard AI work is still going on, even though the AI tag is often no longer used.

Finally, we note that the cognitive modeling community has benefited from the basic work done in hard AI. We noted earlier how the ACT-R and Soar cognitive models (Polk 2002; Laird 2012) rely on ES-inspired procedural rules and declarative memories. Other cognitive models do as well, including the Belief-Desire-Intention (BDI) family of models (Jennings 1993; Rao and Georgeff 1995; Georgeff et al. 1998), the GOMS (goal, operator, method, selection) model family (John and Kieras 1996), the Executive-Process/Interactive Control (EPIC) model (Kieras et al. 1997), and Connectionist Learning with Adaptive Rule Induction ON-line (CLARION) (Sun 2002; Sun and Zhang 2006).

The **"soft" AI stream**, also shown in the middle of figure 4.1, stands in contrast to the hard AI approach just discussed, with its focus on reasoning in a nonsymbolic fashion and addressing issues of uncertainty. In addition, it is identified with "biologically inspired" approaches separate from those associated with language.

Predating the Dartmouth Conference noted above, McCulloch and Pitts developed computational models of idealized or "artificial" neural networks and described, in 1943, how a Turing Machine[11] might be implemented with such a network, thus providing broad motivation for the eventual development of an AI that was inspired by the brain's basic neural components (McCulloch and Pitts 1943). In 1947, they demonstrated how ANNs might be used to model specific perceptual functions, namely audition and vision, in their groundbreaking paper (Pitts and McCulloch 1947), which was theoretically significant but of limited use in solving practical neural processing problems. Their work did, however, spur a number of efforts through the 1950s, including work by Hebb explaining "associative" or unsupervised learning in

---

11. An abstract model of computation based on symbolic manipulation, invented by Turing in 1936.

neural networks (1949), which led to simulation-based analyses of this type of "Hebbian learning" by Rochester[12] and Holland[13] (Rochester et al. 1956). The seminal McCulloch and Pitts ideas also inspired many other works such as that of modeling frog vision in Wiener's group at MIT (Lettvin et al. 1959), notions for finite automata (Piccinini 2006), and many other efforts that used ANNs and their ability to learn patterns over time to model human perception and cognition.

Rosenblatt's simple perceptron (1958) was also a significant development as it was the first well-formed computational-oriented ANN. The simple perceptron learned by associating a stimulus with a response—an associative model, unlike most of those proceeding it. Significantly, the learning approach with Rosenblatt's two-layer model modified the synaptic weights in a connected network to minimize the difference between the desired output and the achieved output, dividing the error among the weights proportional to their size—this is the *credit assignment problem*. However, there were limitations in this approach. Predicting the correct response from a stimulus was correct only when the responses were correlated. In 1962, Block published two specific findings on Rosenblatt's simple perceptron that were important (Block 1962): (1) the simple perceptron required linearly separable classes if it were to achieve perfect classification (the *perceptron convergence theorem*); and (2) if two classes were linearly separable, Rosenblatt's simple perceptron would find the solution. Despite the progress and the findings by Block, MIT researchers Minsky and Papert (1969) published findings that demonstrated there were several classes of problems *single-layered* perceptrons could not address and then extended their (incorrect) judgment to the multilayer perception (as well as others). These conclusions were likely the cause of a significant reduction of funding for ANN research for many years following (Anderson 1988).

Despite the conclusions published by Minsky and Papert (1969), other significant works in the early 1970s occurred. Anderson (J. Anderson 1972) and Kohonen (1972) developed generalizations of Rosenblatt's perceptron that became known as "linear associative neural networks." Van der Malsburg (1973) demonstrated self-organized learning, which also was the first effort to directly compare computer simulation to physiological data (Anderson 1988). As a result, self-organized learning became an important area in neural networks, which took off in the early 1980s. Additionally, Brodie and colleagues (Brodie et al. 1978) successfully modeled the eye of the Limulus polyphemus

---

12. Designer of the IBM 701 computer.
13. Developer of genetic algorithms, which we discuss later in this section.

(horseshoe crab) as a linear system, a significant success in modeling a neuro-physiological system. However, such a model did not scale to larger problems.

Feed-forward ANNs saw an important breakthrough in updating the internal (or hidden) weights in a network with arbitrary layer depth, using the error backpropagation algorithm ("BackProp"), a recursive solution to the structural credit assignment problem for multilayered networks. As noted by LeCun (1988), Bryson and Ho (1969) had figured out how to do this using LaGrange theory applied to layered control problems. The basic approach was rediscovered by Werbos (1974) and then again by Parker (1985). But Back-Prop did not become mainstream in the ANN community until Rumelhart and colleagues applied it to pattern recognition problems and demonstrated its significant utility (Rumelhart 1986). In 1989, Cybenko demonstrated that ANNs were universal function approximators so long as a nonlinear activation function is used (Cybenko 1989). This is a significant insight validating the utility of ANNs in a broad variety of applications.

Other advances in feed-forward ANNs occurred with breakthroughs in understanding the "internals" of neural activity. The so-called "spiking neuron," originating with work by Hodgkin and Huxley (1952), used the precise firing times associated with neurons to encode information (Paugam-Moisy 2006). Previously, temporal data was not well represented in feed-forward ANNs, but spiking neurons are an active area of research, even with their significant computational challenges.

In 1982, Hopfield published a significant work on neural networks (Hopfield 1982) often credited with the renaissance of research in ANNs (Anderson 1988). Significantly, Hopfield treated the network as a state space of activations for each neuron in the network. The goal was to change the activation pattern to minimize the gradient toward an "attractor" state; Hopfield achieved this with a recurrent network architecture and a form of Hebbian learning, consistent with previous works by others. The significance of Hopfield's approach was that he observed that the energy equation he minimized was analogous to an Ising model used in statistical mechanics models of magnetic spins (Fletcher 2016). This observation had two significant impacts: (1) it provided a way to apply theoretical mechanisms to ANNs; and (2) made ANNs a legitimate field of study once again, opening the door for participation in theory development by the physics community.

Ackley et al. (1985) extended the Hopfield network model by using a stochastic state transition approach instead of the traditional gradient descent approach.[14] The work by Ackley and colleagues turned to simulated annealing (SA) to minimize the error function, an algorithm modeled after the namesake statistical mechanical process. Under certain conditions, SA can be guaranteed to find a global minimum of the objective function and may obtain a lower-error network than the gradient descent form, which only guarantees a local minimum.[15] Fletcher has since extended SA in two areas (Fletcher 2016; Fletcher and Mendenhall 2016); Fast Simulated Annealing (FSA) has been introduced by Szu and Hartley (Szu 1987); and Generalized Simulated Annealing (GSA) from Tsallis and Stariolo (1996) has been proposed for feedforward ANN weight updates. The FSA and GSA algorithms were derived from physical models of annealing, which incorporate both a thermal and a quantum mechanical phenomenon.[16] Both FSA and GSA have been shown to reduce training time required and to find lower-error solutions, compared with gradient descent. Additionally, GSA brings stronger guarantees of converging to a minimum energy state (Dall'Igna et al. 2004).

Kohonen introduced the Self-Organizing (Feature) Map (SOM) in 1981 (Kohonen 1981). Kohonen's SOM (also called the "K-SOM" or the "K-SOFM") is motivated not by the interworking of a cell body and the process by which cells fire; rather it is motivated by the projection of the organization of sound frequency in the tonotopic map in the auditory cortex and the spatial organization of sensory control in the somatopic portion of the mammalian brain. The SOM is used primarily in data clustering where its supervised learning dual, Learning Vector Quantization (LVQ), is used for classification problems (Kohonen 1988). Although seemingly not as popular as feed-forward networks, theoretical and application-based research in SOMs and LVQs is ongoing with a strong following (Kohonen et al. 1997; Merenyi et al. 2016).

In the 1990s and 2000s, many researchers pursued the development of other network forms. These include hybridized approaches using genetic algorithms (GA) to create topology and weight evolving ANNs (TWEANN). This includes significant works by Stanley and Mikkulainen (2002) at the University of Texas in Austin, on the neuroevolution of augmenting topology (NEAT). Concurrently, researchers have been working with larger and "deeper" multilayered ANNs—with many applications focused on audition and vision—enabled by three major growth areas in ANN learning: (1) underlying computational and memory increases associated with Moore's law; (2)

---

14. Under certain forms of the state transition probability, it is analogous to the Boltzmann distribution and is referred to as a "Boltzmann machine" (Fletcher 2016).

15. When those conditions are not met, only a local optimum can be achieved.

16. The quantum phenomenon included in the models enables the algorithms to tunnel through the barriers on the potential surface instead of "hopping" over it as is done in SA.

vast increases in datasets available for training (e.g., images of cats on the web); and (3) algorithmic improvements, especially the ability to train multi-layer networks layer by layer. Three researchers are notable for developing methods of "deep" learning across large datasets: Hinton of the University of Toronto, who is now a distinguished researcher at Google (see, for example, Hinton et al. [2006]); his (former) student LeCun of New York University, who is now leading the AI effort at Facebook (see, for example, LeCun et al. [1998]); and Bengio of the University of Montreal (see, for example, Bengio [2009, 2016]). An excellent recent summary of the state of the art has been provided by these three in LeCun et al. (2016). The many successes of deep learning ANNs operating over big datasets over the last few years has re-energized the community (Edwards 2015; Krakovsky 2015; Anthes 2017). As noted by Hinton in *Wired* magazine (Hernandez 2014):

> *We've ceased to be the lunatic fringe. We're now the lunatic core.*

A very different approach to dealing with reasoning under uncertainty was pioneered by Zadeh in the 1960s with his introduction of *fuzzy logic* (Zadeh 1965), basically an extension of classical logic (of hard AI) to deal with sets that are subjectively defined, as humans do in their day-to-day activities. Thus, in a classical sense, a day might be considered warm or cool if the temperature is above 70°F or below that, respectively, so that a 68°F day would be a member of the cool set and not the warm set. The respective *membership function*—the relative "strength" of a day being in one set or the other—is a simple binary function, zero or one. In fuzzy logic, a membership function is defined over an interval, say, in this example, from 60° to 80°F, where, as the day's temperature rises, its membership strength in the warm set rises from zero to one, while membership strength in the cool set correspondingly decreases. This conceptual framework then allows us to deal with continuous variables characterized by uncertainty ("noise") in the following three-stage fashion:

- *Fuzzify* these "input" variables into different sets as just described, generating one or more linguistic variables (e.g., cool, warm) with associated membership strengths

- Use some form of a propositional logic to operate on those linguistic variables, for example, an expert system with a procedural rulebase and inferencing engine, to generate additional linguistic variables desired as the linguistic "output" of the system

- If there are desired continuous "output" variables, recover them via a process of *defuzzification*, a process analogous to the inverse of the initial fuzzification

Fuzzy logic also allows for the combination of other variables (e.g. purely linguistic ones) and with extensions can deal with dynamic time series and dynamic systems. Although theoretical development and practical application of fuzzy logic—also called "fuzzy expert systems"—probably peaked around 2000, there is still a significant community working in this area, based on the publications of the IEEE Computational Intelligence Society.[17]

Alternatives to this type of fuzzy or, more generally, probabilistic reasoning have grown significantly, as application areas initially dealt with in the "hard" AI community (i.e., "toy" microworlds) have graduated into domains characterized by uncertainty, both discrete and continuous variables, and complex system dynamics. The very term *probabilistic reasoning* gained significant visibility with Pearl's (1988) publication describing Bayesian belief networks (BBN). BBNs are composed of nodes that effectively represent a system's states (which may be discrete or continuous), and the links between the nodes represent the probabilistic, and usually causal, relationships between the node variables. These causal relationships are, in turn, represented by *conditional probability tables*, which, when used with Bayes' theorem,[18] provide a means of updating the variables associated with one node when another node's variables change. In effect, probabilities or likelihoods can be propagated throughout the network when any single node variable is updated, in a fashion that follows the rigorous rules of probability theory described by Bayes' theorem and that supports a sparse decomposition of high-order joint probability distributions via limited network nodes and links. More practically, the graphical representation of the BBN allows for rapid construction of network diagrams by domain experts that understand the underlying causality of key variables in a system. Alternatively, methods have been developed to "learn" both the structure of the networks and the values of the parameters, given a large enough and sufficiently rich dataset to drive the learning algorithm (Cheng et al. 1997). BBNs have been used primarily for estimation/diagnosis of a system's state (i.e., "beliefs") in a wide number of domains, including genetic counseling (Uebersax 2004), behavioral modeling (Hudlicka and Zacharias 2005), and failure analysis (Weber et al. 2012). Dozens of software packages exist to ease implementation efforts as well.[19]

Because BBNs deal with probabilistic variables and not with their associated objects or entities in a specific domain, BBNs can be limited in their ability to

---

17. *See* IEEE Computational Intelligence Society, https://cis.ieee.org/.

18. A method for using new evidence to update old beliefs, based on probabilities. See, for example, "Bayes' Theorem" entry in Wikipedia, accessed 9 February 2019, https://en.wikipedia.org/wiki/Bayes%27_theorem.

19. See, for example, "Bayes' Network" entry in Wikipedia, accessed 9 February 2019, https://en.wikipedia.org/wiki/Bayesian_network.

represent complex domains with multiple and possibly varying entities. Probabilistic relational models (PRM) were developed to overcome these limitations via the specification of two templates: a relational component defining the object schema (as one would do with a relational database) and a probabilistic component, defining the probability distribution of the attributes of any given object in the schema (Friedman et al. 1999). This greatly facilitates the "mining" of large relational databases to learn underlying patterns or relationships across objects, thus supporting, among other efforts, "big data" processing that goes beyond simple object recognition, for example, toward the understanding of probabilistic relationships among entities in a given dataset.

One last soft AI technique of note has its beginnings in biologically inspired evolutionary theory. Early work in the 1950s focused on the development of a variety of computer simulations of evolution, many of which eventually became abstracted, formalized, and popularized as GAs by Holland in the mid-1970s (Holland 1975). The basic notion is that we have a population of individuals, each having a different fitness relative to some objective function we are trying to optimize. An individual's fitness, in turn, is determined by its gene, which is essentially a vector of parameters to be adjusted for maximum fitness. A GA-based solution consists of "evolving" the population over a number of discrete time steps, where, at each time step: (1) sections of genes are "crossed over" to produce new genes, mimicking the process of sexual reproduction, and population mixing; (2) some genes are subject to "mutations" that change a small number of parameters in its genes, to mimic the random introduction of good and bad traits in a population; and (3) based on the gene pool, new individuals are created that live or die based on their fitness, so that the more fit individuals are available to propagate their genes at the next time step. This continues until a sufficient number of individuals are sufficiently fit, effectively finding a "soft" optimum in the population space. A wide variety of GA variants have been developed beyond this basic approach, including evolutionary algorithms, which provide more structure to the underlying genes (Back 1996) and evolutionary programming (McDonnell et al. 1995) and many application domains have been explored for optimization (Goldberg 1989).

Holland continued to generalize the GA approach and push it beyond simple "soft" optimization by focusing on populations of individuals interacting, not just at the genetic level but as individual "cognitive" agents interacting with one another (Forrest and Mitchell 2016). The result was the development of the research area of complex adaptive systems (CAS) (Holland 1996, 2006) in the 1980s, which evolved into an area generally labeled as "Complex Systems Theory," where large numbers of possibly "simple" entities interact to give rise to complex patterns of group behavior, such as one sees in insect swarms (Bonabeau et al. 1999).

The discussion in this section is not meant to imply that there is a clear dichotomy between "hard" and "soft" AI techniques. On the contrary, in several instances the boundaries blur, going back many years, including attempts to "soften" hard AI approaches with the development of Gaifman's (1964) probabilistic propositional logic, Zadeh's (1965) fuzzy logic, Shapiro's (1983) certainty factor, and Nilsson's (1986) probabilistic logic, among others. On the converse side, attempts to provide some underlying semantics to the "soft" approaches are more recent, with a key publication by Andrews, Diederich, and Tickle (1995) that surveyed attempts at merging ANNs with ES-type rule sets in three areas: (1) using rules to insert knowledge into ANNs; (2) extracting rules from trained ANNs; and (3) refining existing rule sets using ANNs. This has motivated significant effort in ruleset extraction of high-performance of ANNs of today: for example, DARPA's current Explainable AI program is aimed at providing a human with the "why" behind a decision or recommendation made by "soft" approaches, particularly ANNs that have undergone learning beyond the knowledge given to them by their original developers.[20] Strong interest in unifying hard and soft approaches has motivated a text in neural-symbolic learning systems (Garcez et al. 2012) and, more recently, a review by Russell[21] (2015) aimed at "unifying logic and probability" and a plea by Booch (2016) for "hybrid AI":

> So, whereas Watson was symbolic and Alpha Go was neural, I'm a proponent of hybrid AI, involving the coming together of symbolic computation and neural networks. (Booch 2016)

4.4 Basis of a Common Framework

Referring back to the original figure that began this discussion, figure 4.1, we believe that the different research and development communities shown can provide the impetus for the development of one or more "common frameworks," each having dual uses: one to describe human perception, cognition, and action at a *computational* level (i.e., via modeling and simulation), and one to prescribe engineering systems designs, that, in limited domains and tasks, can replicate human perceptual, cognitive, and motor performance with levels of fidelity adequate to declare them "autonomous" in the sense described earlier in section 1.2.

More specifically, the long histories of robotics and cybernetics have emphasized the importance of situated agency and human-system teaming, both

---

20. *See* DARPA, http://www.darpa.mil/program/explainable-artificial-intelligence.
21. Of hard AI fame with the co-authorship of Russell and Norvig (1995).

from a practical and theoretical standpoint. Both are clearly critical to the development of usable and reliable autonomous systems. But in addition, active work in the robotics area has identified the need for internalized "mental models" of the outside world, a trend that is likely to lead to the development of more cognitive structures to move beyond simple sense-and-locomote strategies that have dominated many recent autonomous robotic efforts. In parallel, the cybernetics stream has, as noted, evolved into a community focusing on the development of cognitive architectures and models that can simulate human behavior in a wide variety of tasks and environments.

In parallel, the cognitive psychology and neurosciences communities have also driven the evolution of cognitive architectures, the former from a "top-down" behavioral point of view and the latter from a "bottom-up" understanding of the neural substrate. Both are advancing the state of the art in developing cognitive architectures and computational models that can be used to: (a) in a descriptive mode, compactly describe empirical findings; and (b) in a prescriptive mode, support hypothesis-driven experimentation into new areas. As a side benefit, discoveries of both communities help drive both the hard and soft AI communities, who often look to psychological or neural inspiration for their constructs (e.g., expert systems, multilayered neural networks).

Finally, the hard and soft AI communities are rapidly evolving, especially in domains that were once considered the province of human intelligence—specifically, competitive games.[22] They are also likely to start converging more rapidly, as each community sees potential advantages in the other's capabilities as we have just described. However, the applications are still very narrowly focused (e.g., image classification, financial trend predictions, "perfect information" games, etc.), and some recent successes may be more a function of growth in computational capability and training set size (in the case of ANNs) than of the introduction of truly revolutionary AI concepts, hard or soft. Nevertheless, these techniques, in conjunction with the more "traditional" computational techniques and methods (e.g., dynamic state estimation and control theory, systems identification, quadratic optimization, constraint-based programming, etc.), are likely to form the key to effective and efficient computation of any behavioral function (human- or machine-based models) that we choose to implement in the foreseeable future.

We see all of these communities as key to the development of *proficient* and *trustworthy* autonomous systems, in the sense that we have discussed those terms earlier in chapters 2 and 3. Bringing them together will necessitate

---

22. For example, IBM's Deep Blue for Chess (McPhee 1997), Deep Mind's AlphaGo and AlphaGoZero for Go (Silver et al. 2016, 2017), and Libratus for Texas Hold'em (Maetz 2017).

common frameworks to bridge the gap across communities, support the development of a common language to describe similar and related concepts in the different fields, and provide the foundation for developing autonomous systems in the future. As we noted earlier, we believe the time is right to reach across these communities and begin the effort to develop a common framework.

Chapter 5

Framework and Functions

In the previous chapter, we described the convergence of a number of diverse research and development communities in the cognitive sciences and engineering application areas that have far-reaching implications for the development of future ASs. As we noted, all of these communities could benefit from the development of one or more "common frameworks" that would facilitate not only better communications across communities but also a transfer of concepts, models, and computational representations, accelerating our understanding of existing systems and supporting the development of new ones. More specifically, if we are successful, a common framework could help us accomplish the following:

- Identifying the fundamental structure common to most or all autonomous systems in terms of the internal component functions, their relationship to each other and the environment, the principles governing their design, and overall control-flow and data-flow

- Finding a place in the autonomy "universe" for those working subsets of the general problem (e.g., data fusion, image classification, path planning, motor control, etc.) and providing connectivity to others working complementary subsets of the problem

- Helping develop a unifying "science of autonomy" underpinning the thousands of "one-offs" we now have in the engineering community

- Separating functionality from enabling technologies so that architecture design can go on in parallel with technology development

- Pointing to where the S&T community needs to invest to develop "missing" functionalities and/or improve technology capabilities

- Dealing with the issue of meaning making and the need for a common frame or context

And, in the longer term:

- Serving as the foundation of a common open systems architecture (OSA) to encourage reuse of developed software modules across applications and domains

- Supporting interoperability across DOD[1]

As we noted earlier, we believe the time is right to begin developing a common framework.

We note that we have used the word "framework" as the title of this chapter to indicate a rather looser notion of an architecture—one that is focused more on broad functions rather than on detailed component diagrams, such as one might think of in terms of, say, software architectures. The goal is to think in terms of the functionalities needed to provide the autonomous system with the task proficiencies, elements of trust, and behavioral flexibilities described in the earlier chapters, particularly if those functionalities/attributes can be generalized across domains and applications. We begin with a brief discussion of some broad considerations for a framework (section 5.1) and then illustrate this notion with a "composite" framework based on concepts drawn from a number of the communities just discussed (section 5.2). We then describe detailed component functionalities within the framework, along with promising approaches to implementation (section 5.3). We close with a brief discussion of "what's missing" from this given framework to motivate further research (section 5.4).

5.1 Considerations for a Framework

There exists a number of architectures, models, and even development toolsets that could serve as a starting point as a common framework for developing autonomous systems across a number of communities. However, we believe (and have argued in the previous chapter) that the best starting point is one grounded in the many cognitive architectures that have been developed to describe human behavior and performance, simply because if they are sufficiently accurate in a descriptive sense, they are likely to also be very useful in a prescriptive sense in guiding the scientific and engineering development of synthetic autonomous systems that perform on par with humans, likely satisfying our *proficiency* considerations at a performance level. Equally important, the potential similarity provided by common human and engineering frameworks motivates the development of similar underlying behaviors, likely satisfying many of our *trust* considerations at a behavioral level. We are thus motivated to look at cognitive architectures and models as inspiration for a framework.

---

1. For example, Air Force UAVs conducting ISR missions for Navy Attack unmanned undersea vehicles (UUV).

Even when we limit our scope to cognitive architectures that can be instantiated as models in a computational environment—as opposed to qualitative models that provide little guidance to the engineer trying to develop an autonomous system—there still exists a wide variety of architectures and a large number of computational models, given the long history of research in this area cited in the previous section. Extensive surveys have been provided by Pew and Mavor (1998), Sun (2006), Ritter et al. (2003), and Zacharias et al. (2008), including a comparative table of a number of cognitive architectures hosted online by the Biologically Inspired Cognitive Architecture (BICA) Society (http://bicasociety.org) and reviewed by Samsonovich and colleagues (Samsonovich et al. 2010). A more recent listing of about 40 cognitive architectures is given at https://en.wikipedia.org/wiki/Cognitive_architecture. Architectures can provide "end-to-end" functionality covering perception, cognition, and action (Thelen and Smith 1996; Cassimatis et al. 2004) or they can focus only on cognition (e.g., early ACT-R models; Anderson 1983); they can be functionally structured (e.g., SPAUN; Eliasmith et al. 2012) or unstructured (e.g., NARS; Wang 2006); they can be symbolic (e.g., ACT-R; Anderson 1983) or subsymbolic, that is, connectionist (e.g., CLARION; Coward and Sun 2004) or a hybrid of the two (Sun 2006, 2010); they can incorporate learning (e.g., Soar; Laird 2012) or not (e.g., 4D-RCS; Prokhorov 2008); and finally, they can control processing (or attention or activation) in a centralized fashion (e.g., Soar, ACT-R) or in a decentralized fashion (e.g., Google DeepMind; Graves et al. 2014).

What we propose here is an abstract framework that provides "end-to-end" functionality that is functionally structured rather than unstructured, that makes no commitment on symbolic vs. subsymbolic processing, and which incorporates learning. It is deliberately engineering focused with a strong "dataflow" orientation that has its basis in a cybernetics view of the world; processing control is not explicitly represented, nor, in fact, is "goal generation," likely a key consideration if we are to eventually develop autonomous systems with initiative and inventiveness. There are many other arguments that could be made about the framework's shortcomings,[2] most notably about its functional dataflow orientation, and the difficulty of reconciling that choice with a looser, nonfunctional approach provided by, say, nonpartitioned rule-based systems[3] or distributed ANN structures.[4] Nonetheless, we put this forth

---

2. And we address some of these in greater detail in section 5.5 below.

3. For example, where declarative knowledge is not even partitioned from procedural knowledge.

4. For example, where a layer's "function" in a multilayered network is not even defined, if ever, until after learning stops.

as one *example of a framework* that may be able to bridge the gaps across the several communities we noted in the previous section and serve as a basis for further development of truly autonomous systems. At the very least, it provides us with an integrated framework for discussing the functionality an autonomous system needs to be *proficient, trustworthy,* and *flexible* in future operational scenarios.

5.2 Example Framework

Figure 5.1 illustrates one of many frameworks that we might consider. Figure 2.1b earlier showed a single autonomous system interacting with an "outside world" and a human teammate via an HCI; the figure below effectively provides an "under-the-hood" view of the autonomous system central to that earlier figure. Thus the "outside world" cloud elements of figure 2.1b drive the input sensors shown on the left side of the figure below, and the effectors on the right-hand side of the figure below drive the cloud elements of figure 2.1b in a closed loop, providing us with the desired attribute of "situated agency" described earlier. There is a similar correspondence between the two HCI blocks of both figures,[5] providing two-way interactions between the autonomous systems and the human teammate, who is not illustrated in this "under-the-hood" view.

Proceeding in a *left-to-right* fashion, we see that sensors and databases drive (i.e., provide inputs to) a sensor/data fusion block,[6] which drives a "layer" of the framework called the situation assessment and decision-making (SADM) layer, whose outputs eventually drive an effectors block so that the autonomous system can effect or take action upon the outside world, again closing the loop on the outside world "cloud." A quick look inside the SADM layer shows a very stylized, staged set of processes, where one block's output serves as another block's input; we will discuss this shortly, but for now it serves as a simplified representation of many of the basic functions that we believe an autonomous system must accomplish to be proficient in its performance.

Proceeding in a *right-to-left* fashion, we see that the SADM layer also drives information-seeking functions via sensor management and data mining ("collection management" in ISR terms), learning activities, and HCI functions—relations we will expand upon shortly.

---

5. Figure 5.1 expands on the HCI theme by including "collaborative environments" (CE) to allow multiple humans to interact with the autonomous systems (and with each other).

6. "Block" is used interchangeably with "function" in this section, given this framework's inherent "block diagram" rendition.

Proceeding in a *top-to-bottom* fashion, the HCI/CE block supports interaction between human teammates (at top of diagram, not shown) with the sensor/data fusion block as well as with all of the blocks in the SADM layer, providing for a closer look into the inner workings of the autonomous system by its human teammate(s) at all stages of processing. Also shown are interactions between blocks in the SADM layer and a learning/adaptation layer, which provides for learning across all functions comprising this layer. The learning layer also drives sensor/data fusion and serves to update the domain knowledge layer. The domain-specific knowledge base layer contains knowledge needed by the autonomous system, whether it is structural/parametric knowledge needed by an ANN for image processing, dynamic model-based knowledge about a specific platform, "declarative" knowledge about a given mission, "procedural" knowledge about a specific role/task, etc. Finally, as illustrated, toolsets and technologies underlie all of the other layers, supporting a variety of implementations of the functions/processes illustrated. We now describe the components illustrated in the figure, what functionality they provide, and possible technical approaches for their implementation.

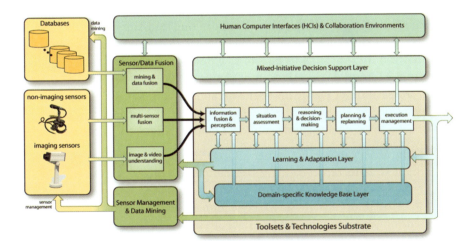

Figure 5.1. Example autonomous systems framework

5.3 Component Functions

In this section, we describe the component functions within the example framework just introduced, block by block, along with promising approaches to implementation. We also note, where appropriate, how the given functionalities can support the three major behavioral dimensions needed for autonomy: *the properties of proficiency, the tenets of trust,* and, especially, *the principles of flexibility.*

We start left-to-right with databases (section 5.3.1) and sensors (5.3.2) and proceed to sensor/data fusion (5.3.3), situation assessment and decision-making (5.3.4) and effectors (5.3.5). We then go top-to-bottom, starting with human-computer interfaces and collaborative environments (5.3.6) and proceed to learning and adaptation (5.3.7), domain-specific knowledge bases (5.3.8), and toolsets and technologies (5.3.9).

5.3.1 Databases

The **database** block represents the long-term recall and real-time working memory of the autonomous system. Across multiple heterogeneous sources, using content-optimized data structures and indexes (Manolopoulos et al. 2000), the database function provides the AS with time- and space-efficient information storage and retrieval to support long lead-time and real-time decisions.[7] Data stores may be organic to the AS, that is, resident or on-board; remote from the AS, such as part of a "cloud" (as illustrated in figures 2.1a, 2.1b, and 2.1c and as described by Deptula [2006]); or resident in the on-board stores of other platforms (Baldwin and Talbert 2005).

Operationally, *inputs* to the database function are of three fundamental types: data flows, data requests, and data manipulation:

- **Data flows** (not shown in fig. 5.1) are sensor, user, and internal system input streams representing time-stamped events or object states. In the process of input, data can be clustered, correlated, and stored in highly structured or unstructured collections. Low-loss/no-loss *compressive sensing* techniques applied to data streams enable high density storage (Eldar and Kutyniok 2012), though with added computational cost and operational degradation (Hytla et al. 2012). Beyond compression, autonomous operations require trusted, responsive, predictive models for deciding what data to keep, compress, or let "fall the floor." Research is required to

---

7. The state of the art in heterogeneous data indexing and retrieval is largely attributable to industry leaders IBM, Oracle, and Google and their operationalization of decades of work in academia.

develop these models as a function of time, resources, and mission objectives. Further, research in adaptive interfaces and human-computer interaction conditioned on cognitive loading will be essential for time-critical, human-machine teaming functions in an AS (Blaha et al. 2016).

- **Data retrieval requests** (queries) are data store accesses spanning possibly multiple heterogeneous data stores (Bonnet 2000). Supporting broad flexibility in cognitive load and task partitioning, queries can be user-generated, system-generated, or standing subscriptions. Data accesses may also be triggered by scheduled or anomalous events. For example, in an ISR-supporting system, an AS can query the database(s) for static, dynamic, historical, or graphical products in response to the detection of an unrecognized ground object or activity. In full autonomous modes, an AS may query system logs and event histories to control on-board or off-board system resources, with or without operator intervention. This function illustrates still-to-be-achieved "trusted automation," a research area needed to enable complex autonomous warfighting systems to act with lethality or put humans at risk.

- **Data manipulation** inputs are directives to a database system to, for example, reconstitute indexes, adjust the degree of compression, or cluster contents physically or logically according to mission-relevant, possibly dynamic relationship structures (schemas). Dynamic database manipulation is an area ripe for research to reduce the cognitive load on humans teaming with the system and balance system resource task loading. Employed in an autonomous system, advanced manipulation inputs could invoke the execution of downstream analytical applications, such as extracting object movement tracks from stores of video or imagery sequences. Many such stand-alone imagery analytical functions exist (see, for example, Blasch et al. 2014) and are recognizable in DARPA's Insight program (http://www.darpa.mil/program/insight) and OSD's Data-to-Decisions initiative (Schwartz 2011). However, additional research is required to ensure trusted, platform-specific, mission-scalable performance of these embedded techniques. This is especially critical when they are invoked during closed-loop autonomous system operation.

In the diagram, we show *outputs* from the data stores as serving the downstream functions of the sensor/data fusion block, which could be used to support a variety of AS functions—for example, for a platform-based ISR mission, map data for navigation, or, for a rapid-response cyber defense mission, a catalog of previously observed attack maneuvers. In addition, outputs could include reactive responses to intentional queries—for example, summarized and cor-related system health event trends or time-space-geo-referenced location his-tories of a cluster of related objects. Triggered outputs could also support a rapid-response cyber defense mission by automatically generating likelihood estimates of attack source and path from a model based on previously ob-served attack vectors (Simmons et al. 2014). Intuitively, engagement with a smart database system should enable task, peer, and cognitive flexibility as the mission demands; in the cyber defense realm, system-generated alerts would be accompanied by access to a user-readable explanation of the events that prompted it.

Of course, the outputs from, say, a data-mining exercise conducted by an AS could be the penultimate output of the AS—for example, the use of clustering/classification algorithms to identify "groups" within the data or the use of as-sociation rule learning to identify underlying relations between variables (Ag-garwal 2015). A vibrant community within the ACM focuses on just this area: the Special Interest Group (SIG) on Knowledge Discovery and Data Mining (http://www.kdd.org/), a group that has been partly responsible for the current excitement regarding "big data" mining with machine learning algorithms.[8] Though typically a manually supervised effort, an AS-supervised effort would have a number of interesting applications in the ISR and cyber communities.

As the examples above suggest, the line begins to blur between "smart" data storage/access functions and downstream analytics functions. While a robust body of data analytics R&D supports a broad spectrum of industrial and military applications, lines of inquiry in data analytics scalability, trust-ability, and transparency for autonomous systems remain underserved. DARPA's "Explainable AI" program[9] and Office of the Director of National Intelligence's In-VEST initiatives[10] are positioned to tackle some of these chal-lenges, but follow-on efforts should extend them to autonomous applications.

5.3.2 Sensors

The **sensors** block serves as a general placeholder for three separate classes of sensors:

- Those driven by aspects or entities of the external environment associ-ated with the "cloud" of figure 2.1. On an aircraft platform, say, this

---

8. See, for example, the paper by Levine of Andreessen Horowitz, http://a16z.com/2015/01/22/machine-learning-big-data/.

9. *See* DARPA, http://www.darpa.mil/program/explainable-artificial-intelligence.

10. Intelligence Ventures in Exploratory Science and Technology, https://www.dni.gov/index.php/resources/in-vest.

might include environmental sensors like an altimeter, a passive sensor like an electro-optic camera, or an active sensor like an active electronically scanned array (AESA) radar.

- Those driven by components or subsystems directly associated with the AS itself. Again, on a platform, this might include an engine RPM monitor, a temperature monitor supporting internal system health assessment, or a CPU computational load monitor.

- Those associated with explicit communications from entities outside of the AS. Again, on a platform, this could be a tasking update from a ground control element or a flight formation coordination update from a swarming teammate.[11]

Clearly, there exists a wide variety of sensors spanning many modalities and levels of complexity, ranging from simple transducers of physical aspects of the environment (e.g., an angle-of-attack sensor) to complex subsystems of their own (e.g., an electro-optic/infrared [EO/IR] sensor ball on a UAV. We wish only to note that, at a sufficiently high level of abstraction for AS considerations, a sensor may be regarded as an element that "transduces" one form of energy into a suitably encoded electrical signal that can be subsequently processed by the AS. In addition (and this is shown on the diagram), we assume that some or all sensors can be "managed" in some fashion—for example, a platform-associated EO/IR sensor ball can be aimed or its field of view changed or its sensitivity adjusted to meet AS tasking needs; likewise, a distributed sensor network associated with a cyber intrusion detection system (IDS) might have its configuration changed depending on the operating mode of the IDS. This is not simply a "nice to have" feature in a sensor but can be critical to supporting "active sensing," in which a sensor interacts with its environment over time (e.g., as a camera might maintain track on a moving target) to gain a richer representation of the signal or object being sensed (Bajcsy 1988).[12]

At an abstract level, then, inputs to the sensors block consist of (typically) physical/electromagnetic variables in either the outside environment surrounding the AS or in the internal environment within the AS. Additional inputs come from communications to the AS from outside entities as well as

---

11. We have "bundled" communications channels as part of the sensors (and effectors) blocks for expository simplicity, recognizing that the C2 communications aspects may very well deserve their own explicit representation in the proposed framework at some later point. For the discussion here, however, we will stand with a simpler representation format.

12. This is not to be confused with "active sensors" in which a sensor may generate a signal to interrogate its environment, like RADAR or LIDAR. Active sensing can use active or passive sensors to make rich inferences of its environment.

sensor management signals from inside the AS. Outputs from the sensors block are either transduced signals or communications, both available for follow-on processing by other AS modules.

Sensors are a key autonomy enabler since they provide the foundational information used by an AS to understand and react to the environment, as we have outlined earlier in section 2.1 on situated agency. For example, in platform-associated electronic warfare/electronic attack (EW/EA) context, changing the sensor mode (such as the waveform), the platform flight path, or the EW jamming based on the local environment will provide a core capability for adapting to red force countermeasures. Evolution of sensors—whether RF or optical—has come a long way from the analog-based systems of the past, and the very flexible and fast digitally based systems used today could benefit significantly from AS-based agility in hostile environments (Kirk et al. 2017).

Sensor interaction with the environment can be facilitated by machine learning and information theoretic techniques that can provide high-accuracy closed-loop understanding of the situation. This capability—on either a platform or in a ground-based facility—enables operation of sensing systems to produce ISR products in complex and perhaps contested environments. Jointly interactive estimation and control techniques can enable an AS to determine the quality of the sensor information generated as a function of, say, the sensor mode or terrain geometry and then provide the capability to improve that information through sensor mode/parameter settings and/or or controlled flight path changes. Changing sensor parameter settings—such as sensitivity, field of view, waveform output countermeasures, and selectivity—can broaden sensor performance across operating conditions while adding to the task and cognitive flexibility of the AS (Anaya et al. 2014).

Miniaturization, performance improvements, and cost reductions of offensive, defensive, and internal sensors have been marching forward for decades as a result of new metamaterials for antennas, sensor digitization, and advancements in low-noise amplifiers. Capabilities such as passive geolocation of threat emitters, active radar, and countermeasures systems, which used to require complete aircraft to carry the associated sensors and processors, now occupy a fraction of the physical space and use significantly less power (Langley et al. 2011).

Because of these trends, the deployment of platform-based sensing systems can range from the exquisitely simple to the traditionally complex. Risk can now be transferred to less expensive forward-deployed systems to improve situation understanding of the adversarial environment and serve to cue more discriminatory (but more expensive) systems that are not placed in harm's way. In addition, AS-based techniques for coupled sensor management and platform control for mission execution and weapons delivery can now be better synchronized for greater mission effects—for example, improvements in detection, geolocation, identification, and tracking of mobile target/threats employing concealment, camouflage, and deception (CCD) activities. Likewise, the use of AS-based techniques for coordinating distributed, networked, and multiphenomenology sensors has the potential for revolutionizing traditional single-sensor, single-platform ISR collection practices.

Sensors are also critical to AS health management, detecting drops in subsystem performance and outright failures. Once detected and identified, internal failures can be mitigated by accounting for a subsystem's contribution to overall mission success and deciding on a prioritized reallocation of resources, such as power, cooling, processing, or even mission planning. For example, during a threat engagement, one might balance the use of high-power EW countermeasures against executing a high-g evasive maneuver. Clearly, both internal and external sensor interplay will be critical in the internal resource management of the platform subsystems as well as the external management of the platform's trajectory, its vulnerability to adversary weapons, and its own weapons capabilities (Schumann et al. 2013).

Digital processing has ushered in a new era of selectivity, accuracy, and precision for sensing systems, and new commercial processing hardware and advanced digital software techniques will accelerate the growth of next-generation systems. Flexibility provided by digital architectures also enables the implementation of a suite of sensors ranging from the exquisite but expensive to the simple but expendable. In addition, the maturation of multifunction sensor systems will enable the sharing of significant portions of the processing chain to perform, in parallel, diverse tasks such as radar, data links, electronic support measures and countermeasures, which, in turn, should support distributed and risk-variable sensing. Finally, new digital processing techniques for sensor product formation will continue to remain a rich field of research as new machine learning techniques are incorporated (see, for example, section 5.3.7). Orchestrating the capabilities of these enhancements will call on the many of the AS capabilities outlined in this section (e.g., sensor fusion, SA, planning, execution monitoring) but will also provide the AS with the critical real-time, sensor-based information needed to operate effectively in its environment (Chabod and Galaup 2012).

5.3.3 Sensor/Data Fusion

The **sensor/data fusion** block provides the first step in processing available data from the sensors to generate knowledge and higher levels of information needed by the AS for mission and task understanding.

5.3.3.1 Sensor Processing and Fusion. The **sensor processing and fusion** function refers to a process by which digital representations[13] of the sensed environment, created via multiple sensors or classes of sensors, are processed independent of each other and then are combined or "fused" to infer or estimate information concerning the sensed environment. Figure 5.2 illustrates the overall construct of both sensor processing and sensor fusion, each of which can have a series of processes that can be application-independent or application-dependent. As shown, sensor *processing* can occur internal to the sensor or external to the sensor. Sensor *fusion* can occur either as an "upstream" function working on externally processed sensor data (black lines); or as a "near sensor" function working on internally processed sensor data (gray lines) (Zheng 2015).

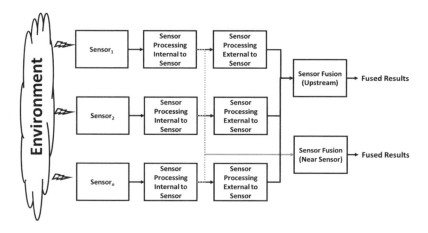

Figure 5.2. Example processing chain for multiple sensors that sense different aspects of the same environment and ultimately are used to produce a fused result

---

13. As described in section 5.3.2 above, the front end of a sensor is analog and typically converts (or transduces) a physical variable into a voltage "signal" that then undergoes an analog-to-digital conversion, sometimes outside of the sensor, but, increasingly commonly, inside the sensor. The result is a digital representation of the sensed environment for that sensor's transduction modality and can be numerical, categorical, textual, or any mixture of the three and can also be dynamically varying over time.

Internal sensor processing is driven by the type of sensor. For example, imaging sensors require different internal processing than audio sensors. For imaging sensors, internal sensor processing could include adjusting the integration time to change the amount of photons the focal plane array can collect, which could be followed by an analog-to-digital conversion, which is then followed by bad pixel mapping (bad elements in the focal plane array) and then a bad pixel signal re-estimated using, for example, bi-linear interpolation. These processes can be chained together to meet the needs of the sensor and the application, and some may be required and some may be optional. In hyperspectral remote sensing, some sensors attempt to determine the center wavelengths for each of the spectral bins by locating known atmospheric absorption features, adjust for propagation loss in the optical chain, stitch the results between the two primary focal plane arrays (one that captures visible-to-near-infrared energy and one that captures near-infrared-to-shortwave infrared energy), and then correct for atmospheric conditions. In this example, atmospheric correction might be optional. These same processing steps could occur off the sensor.

External sensor processing is driven by the application. Whether a series of chained processing steps occurs on the sensor or off the sensor depends on the maturity of the technology and the application. Much of the processing external to the sensor is also chained together, and some processing techniques are common irrespective of the application. Processing off the sensor includes signal denoising, signal compression, and prototypical machine learning tasks such as feature extraction, classification, and clustering. To be more specific, consider change detection, which takes a reference and comparison image and preprocesses them, performs change detection, then performs an assessment that terminates after reporting potential changes between the image pairs (Vongsy et al. 2009). The images are acquired at different times, from different views, and possibly from different sensors and need to undergo atmospheric compensation, coincidence image matching, spatial and spectral interpolation, and illumination correction before change detection. The change detector can be as simple as image differencing but often includes more sophisticated techniques such as the generalized likelihood ratio test that builds in correction for misregistration (Vongsy et al. 2015) and image parallax (Vongsy et al. 2012). Assessment uses the change detector output and applies post processing to identify likely changes from the reference image to the comparison image (Vongsy and Mendenhall 2010). All of this processing is external to the sensor and chained together. In many cases, as with change detection, the processing chain terminates after the sensor processing tasks are finished, and the results are provided to the end user or system. However, improvements can be made by integrating the results from sensor processing in a meaningful way. This is the goal of sensor fusion.

Sensor fusion generically refers to a process by which data from several sensors are combined or *"fused"* in some meaningful way to produce something more than could be produced by any one sensor alone. More specifically, the type of fusion processing of interest is sensor *data* fusion, or combining the *outputs* from the several sensors *and not* combining the sensors. As with many topics, the literature appears to vary on the use of the term fusion. Some use "fusion" and "combining" synonymously. Some use "integration" to define "fusion," and vice versa for others. The action of combining is a special process in the processing chain presented above in figure 5.2 and is more than just a process like denoising and compression. The first definition of fusion was given in 1991 by the Joint Directors of Laboratories (JDL 1991) then revised in 1997, which is the most accepted definition of data fusion (Hall and Llinas 1997): "Combining things for better results."[14] For example, one could combine the video of a person speaking with the audio of what is spoken for better transcription of the person's speech, by combining video-based lip-reading results with audio-based speech-to-text conversion results. The effectiveness of both would depend on the contents and quality of the associated databases (Noda et al. 2015). If video-based affect is also interpreted, then further refinement of the transcript might result by the inclusion of inferred emotional state of the speaker (Picard 1997). This kind of multimodality fusion is very much like what humans do naturally in a variety of situations, including shape perception (Krauthamer 1968), self-motion perception (Zacharias and Young 1981), or when enjoying the illusion of a ventriloquist's act (Soto-Faraco et al. 2002).

Fusion can occur at the signal, feature, or decision level. Regardless of where the fusion occurs, one of the main *challenges* is the need to weight the various inputs and *combine* them appropriately for the task at hand. Hence, either an implicit or explicit performance model is needed for each sensor (or INT) as a function of the current context or set of operating conditions. As a trivial example, the performance model would suggest additional weight on an infrared (IR) sensor during the nighttime condition verses a signal coming from a sensor operating at a visible wavelength. In general, battlefield conditions make these performance models much more complex than the example but nonetheless extremely important for fusion. As can be surmised, the performance model allows the fusion algorithm to weight which sensor is providing the more accurate call given the current situation. The performance model is

---

14. See appendix C for a more formal definition, discussion, and examples.

also central to choosing a sensor and tasking the sensor to make the necessary additional measurements required to make mission critical decisions.

Another challenge of fusion is the need for a common representation and the need to fuse at multiple levels of abstraction, to include control levels. In many cases, the representations are reasonably clear as the sensors are of the same type (e.g., various imaging sensors). In such cases, the simplest approach to fusion concatenates the sensor features, but sometimes this leads to a reduction in performance (Mura et al. 2011) because the features can be redundant, or there is not enough labeled to accommodate the increase in dimensionality, or in fact the representation is not the right one. Such a simple approach likely does not accommodate the fusion of disparate information sources such as text, imagery, signals, cyber, and so on. For these disparate information sources, the common representation is not at all clear; however, representation is central to the ability to fuse key sources of information. A particularly promising approach to fusion is using graph-based feature fusion (Liao et al. 2015), which includes the data mining necessary for the appropriate processing of the streaming sensor data.

Addressing the challenges fusion presents, along with those associated with the use of machine learning and adaptive sensor management algorithms, is one that can be aided by simulation. The number of situations that arise in combat and the number of information sources that need to be combined are far greater than our ability to gather measured data to represent all these combinations. In addition, the fusion processes and the sensors that feed them need to interact with, adapt, and stimulate the environment to perform the mission. Hence, given the combinatorics of the moving parts and the need to interact with the environment including the adversary, simulation is vital to the development and test of any fusion technology. However, there are challenges associated with the simulation technology as well. Fundamental to effective simulation is the fidelity verses computational complexity trade-off in representing and controlling the sensors, platforms, environments, and adversary elements. It is likely that the simulation environment and fusion algorithms will share many of the technical challenges delineated above and, therefore, the concurrent development of both the fusion algorithms and the simulation environment may be the most effective way forward.

The three *principles of flexibility* are directly applicable to sensor processing and fusion; some of the key considerations are as follows:

- *Task flexibility*: A key attribute for the AS will be the ability to change the task at hand depending on the current state and the predicted state, based on existing performance models embedded in the sensor processing block. Simulation will be important as training trials can deliberately expose the AS to conditions outside the bounds of its models to teach it to adapt or, importantly, to abandon the current task to perform a higher payoff task or, perhaps, to multitask to achieve important mission objectives. Common representations will be vital not only for combining the variety of sensor information available but also for achieving task flexibility.

- *Peer flexibility*: Performance models/functions are central to making distributed fusion decisions to achieve mission objectives. In addition, simulation of multiple scenarios will allow AS teams to learn to adapt to different conditions and learn how to rely on each other and adapt to the adversary, the environment, and potential attrition of their teammates. Common representations will be necessary for a common situation awareness, based on an ability to jointly reason and fuse distributed sensor information at the signal, feature, and decision level.

- *Cognitive flexibility*: Different situations will require different representations, different fusion approaches, and different solution strategies. Cognitive flexibility will address such issues as: what sensors should be tasked to provide an actionable decision, at what level should they be combined, and how should they be weighted? These questions will not yield to a single algorithm or single processing strategy, so there will be a need to adaptively change representations and decisions based on internally held performance models. In addition, an ability to simulate multiple scenarios and environments will be needed to develop and test the system's cognitive flexibility under different situations.

The development of fusion systems will build on the underpinnings of statistics and machine learning to combine information rigorously and reliably for key mission decisions. The challenge, however, is to determine how to use these technologies to represent the large space of dependencies driven by the interaction of environment, sensor, and adversary states. Without the ability to model these dependencies, the combination of sensor information, the ability to predict next states, and the ability to prescribe an action will continue to rely solely on human judgment, thus undermining the autonomy vision. In addition to modeling these dependencies via performance models, however, the need to develop robust systems that can react to situations that have not been modeled is also key to developing effective ASs. Common representations that not only combine disparate information but also provide a common situational awareness will be paramount. Finally, the use of simulation is the only viable method foreseen to both develop and evaluate fusion approaches and ASs' ability to achieve task, peer, and cognitive flexibility.

5.3.3.2 Data Mining and Fusion. This block also supports the **data mining and fusion** function,[15] which begins with attention to what actual databases are used to support the mining function. Data mining is a process used to solve problems by analyzing existing data stored in a database through automatic or semiautomatic approaches (Witten et al. 2011; Aggarwal 2016). Data analytics is also a process of examining data to draw conclusions about what is in the data (Rouse 2008). This process relies on, for example, machine learning, pattern recognition, artificial intelligence, etc., as the processing mechanism; for the purposes of this document, the terms *data mining* and *data analytics* are synonymous. The importance of data mining cannot be understated; it is used by high-profile commercial and governmental organizations and companies. It is critical in fraud detection and in high-profile legal cases such as the Enron scandal and is used by some of the world's largest online retailers (e.g., Amazon.com) for recommending products to would-be customers, by Facebook AI to mine imagery data, and by thousands of others for various uses.

Until recently, the key to data mining was selectively choosing what data to store and how to index the data for efficient access when it was needed. But that has changed with the revolution in *big data* processing (also called big data analytics), which, in its most basic form, deals with massive data volumes (e.g., petabytes) useful in executing some task (Reed and Dongarra 2015; Agrawal et al. 2011). It is most often characterized by the 4Vs: volume (how much), variety (how different), velocity (how fast), and veracity (how good). However, it is the actual machine learning and pattern recognition (ML/PR)—the underlying mechanisms in the (semi)automatic "mining" of the data that is seeing the patterns—that has generated the most interest in recent years. The ML/PR literature is vast, and many effective approaches to solving practical problems of interest exist and can be found in various sources (Witten et al. 2011; Duda et al. 2012; Hastie et al. 2009; Schölkopf and Smola 2002; Bishop 2006; Mitchell 1997; Kohonen 2001; Haykin 2009).

The approach to data mining is not one that should be done haphazardly. A sound process should be followed in order to improve the rate of success of the data mining task. One of the most prominent data mining constructs is the Cross-Industry Standard Process for Data Mining (CRISP-DM). CRISP-DM was developed as a joint venture between NCR, DaimlerChrysler (formerly

---

15. We introduce the term "fusion" here to indicate that data mining may occur over multiple heterogeneous databases, to highlight the parallel with sensor fusion across multiple potentially heterogeneous sensors.

Daimler-Benz), and SPSS (formerly ISL) and was funded by the European Commission. The project ended in 1999 with the finalization of the CRISP-DM process (Chapman et al. 2000). According to a 2011 survey, CRISP-DM remains the most popular methodology for data mining (Piatetsky 2014). CRISP-DM is a data-centric data mining methodology that captures business understanding through deployment (Shearer 2000).

Figure 5.3 illustrates the relationship of the six components of CRISP-DM:

- *Business understanding* is where one develops an understanding of the problem and the data mining goals of the organization and then fleshes out any requirements.

- *Data understanding* looks at representative data to increase the data mining professional's understanding of what the data physically looks like and gain any early insights as to what it might be able to inform.

- *Data preparation* includes such tasks as data cleaning, data transformation (e.g., data whitening, application of a wavelet transform or a Fourier transform, Principal Component Analysis, etc.), and attribute selection (e.g., feature selection) prior to being ingested by the modeling engine (e.g., density estimators, classification tools, or regression tools).

- *Modeling* consists of model investigation to determine which models work well for the data and the task and then doing parameter tuning (e.g., selecting learn rates and schedules in a Back Propagation Neural Network).

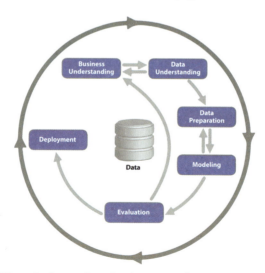

Figure 5.3. CRoss-Industry Standard Process for Data Mining (CRISP-DM). (Image used with permission from Jensen [2012])

- In *evaluation*, one ensures the model(s) actually enables one to meet the original requirements or business objectives; at this stage a determination will be made if the results of the data mining task should be used.

- *Deployment* is as it sounds, deploying the model(s) within the organization so that it serves its intended purpose. This phase may also require customer training/education so they know how to use the model and correctly interpret the results.

Figure 5.4 (next page) summarizes the six component of CRISP-DM and the tasks and outputs associated with each (Chapman et al. 2000).

A critical USAF capability gap that data mining is particularly well suited to address exists in the USAF core function of global integrated ISR (GIISR). The current generation of intelligence products generated from ISR data is dominated by a process of manual data analysis, occasionally augmented by automated "event" detections. This process can often take hours. Autonomous systems incorporating data mining and fusion could, instead, automatically extract patterns of interest and generate products in minutes (Bomberger et al. 2012). If these patterns were extracted using only predefined signatures, they would be valuable simply because of their potential to reduce the human analyst's drudgery work and free him/her to accomplish "higher" cognitive analytical tasks.

A significantly greater potential could accrue with a truly autonomous data mining solution demonstrating cognitive flexibility, which could develop an understanding of novel yet still relevant patterns that need to be recognized, captured, and disseminated. These ISR-focused AS decision aids, driven by extracted information, would be able to change their tasking in a variety of ways, such as initiating new ISR captures, engaging in peer-to-peer flexibility as task demands change, and reprioritizing tasking based on changed mission situations and objectives. These ASs could learn patterns hierarchically in a graph structure by adaptively clustering track reports in a fused feature space merging similar paths, as has been demonstrated in field exercises (Bomberger et al. 2012). Another midterm application of AS-enabled ISR includes link analysis, that is, identifying the relationships between nodes that represent people, places, or events (Picoh et al. 2004). The current manual approach includes taking fragments of evidence from multiple databases and "linking" them together using inductive reasoning and heuristics; but because of human limitations in searches and making connections, the analysis is limited to small datasets. In contrast, an AS-enabled data mining approach could deal with a large number of heterogeneous datasets encompassing a vast array of data, taking fragments of confirming and conflicting evidence from multiple

Business Understanding	Data Understanding	Data Preparation	Modeling	Evaluation	Deployment
1. Determine Business Objectives - Background - Business Objectives - Business Success Criteria	1. Collect Initial Data - Initial Data Collection Report	1. Select Data - Rationale for Inclusion/Exclusion	1. Select Modeling Techniques - Modeling Technique - Modeling Assumptions	1. Evaluate Results - Assessment of Data Mining Results with respect to Business Success Criteria - Approved Models	1. Plan Deployment - Deployment Plan
2. Assess Situation - Inventory of Resources - Requirements, Assumptions, and Constraints - Risks and Contingencies - Terminology - Costs and Benefits	2. Describe Data - Data Description Report	2. Clean Data - Data Cleaning Report	2. Generate Test Design - Test Design	2. Review Process - Review of the Process	2. Plan Monitoring and Maintenance - Monitoring and Maintenance Plan
3. Determine Data Mining Goals - Data Mining Goals - Data Mining Success Criteria	3. Explore Data - Data Exploration Report	3. Construct Data - Derived Attributes - Generated Records	3. Build Model - Parameter Settings - Models - Model Descriptions	3. Determine Next Steps - List of Possible Action Decisions	3. Produce Final report - Final Report - Final Presentation
4. Produce Project Plan - Project Plan - Initial Assessment of Tools and Techniques	4. Verify Data Quality - Data Quality Report	4. Integrate Data - Merged Data	4. Assess Model - Model Assessment - Revised Parameter Settings		4. Review Project - Experience Document
		5. Format Data - Initial Assessment of Tools and Techniques			
		6. Dataset - Dataset Description			

Figure 5.4. Tasks and outputs for each of the six components of the CRISP-DM model. Bold represent tasks, standard text are outputs. (Image recreated from Chapman et al. [2000])

databases and autonomously accomplishing the link analysis. A big challenge here is what is called the graph association problem (Tauer et al. 2013): the generation of separate graphs that capture a description of possibly the same entities and relationships requires merging into a single graph. One AF-relevant application of correct association is the assigning of labels to unlabeled individuals (actors) in a large heterogeneous social network (Bui and Honavar 2013).

Recent advances in AI, for example IBM's Watson and Google's AlphaGo (Ferrucci et al. 2010; Silver 2016), rely heavily on their ability to make inferences across large datasets. The accuracy of these inferences is related to the model's accuracy. In lieu of having access to the underlying processes that generated the data, data mining algorithms need significant amounts of varying data types to build accurate model estimates (inferences of the underlying processes), such as with deep-learning approaches. Big data environments and the analytics that sit on top have served and continue to serve a pivotal role in the advancement of data mining and machine learning. Advances in this area will have significant potential for advancing AS capabilities in dealing with large datasets in the future as well as enabling more efficient and effective machine learning to improve AS behaviors over time (see below).

5.3.4 Situation Assessment and Decision-Making

The **SADM** layer is composed of five serial blocks as illustrated in figure 5.1.[16] We provide a brief overview of each, in the following.

5.3.4.1 Perception and Event Detection. The **perception and event detection** block provides the basic function of transforming the products of sensor processing and data mining into higher level "percepts" or "events" that are important at some level to the tasking and performance of the AS.[17] A percept or an event is any part of the representation of the environment comprehended as a whole; defining how a percept or an event is related to or can interact with other percepts or other events is comprehension of that percept or event (Endsley 1995b).

Variables associated with perception can be continuous variables, such as the six coordinates defining a platform's location and orientation (x, y, z, roll, pitch, yaw), or they can be discrete, such as presence or absence of a user's log-on. Providing good estimates of these variables is the function of percep-

---

16. Again, we recognize that these functions are most likely more interconnected, but we have deliberately simplified the flow for exposition purposes.

17. This is the equivalent of the Level 1 function of Endsley's model of Situation Awareness, "Perception of Elements in Current Situation" (Endsley 1995a).

tion at its lowest level, and engineers have accomplished this in a variety of fashions: using the sensor signal itself as the estimate, assuming the sensor is relatively "noise free," or using multisensory, multivariable optimal estimators, such as the Kalman Filter frequently used in flight control systems (Kalman 1960; Gelb 1974). The Optimal Control Model of the human operator/pilot incorporates a version of the Kalman Filter for its perceptual "front end" and has been shown to replicate human behavior in a wide variety of "human-in-the-loop" tasks of the type noted in section 4.1 (Kleinman et al. 1970).

Variables associated with *event* detection can be thought of as semantic variables or states such as "high altitude" or "illegal log-on." As such, detected events can also be thought of as a generalization of recognized objects at some point in time. A variety of methods might be contemplated for mechanizing the event detection function. For example, to transform perceived (estimated) continuous variables, one could simply "bin" values, for example, so an altitude could be categorized as either low, medium, or high. That binning function, in turn, can be accomplished with simple "hard-edged" threshold functions or via the use of "softer" fuzzy logic categorization and event set membership strength (Zadeh 1965, 1996).

Perception and *event detection* functions do not need to be separated and sequential as just described. For example, in the application of ANNs to image recognition, we might have a sensor's pixelated array as inputs to the block we are describing and a single output node of the ANN driving the output of the block, declaring the presence or absence of a cat video, based on a simple scalar value between zero and one. The net result is the same: sensor inputs in and a semantically categorized event/object out. Naturally, with the appropriate mechanisms to accomplish the task, this block might also combine several events into a higher-level event—for example, a simple rule-based system or a more complex fuzzy logic inferencing engine. This is discussed shortly.

Three key issues in developing an effective perception and event detection block are dealing with: (1) dynamic changes in the input over time and maintaining proper semantic categorization; (2) noise in the inputs; and (3) unexpected events, that is, events that the system was not designed to detect (e.g., a dog video) but that are still important to handle to complete the task or mission assigned the AS. Current approaches that use fixed dynamic models to estimate continuous variables buried in noisy signals, and that use predefined "templates" or "detection rules" to categorize the input signals from the sensor and data mining blocks, may run into trouble when systems fail and/or the unexpected or the previously undefined situation occurs.

To achieve the cognitive flexibility in the area of detection of novel events, solutions for the "zero-shot" learning problem have to be developed (Burlina et al. 2015). Direct classification is not normally achievable without training exemplars for all the classes of objects to be encountered. One approach to solving this problem is the extraction of semantic attributes that can be associated with attribute relationships. Another extremely promising approach to zero-shot learning is via cross-modal transfer (Socher et al. 2013). By extracting the distribution of words in texts as a semantic representation, generating knowledge about the visual world through natural language, and applying that to classify untrained objects, a reasonable level of performance is possible on unseen classes of objects or events, where accuracy for known classes is traded for accuracy in novelty detection.

5.3.4.2 Situation Assessment. The **SA** block serves to transform the events (and associated perceived states as needed) into SA that supports the successful accomplishment of the task/mission assigned to the AS. In effect, this block supports the comprehension of what the events *mean* with respect to accomplishing the task/mission, that is, understanding the current situation and how the current situation is likely to evolve into the future, that is, the projected situation.[18]

As described in section 3.2, the first component, "current SA," refers to awareness based on signs denoting elements in the environment *external* to the AS, including the AS mission and task (which may change over time via the associated C2 system with which it interacts), the overall environment/context in which the AS is operating (especially if it is a platform), the basic task/mission SA elements needed for success, the status of its teammates (if any), and, if available, the status of its adversaries. SA also depends on signs from the environment *internal* to the AS, including self/health awareness (including the health of all associated subsystems that the AS "owns"), the behavior and performance relative to what is expected normally, and the AS capability margins based on a knowledge of the AS's self-knowledge of its nominal operating envelope and where it is currently operating within that envelope. Ideally, all or most of these components of both external and internal SA would have associated confidence assessments to provide indications of uncertainty in the assessments, so that future decisions based on these assessments can fold uncertainties into the risk calculus underlying decision-making. How those confidence assessments are calculated will obviously depend on the scheme used for generating SA, which we discuss shortly.

---

18. This is the equivalent of the Level 2/3 functions of Endsley's model of Situation Awareness, "Comprehension Current Situation" and "Projection of Future Status" (Endsley 1995a).

The second component, "projected SA," refers to an understanding of how the current situation[19] is likely to evolve into the future, based on relevant signs, over the mission/task time horizon of interest. This is relatively straightforward for, say, simple physical systems in which the dynamics can be well-modeled (e.g., an orbital flight path of a satellite), but as the underlying systems dynamics become less determined by well-known physical laws (e.g., individual adversarial decision-maker behavior) and/or the systems become more intentional and complex (e.g., teams of adversaries and defensive systems), predictions become increasingly difficult to make with confidence for a given time horizon. Nevertheless, projections are often made, at least by humans, based on experience, qualitative "mental models" of the underlying drivers of situational evolution based on that experience, and an ability to "run" those mental models on the current situation to generate one or more likely outcomes, to project the current situation into the most likely one, some period of time into the future (Klein 1998, 2008). Naturally, as the time horizon grows, predictions also become more difficult to make, as happens with even well-grounded models of purely physical systems.

Various techniques have been used in developing quantitative models of *human* SA, as described in the Situation Awareness chapter of Pew and Mavor (NRC 1998). For example, *production rule systems* have been used to match a current set of events (e1, e2, e3, ...) to a predefined *situation* S_1, when all the events are deemed to be "true" or "sufficiently true." Closely related approaches that attempt to account for close but not perfect matches between event sets and situations include *expert systems* (with confidence factors; Nikolopoulos 1997), *complex event processing* (CEP; Luckham 2002), and *case-based reasoning* (CBR; Kolodner 1993). More sophisticated approaches taking a probabilistic approach to "matching" event sets with situations, where the events have associated likelihoods or probabilities of being true, include *Bayesian belief networks* (Cooper 1990), in which a node in a BBN network may define a situation and assign it a probability of "trueness, depending on the event states and the network model employed by the SA model. Such models or approaches could serve as the basis for an AS's implementation of the "current SA" component block. Comparable models of how humans develop projected SA are less well developed, most likely because of a fundamental lack of understanding of how humans actually project current situations into the future—for example, whether they rely on temporal projections or simply base future forecasts on current/future pair matching based on past experience.

---

19. Actually, the assessed current situation, which, for any number of reasons (e.g., poor sensors, adversary deception, etc.) may significantly deviate from the "actual" or "ground truth" situation.

This may be less of a constraint in developing a capability for "projected SA" in an AS, however, since there exist well-understood methods of simulation-based temporal projection (extrapolation), assuming the existence of quantitative models underlying the projection (Ziegler et al. 2000).

Whatever method is used for implementing an SA function, present or future, two key features already exist that can extend the basic functionality of assessment based on simple component events/objects. First is the capability to reverse the assessment function so that, given a hypothesis of the current situation, one can infer "missing" events that were either not detected or were associated with low probabilities of occurrence. This capability can then provide guidance to the perception and event detection function to go look for evidence confirming the missing events and the overall assessment (or, conversely, confirming that the events have not occurred and that the hypothesized situation is not true). In an ES mechanization, this might be implemented via backward chaining, starting with the hypothesized situation and inferring which antecedent events must be true, given the ruleset; in a BBN implementation, this might be accomplished via a sensitivity analysis of the network, to effectively generate the first partial derivative of a given situation relative to a given event (or evidence, in BBN parlance). Related to this capability to reverse the *current* assessment function is that of reversing a *future* hypothesized assessment to generate an associated current hypothesized situation that could evolve into the future situation. In a current/future pair matching scheme this would be straightforward; with a simulation-based temporal projection scheme, this could be considerably more difficult, since the projection models would have to work in reverse. The second key feature that could extend functionality is the ability to provide some associated probabilistic assessment of current and projected SA. For continuous variables, this could be simple statistical measures of variance or covariance, either static or growing over time; for discrete variables, this could be associated likelihoods or probabilities of "trueness." This kind of information associated with one or more assessed situations could then be used "downstream" in subsequent decision making and planning activities, where decisions and plans can be informed by the likelihood of—and possibly the confidence in—one assessed situation over another. How these probabilistic measures are generated will depend strongly on the underlying method employed for generating the SA estimates themselves.

5.3.4.3 Reasoning and Decision-Making. The **reasoning and decision-making** block serves to transform the assessed situation(s) into decisions for planning and replanning as well as for current and/or future actions. The basic problem can be broken down as one of how does one "reason" to an appropriate decision. Reasoning can occur in one of several ways. For example, reasoning can involve judgments that maintain *coherence* among sets of symbols (linguistic, numerical), such as when rules of logic are employed or a mathematical equation is solved. The coherence approach can be found in the literature on rational choice theory (e.g., von Neumann and Morgenstern 1944; Bernoulli 1954). As another example, reasoning can entail judgments that establish *correspondence* with elements in the real world when recognizing situational patterns (Klein 1997, 1998, 2008), such as when inferring the occurrence of a robbery when entering a restaurant based on a complete absence of staff.

With increasing automation, many tasks that were solely correspondence-driven for the human—such as pilots navigating using an out-the-window view—become supported by coherence-driven approaches—such as pilots navigating by looking at numerical displays that represent aspects of an out-the-window view (Mosier 2009). Either reasoning approach can then be taken to render a decision.

If one is to base reasoning on the coherence approach, one would consider issues such as probability, rules, and utility. Accordingly, one would consider:

- The current and projected situation or the set of situations and their likelihoods

- The mission/goal/task objectives

- The resources available

- The constraints, either physical or those imposed by rules of engagement (ROE)

- The likely outcome of the decision taken and its associated value or "utility" (see below)

The answer might be in the form of a simple and single decision or in the development of a multistep plan (see section 5.3.4.4, below).

A range of decisions can be considered, but here we make the simple distinction between fast/reflexive decisions and slow/deliberative reasoning, based on findings in the human decision-making literature. In the former category, a leading model of human decision-making is *recognition primed decision-making* (RPD), which describes how seasoned experts make timely decisions in complex situations that are often characterized by uncertainty: they identify the salient features of a problem, quickly assess the situation, and then decide, based on their expertise and experience (Klein 1998). In the latter category, a number of "rational actor" models exist that are based on work conducted by the operations research, game theory, and economic analysis communities, which focus on expected utility to determine an optimum decision. Expected utility is based on the situation, its likelihood, and the utility (or payoff) for making a given decision. The optimum decision is then chosen on the basis of maximizing the expected utility across the set of all possible decisions (von Neumann and Morgenstern 1944).

For both general approaches, there are several variants. For example, under RPD theory, there are three basic variants: the most basic, in which an assessed situation leads directly to a recommended decision and action plan; one of an intermediate level of complexity, in which a given situation leads to a recommended "family" of decisions/actions, which then need to be more closely evaluated and selected from based on the specifics of the given situation; and a more complex mode in which humans effectively conduct mental "simulations" of potential decisions/actions to select a "best" decision based on the projected outcome (Klein 1998; Pew and Mavor 1998). Likewise, there exists a number of variants under utility-based decision theory: utility itself can be a complex function of an individual's personal preferences (Kahneman and Tversky 1979); the selection of the "optimum" decision need not actually be optimizing but rather simply satisficing, as in rank-dependent utility optimization (Tversky and Kahneman 1992); incorporating multiple dimensions of preferences via multi-attribute utility (Keeney and Raiffa 1993); and using game theory to project future actions by competing entities over a number of "moves" or a specific time horizon (Aumann 1989). With a utility-based decision-theory approach, several optimization techniques can be applied, and the selection of the best will depend on the characteristics of the utility function being optimized, such as convexity, linearity, and so on (e.g., Aoki 1971; Onwubolu and Babu 2004).

If one is to base reasoning on the *correspondence* approach, situational pattern recognition would play a key role. One reasoning method involving situational pattern recognition is case-based reasoning (Russell and Norvig 2010). Generally, pattern-recognition-based reasoning can be implemented with straightforward production rule models if a single situation is plainly evident. More sophisticated methods need to be brought to bear for situational uncertainties or if a single decision or course of action needs to be selected from several candidates, via optimization, "fast time" simulation to assess potential outcomes, or alternative methods for elaborating on basic decisions and outcomes.

For either of the approaches, if the decision is not a simple single-step decision yielding a single outcome, a more complex series of sequential actions—and thus planning—may need to be initiated.

Future technology development will need to mature decision-support tools based on an assessment of the advantages and disadvantages of coherence versus correspondence approaches to reasoning—for example, the inability of coherence approaches to handle unforeseen circumstances and the lack of statistical relevance found in correspondence approaches. These disadvantages may be overcome, however, if the AS has the cognitive flexibility to choose the appropriate approach or combination of approaches and capacity to carry out the more complex tasks generated by this "hybrid" reasoning approach.

With respect to hybrid design, Murphy and colleagues describe a hybrid robotic design in which both reactive and deliberative processes are present in the system (Murphy et al. 2002). Reactive processes refer to robots operating in real-time using sensors, actuators, and processors with no memory. Deliberative processes refer to more cognitive-oriented functions, such as planning, remembering, monitoring self-performance, or reasoning about the state of the robot relative to the world. Because planning covers a long time horizon and requires global knowledge, it is assumed that planning should be decoupled from real-time execution. Thus, one would first PLAN—involving deliberation, global world modeling, task or path planning—and then SENSE-ACT—involving innate and learned behaviors. The hybrid architecture may be one of the best general architectural solutions (Murphy et al. 2002). An example of a hybrid robotic architecture is the Autonomous Robot Architecture (Arkin 1989), which is based on schema theory and is the oldest of the hybrids.

The hybrid reactive/deliberative design appears analogous to human cognition: human reasoning and decision making are governed by two sets of cognitive processes that can be dissociated experimentally and neurologically (Patterson 2017). One system is called *Analytical Cognition*, which entails conscious deliberation drawing on limited working memory resources: it is effortful, rule-based, symbolic, limited in capacity, and slow. The other system is *Intuitive Cognition*, which entails unconscious situational pattern recognition unconstrained by working memory limitations: it is independent of conscious "executive" control, large in capacity, and fast (Patterson and Eggleston 2017). It is tempting to draw a comparison between deliberative robotic design and human analytical cognition on the one hand and between reactive robotic design and human intuitive cognition on the other. Whereas the cognitive-oriented functions of deliberative design do seem to match up well with the characteristics of human analytical cognition, the reactive design does not match up well with the properties of human intuitive cognition. In the latter case, reactive design is relatively low-level whereas human intuitive cognition is surprisingly high-level and involves situated meaning making (Patterson and Eggleston 2017). Thus, to approach the level of sophistication offered by human intuitive cognition, reactive robotic design would have to be refashioned as a form of anticipatory, proactive design based on situated meaning making.

5.3.4.4 Planning and Replanning. The **planning and replanning** block provides a means of elaborating on a decided course of action, in which multiple steps comprise that course of action. This block is composed of the following subsidiary functions: (a) planning, which is essentially determining the sequence of actions that need to be taken to move from an initial situation or "state" to a desired goal condition that may exist in one or more states, to satisfy the objectives of the previously described reasoning and decision-making activities; (b) scheduling, the temporal assignment of the action sequence to a timeline, overlapping actions depending on their interdependence; and (c) replanning, which occurs to adjust the plan based on changed objectives, failed plans, changed circumstances (as seen in the assessed situation), etc.

A broad survey of *human military planning activities* is provided by the Planning chapter in Pew and Mavor (1998), covering information collection, SA, course of action (COA) development, COA analysis and selection, plan monitoring, and replanning.[20] A dozen different computational models of these planning-specific activities are also reviewed, with underlying methods that fall into one of four general approaches: (1) production rules or decision tables that closely follow written planning doctrine; (2) combinatorial search or genetic algorithms, to develop, respectively, sufficing or optimal plans; (3) planning templates or CBR to support plan elaboration; and (4) simulation-based planning to quickly generate, evaluate, and iteratively modify plans.

Pew and Mavor (1998) also review more generic planning methods developed by the AI community, based on means-end analysis,[21] in which a plan is viewed as simply "a sequence of operators transforming a problem's initial state into a goal state." Work since then has attempted to address these shortcomings that limit applicability to "real world" problems (e.g., real-time sensor-driven path planning for robotics), with approaches including *reactive planning* to deal with a changing environment, short time horizons, and continuous feedback on plan success/failure; *refinement methods* for plan elaboration at different abstraction levels; incorporation of *domain heuristics* to speed planning; *temporal modeling* for dealing with plan dynamics; and *nondeterministic*

20. Which is a broader range of activities considered here, namely on COA development, analysis/selection, and replanning as required.

21. Starting with the General Problem Solver means-ends analysis paradigm introduced by Newell and Simon (1963).

methods for dealing with noisy partially observable information and nondeterministic systems (Ghallab et al. 2004, 2016). Underlying many of these approaches are several different computational techniques and methods, too numerous to list here but well described by Ghallab et al. (2004, 2016).

Planners can exhibit task *flexibility* through the generation of plans or policies that better cover all possible contingencies. This occurs as contingent, conditional, or conformant planning or as a form of planning under uncertainty (probabilistic planning) (Ghallab et al. 2004). A contingent plan is a plan that includes action sequences for situations in which the goal state is not reached and how to correct to meet the goal condition. A conditional plan is one that includes branches in the action sequence solution for actions that may result in multiple outcomes. Executing a conditional plan requires that the system be able to determine which state the system is in (observable) to execute the correct action. A conformant plan is a plan that includes actions that will reach a goal state when the actions may have multiple outcomes but does not require observability of the state.

Conformant planning can be conducted through model checking or through classical techniques operating on transformations of the problem space into other representations such as Binary Decision Diagrams (Ghallab et al. 2016). The output of a conformant planner is considered a strong plan. A strong plan is one in which the resulting plan includes all nondeterministic branches and has solutions that lead to the goal state or a set of goal states that all meet the goal condition. This is in contrast to the "classical" weak plans that are a single shot plan from initial to goal conditions and rely on execution management in the face of nondeterminism.

The planning community has not conducted a conformant plan planning competition within the last 10 years. However, there have been significant planning advancements that are related to cognitive flexibility. These advancements, discussed in the next section, can be directly applied to this domain through recent advancements in polynomial time algorithms that convert contingent plans to problems that can be solved with classical planning algorithms (Palacios et al. 2014).

Planning under uncertainty often formulates the problem as a Markov decision problem (MDP) or partially observable Markov decision problem (POMDP; Kaebling et al. 1998). Both formulations leverage the Markov assumption, that the current state captures the history of past states and actions, as well as solve for some form of a policy that captures which action should be taken in every state to maximize future expected utility.

The most recent developments have included the development of the Relational Dynamic Influence Diagram Language (RDDL) used in the 2014 International Probabilistic Planning Competition (IPPC; Vallati et al. 2015). The RDDL extends research on the abstract decision diagram (ADD; St-Aubin et al. 2001) and binary decision diagram (BDD; Hoey et al. 1999). These approaches use influence diagrams to represent more concisely the domain and increase the speed of solution. The current best-performing planners, PROST (Keller and Helmert 2013) and G-Pack (Kolobov et al. 2012), leverage either a variant of Upper Confidence Bound 1 applied to trees (UCT; Kocsis and Szepesvari 2006) or modification to the Labeled RTDP algorithm (Bonet and Geffner 2003). The UCT algorithm performs a search by sampling paths from the start to the goal and updating policy values based upon the rewards received. The Labeled RTDP algorithm leverages heuristics and learns the policy through repeated greedy searches of the policy space, labeling states as solved as they converge to improve convergence.

Peer flexibility with planners has tended to focus on multiagent and distributed planners. It does introduce several difficulties, such as interagent constraints, distributed planning techniques, preferences and resource demands, human interaction, and privacy (de Weerdt and Clement 2009). Several of these interactions are formalized in the coalition formation (Chalkiadakis et al. 2009) and multirobot task allocation (Gerkey and Mataric 2004) literature. More applicable to AS peer flexibility is dealing with failures in the plan, the peer, or interacting with human participants. Failures in the plan and peer require an execution monitor and replanning and repair mechanism (Komenda et al. 2012). Alternatively, having the agents be entities in an emergent system leads to a very robust implementation but due to homogeneous implementations lacks a leader that can set intent (Genter and Stone 2016). Work on understanding how social leader emergent and goal setting can be done is an area of future research.

When working with humans, recent work leverages probabilistic plan recognition that identifies possible plans of the human and then looks at compromise, opportunism, and negotiation as possible deconflictions (Chakraborti et al. 2016). This does require models of the domain and of the user.

Cognitive flexibility–related research has focused on systems that consist of several search algorithms, called planning portfolios (Nunez et al. 2012; Helmert et al. 2011; Roger et al. 2014; Valenzano et al. 2013), and heuristics (Torralba et al. 2014). These planners perform a computational load balancing during the search to identify the algorithm or heuristic subset that appears to be solving the problem quickest.

The planners that won the most recent International Conference on Automated Planning and Scheduling competition in 2014 tended to leverage this strategy. For example, in the sequential optimal category, SymBA* (Torralba et al. 2014) alternated between searching using heuristics and performing a reverse direction search to construct abstract heuristics to improve the search results. Similarly, IBACOP2 (Cenamor et al. 2014) leveraged 11 planning algorithms and learned heuristics to identify the subset of planning algorithms to use on a domain based on features in the domain description.

The organizers of the competition did note that "we observed a worrying small number of planners able to deal with preferences and temporal models."[22] We would like to add to this by noting that the planning portfolios also tend to be missing algorithms exhibiting the latest developments from the evolutionary computation community (Kim et al. 2003). There are also significantly fewer planners competing in the uncertainty tracks.

In summary, although Ghallab et al. (2014) discuss the broadening and crossover between different subcommunities, there is still need for further encouragement in this area. Example areas include:

- The multiagent systems research that focuses on peer flexibility does not appear in the planning research that is predominantly deterministic; there isn't a planner submitted spanning the gap between deterministic and uncertainty planning domains.

- The developments in the evolutionary computation community on satisficing algorithms do not seem to be being incorporated into the satisficing planners and planning portfolios.

- Portfolio planners have not migrated into the stochastic planning domains.

A way to encourage this cross flow of ideas is to encourage the planning community to tackle all three of the flexibility principles in their competitions, rather than specialized pipelines.

5.3.4.5 Execution Management. Execution management serves two purposes. The first is as an *effector manager*, a method that converts planned actions into commands and reacts to events. It serves as a manager and coordinator of the effectors that execute the plan. The second is as a *plan execution monitor*, monitoring the situation, the status of the system, and the success or failure of the steps of the plan being executed.

As an *effector manager*, this block issues a range of commands, from very high-level subsystem commands (e.g., locomote to a specified waypoint) to very low-level component commands (e.g., set the arm's joint angles as specified). The level of specificity will depend on the AS and the level of effectors being

---

22. *See* ICAPS 14 results at https://helios.hud.ac.uk/scommv/IPC-14/repository/slides.pdf.

managed. In hybrid robot architectures, this is often termed a sequencer (Murphy et al. 2002).

As a *plan execution monitor* this block depends on inputs from the other blocks in the SADM layer. For example, it relies on detecting events that signal success or failure of a step in the plan, a change in the current state that may change one or more of the broader conditionals upon which the plan was based, a change in the goals that motivated the generation of the plan, a reconsideration of plan step sequencing, or some combination of these factors. Murphy et. al (2002) groups the large number of approaches available into a performance monitoring and problem-solving component. Often, in three-layer architectures the sequencer includes these capabilities (Gat 1998; Pettersson 2005).

Effector management and plan execution monitoring have been key functions in many AI planning and autonomous robotics efforts, and there is a mature technical basis for these functions. For example, the System for Interactive Planning and Execution (SIPE; Wilkins 1988) is a goal-oriented hierarchical planner that can deal with different levels of abstraction, so it can effectively decompose higher abstractions into lower level effector commands, serving as an effector manager. SIPE also replans during execution, based on its monitoring of how well the plan is being executed, so that it can conduct "plan repair." Building on SIPE is the Continuous Planning and Execution Framework (CPEF; Myers 1998) developed for planning and replanning in a highly dynamic and unpredictable environment—for example, one supporting a "Joint Forces Air Component Commander (JFACC) in the execution of realistic air campaigns" (Myers 1998). CPEF uses a *plan manager* to oversee execution of the planning, execution monitoring, and plan repair functions. O-Plan (Tate et al. 2000), another hierarchical planning system, similarly provides for goal-oriented plan generation and plan repair in the face of dynamically changing situations that were not anticipated in the original plan, via monitoring of plan execution. A multiagent system approach to plan monitoring is discussed by Dix et al. (2003), in which the plan generated by a declarative planner is compared with one achieved by the MAS, to detect possible agent collaboration failures.

In addition to these direct plan manager and monitor functions, this block can also serve as a more general monitor of higher-level assessment of AS behavior and performance. On the behavior side, the block could serve to monitor, for example, where in the "design operating envelope" the AS is currently operating—specifically, how close it is to an envelope boundary (in terms of capability, expertise, training, etc.) and the likelihood of violating that boundary over some future temporal window. On the performance side, the block could assess higher-level mission/task performance (not just plan success) to provide the AS with self-awareness of overall performance for potential correction or remediation, via, say, reconfiguration or repair. How these self-assessment functions are divided up between this block and that of the SA block is clearly a design decision that needs to be made for the specific domain and application of the AS, but it should be clear that self-awareness and behavior/performance awareness need to be closely linked.

Recent surveys specific to execution management include Pettersson (2005), which leverages the industrial control application taxonomy for execution management to robotics and autonomous systems related work. The more recent survey by Ingrand and Ghallab (2017) finds that even in systems since Murphy et al. (2002) there are three primary functions of planning, acting, and monitoring and that future work needs to include additional design dedicated to integration that allows for more flexible relationships as well as sharing and comparative testing.

5.3.5 Effectors

The **effectors** block provides for the major "output" of the AS, here serving as a general placeholder, based on the situation and the desired outcome, for three separate classes of effectors:

- Those driven by the execution management block as just described, covering a range of effectors "internal" to the AS itself,[23] from high-level subsystems to low-level electromechanical components. On an aircraft platform, for example, this might include a flight control subsystem (FCS) given "locomotion" commands to achieve higher level goals by the AS, an auxiliary power unit (APU) given start up commands under a power failure scenario, or low-level actuator commands given to a bomb bay door opening component. Alternatively, in a supervisory control and data acquisition (SCADA) environment, this might include networked SCADA subsystems; and within a SCADA subsystem, this might include lower-level programmable logic controllers (PLC) or digitized proportional-integral-derivative (PID) controllers that serve to control the SCADA process.

- Those effectors "external" to the AS, directly interacting with elements or entities outside of the AS (as shown in the "cloud" portion of figures 2.1a, 2.1b, and 2.1c). Again, in a platform application this might include an

---

23. We exclude here actively controlled sensors, even though they are actively controlled subsystems, since sensor management is intimately tied to sensor processing, "active" sensing, and situation assessment. We provide for this separate loop closure via the feedback loop in figure 5.1, from the execution management block directly to the sensor management block.

EW subsystem (for example, a defensive chaff/flare subsystem or an active radar countermeasure subsystem) or a more traditional kinetic weapons resource management subsystem that serves to prioritize targets and conduct weapons-target pairing to maximize probability of target kill at minimum expense of weapons (e.g., missiles). Alternatively, in offensive cyber operations, this might include hundreds of thousands of infected computers to launch a distributed denial of service (DDOS) attack, or a single well-crafted packet to analyze and exploit vulnerabilities in a network, or new zero-day exploits that have not been patched yet.

- Those associated with explicit communications to entities outside of the AS. Again, on a platform, this could be an ISR download to a ground control element or a flight formation coordination update to a swarming teammate.[24] In a networked C2 environment, this could include battle management communications from commanders, between peers, and to subordinates. A wide variety of machine-to-machine protocols already exists to support these communications, from standards associated with the internet (TCP, UDP, TCP/IP, etc.) to specialized application protocols serving particular services and devices (MQTT, CoAP, LWM2M, etc.). However, if these standards and protocols are to be used, then they will need to be encrypted and secured to prevent eavesdropping and/or data tampering during transmission.

As we noted in section 5.3.2 on sensors, a wide variety of actuators exists spanning many modalities and levels of complexity. However, at a sufficiently high level of abstraction for AS considerations, an actuator may be regarded basically as an element that translates an output "command" generated by the AS to an entity, subsystem, or component existing in either the outside environment surrounding the AS or in the internal environment of the AS itself. Additional outputs come from communications from the AS to outside entities.

We should also note that with the use of all kinetic and nonkinetic effectors, timely and reliable BDA is called for if the AS is to efficiently execute its mission. The situated agency provided by the AS's closed-loop sensor feedback can provide the needed measurements for this assessment.

We now discuss four layers in the AS diagram, going from the top down: The **Human-Computer Interfaces and Collaborative Environments** layer,

---

24. Again, we have "bundled" communications with the effectors block, this time outbound, as we did with the sensors blocks for expository simplicity, recognizing that the C2 communications aspects may very well deserve their own explicit representation in the proposed framework at some later point.

the **Learning and Adaptation** layer, the **Domain-Specific Knowledge-Base** layer, and the **Toolsets and Technologies** substrate.

5.3.6 Human-Computer Interfaces and Collaborative Environments

The **HCI** and **CE** layer provides a multipath interface between the AS and one or more human operators, where the HCIs provide multiple human-AS interfaces and where the *collaborative environment* provides multiple human-human interfaces needed for collaboration among human members of a team. Because the CE portion is basically outside the scope of AS functionality, we will focus on the HCI portion in the discussion here.[25] This combined block is basically an elaboration of the simpler HCI block shown earlier in figure 2.1b, extended to support multiple interfaces with multiple human operators, serving as supervisors, peer-to-peer teammates, or subordinates.

The importance of well-designed AS HCIs was discussed in 2015 by the Air Force Office of the Chief Scientist in *Autonomous Horizons*, volume I (Endsley 2015c), a work based on decades of experience with automated systems. Issues that arise with poorly designed systems include the following:

- Difficulties in creating autonomy software that is robust enough to function without human intervention and oversight

- The lowering of human situation awareness that occurs when using automation, leading to out-of-the-loop performance decrements

- Increases in cognitive workload required to interact with a greater level of complexity associated with automation

- Increased time to make decisions when decision aids are provided, often without the desired increase in decision accuracy

- Challenges with developing a level of trust that is appropriately calibrated to the reliability of the system in various circumstances

As noted in chapter 3, these issues can be addressed and ameliorated by developing robust and proficient ASs, providing cognitive congruence and transparency, enabling mutual situation awareness, training for human-

---

25. Although one might consider the need for a full-scope CE—including the AS—in situations in which collaborative AS-dependent teaming is critical for task success, for example in multi-operator machine-aided planning. The general design requirements of such a CE is beyond the scope of this report, however.

system teaming, and ensuring the incorporation of "good" human factors engineering when developing HCIs. Here we focus on HCI requirements.

Given that the development of effective HCIs, in general, has received considerable attention for many decades and in several professional organizations (e.g., ACM, HFES, IEEE, NDIA, etc.), we focus here on the design requirements for *AS-specific HCIs*. Based on the *general* recommendations of chapter 3, AS HCIs should accomplish the following:

- Because AS proficiency is key to engendering human trust, an AS HCI needs to provide a sufficiently broad range of behavioral and performance metrics (as noted above in the execution management discussion above), associated with both internal system "health" and external mission-related accomplishment. The HCI needs to provide both a broad overview of AS health/mission status and a means for "drill down" by the operator/teammate to support diagnosis, reconfiguration, or other activities needed to deal with anomalies, system failures, or the like.

- If cognitive congruence and transparency (explainability) are built into the system as recommended in section 3.1, then the HCI should support both. For example, providing *congruence* may be a simple as providing, at a high level, a graphic depiction of the framework or architecture of the AS, with tags of current activity occurring at each "block" in the graphic, with drill-down capability supported at each block. Providing *transparency* via the HCI may require higher levels of interaction between the AS and the human teammate, in which "conversations" might address questions like "Why do you assume that?" or "Why do you want to do that?" To allow some level of inspection into any functional block of the autonomous system would also likely be desirable operationally.[26] For example, an operator may request to know all of the events that were detected by the AS, what operating assumptions the AS is holding, what its SA is given those events and assumptions, how confident the AS is in terms of assessing alternative courses of action, and then perhaps how well the AS thinks the execution of a plan is going, relative to the desired outcome. This detailed look into separate functional blocks of the AS is indicated in figure 5.1 by the several arrows connecting the HCI/CE to the blocks in the SADM layer.

- Because AS situation awareness of self and environment is so critical for proficiency, the HCI needs to be able to support accurate displays of

---

26. And likely during development for troubleshooting, and during test and evaluation, for verification and validation.

same. As noted earlier, SA spans a range of key knowledge, from health of individual AS subsystems, to overall AS health and location in the AS design envelope, to task- and mission-related performance. How this information is portrayed to the human teammate via HCI displays is a significant design problem in itself, with a considerable literature beginning with alarms and warnings (Fitts and Jones 1947) and evolving to more insightful status displays of system operation (Card et al. 1983; Kieras and Meyer 1997). Going the other direction, an HCI would ideally provide the AS with any relevant and potentially useful SA held by the human and not readily available to the AS. How this would be visualized or verbalized by the human, through the HCI, is a matter of design and probably highly domain dependent. However accomplished, the design goal of the HCI would be to enable the passing of relevant SA information in both directions, to facilitate the holding of mutual SA needed for team proficiency, as discussed earlier in section 3.2 and illustrated in figure 3.7.

- To support timely and effective human-system teaming, the HCI needs to enable straightforward and relatively effortless (on the part of the human) bi-directional communication during task planning and execution. This will likely involve a combination of verbal and visual cues generated by the AS, and verbal and graphical user interface (GUI) activities on the part of the human, necessitating speech synthesis and understanding on the part of the AS. The HCI will also need to support more abstract communications associated with the explication and adaptation of common and complementary roles and goals, especially in dynamic retasking situations when C2 relationships may change. Finally, because the HCI needs to be able to support virtual and live human-system training prior to fielding, considerations should be given to augmenting the HCI with information that can accelerate developing proficiency, such as augmented displays providing "truth" data, decision aids that can make suggestions, planning aids that can guide retasking decisions, etc.

These are desired generic HCI functions and characteristics. Ones that are more *specific to "in-motion" AS applications* include the following (Endsley 2016):

- Pilot control interfaces that adhere to military standards for human factors of vehicle control systems

- Data integration to reduce workload

- Multisensory cues to compensate for loss of haptic and auditory information

- Improved spatial awareness of the environment and relevant objects in the environment to include self-orientation, wayfinding, contextual awareness, and an understanding of SA limitations

- Predictive displays to help compensate for time lags

- Displays to support understanding and projection of AS operations (including monitoring, diagnosis, and mission/payload management), rapid shifts in level of control between the pilot and the AS for various functions, and real-time assessments of trust in the AS

- Displays to support coordinated action with manned platforms, multiple AS platforms, and other teammates (e.g., analysts and commanders)

HCI requirements *specific to "at rest" AS applications*, are, unfortunately, not nearly as well characterized. For example, in applications supporting air operations in general (e.g., ISR collection management and analysis, air tasking order [ATO] generation and updating, BDA, etc.) and in non-air operations (e.g., cyber defense/offense, space operations) HCI designs are often "one-offs" with little standardization across applications and systems. However, there are still several HCI design principles—for visual displays at least—to be followed to support intuitive understanding and efficient use of screen space; as noted by Gouin et al. (2004), these include:

- Interface design. Careful design of the screen real estate and interface widgets to ensure an efficient interface

- Hierarchical representation. Outline/tree views to present information using a hierarchical representation, with the ability to expand or collapse certain hierarchies selectively

- Object explorer widget. Also used to present a hierarchical view of objects, but selecting an object leads to a different visual representation

- Information categories. Subdivision of the information into meaningful categories, using sub-areas and tab folders

- Multimedia information. Use of multimedia (video, imagery, alarms) such as TV feeds, reconnaissance video, and collateral imagery to enhance situation awareness

- Hyperlinks. Use of hyperlinks to provide association between information elements and a capability to drill down into the information

- Multiple views. Information must be presentable in multiple views

- Drag and drop. The user can pass information easily between two applications or tools using drag-and-drop operations

- Animation. Use of animation to display temporal information, for example, animating a plan as it evolves over time

More recent efforts in more abstract data visualization have not only incorporated good HCI design but also taken advantage of an understanding of human visual perception (Ware 2013), the use of good design principles (Steele and Iliinsky 2010), and the availability of faster, cheaper digital graphics. A wide variety of texts guides designers of more complex displays in which the data is more abstract, multidimensional, interrelated, dynamic, and stochastic; in addition, there are a number of professional societies and conferences devoted to data visualization, including the IEEE Visual Analytics Science and Technology (VAST) conference, and the ACM Special Interest Groups on Computer-Human Interaction and Computer GRAPHics (SIGCHI and SIG-GRAPH). Designers of modern displays of an increasingly information-centric Air Force, with or without deployed ASs, would do well to be guided by these texts and to be active participants in these professional societies.

The sister function to data visualization, direct data manipulation, has progressed little from the basic desktop object functionality (selecting, dragging, changing size, etc.) developed in the 1970s by Xerox PARC among others, commercialized in the 1980s by Apple, and described by Shneiderman (1983), and it certainly has not reached anything near the envisioned functionality portrayed in the science fiction movie "Minority Report" (2002) where an operator can directly manipulate "swarms" of data objects in three dimensions while simultaneously visualizing it. Significant opportunities therefore exist for innovative developments, especially as big data starts to dominate areas such as multi-INT fusion, logistics trend analysis, and so on.

Complementing data visualization and manipulation is natural language interaction between human and machine. Clearly, a wide variety of speech UI systems exists, from limited implementations like text-to-speech generators for verbal alarms and word recognition systems that can support menu selection, to more advanced implementations that can support an approximation of a conversation between human and machine, like the original ELIZA program developed by Weizenbaum (1966), to more recent efforts aimed at providing semantically and syntactically correct conversational responses (e.g., Litman et al. 2000; Shang et al. 2015) or interactive machine translation (Green et al. 2015). Such systems can support limited speech interactions with ASs, but progress still needs to be made if the conversations are to be made more "natural"—that is, for the human—and "deeper," in terms of the machine truly understanding what's in an image and describing it accurately or providing a rationale for the concepts that lie below the surface of simple statements in a conversation. Significant progress is being made, however, via the use of vector spaces to characterize individual words taken from a large dataset of sentences, followed by the use of deep neural network learning on those vectors to cluster and classify entire sentences, discovering analogies, and what might be considered *concepts* along the way (Goth 2016). This may bring machine-based *speech understanding* closer to reality, which is certainly a prerequisite for AS explainability and, eventually, trust (Knight 2016).

5.3.7 Learning and Adaptation

The **learning and adaptation** layer in our AS framework facilitates both initial development of AS behaviors (offline) and flexible or adaptive behavior once in the environment (online). This function, in general, requires a knowledge of "ground truth" and an assessment of a function's (or an entire AS's) performance with respect to that ground truth, relative to the overall goals of the AS. It also requires some reasoning and manipulation of the representation being used for the accomplishment of the tasks associated with those goals. The advantage of learning and adapting either before or after fielding is threefold:

- Reduction of "knowledge engineering" on the part of the designer, since all possible contingencies need not be anticipated nor engineered (hand coded) for

- Matching the information structure of the environment, or adapting to environment, after initial fielding

- Acquiring/integrating new knowledge with "experience," thereby becoming more skilled/expert over time

This approach does, however, require the system being designed to *acquire* new knowledge and to be able to *integrate* that knowledge so it can effectively be used to drive the learning/adaptation function. The subfunctions of acquisition, integration, and learning/adapting are tied to whatever representation the AS uses. That choice of a cognitive approach determines how the initial knowledge is provided to the system and how during learning over time that knowledge can be modified.

When building a *learning* AS, there are three key phases. First, the designer must choose a technical implementation to support a given functionality. For example, the designer might decide to use an ANN as a means of detecting specific events (in the perception and event detection block, for example). Second, the designer must encode the function in the representation language of the technical implementation, deciding on key architectural parameters and the *learning* mechanism. For example, if the model is a feed-forward ANN, then the designer would specify the architecture (e.g., number of neurons per layer, the number of layers, interconnectivity density, and node characteristics) and the process used to set the weight values within a given node (e.g., backpropagation using gradient descent with an error-based objective function, the specification of a "training" dataset, and any preprocessing such as vectorization). This phase could be considered the initial "knowledge acquisition" phase of learning by the AS. The last phase is the use of the resulting function to accomplish one or more tasks assigned to the AS. If additional adaptation is called for ("with experience"), then learning will continue as the AS performs its assigned functions after initial fielding. This phase could be considered the ongoing "skill refinement" phase of learning by the AS.

Note that in figure 5.1 we show the learning/adaptation layer (potentially) interacting with all blocks in the SADM layer as well as the Sensor and Data Fusion layer so that learning can occur almost anywhere in the AS structure. This supports both initial knowledge acquisition as well as experience-based skill development across a broad range of AS functions, but to accomplish this requires learning structures and algorithms tuned to the particular implementation technology used for a given block. We described some of the learning features above in terms of ANN learning, but there are clearly others. As described by Domingos (2015), there are at least five fundamentally different approaches to learning (and associated "tribes"), based on the underlying approach to knowledge representation and reasoning:

- Symbolic representation and reasoning via deduction, and learning via induction

- Neural networks and learning of node weights via backpropagation

- Genetic/evolutionary algorithms and learning by evolving

- Bayesian networks inferencing, and learning of network structure/parameters via Bayesian inferencing

- Case-based (or analogical) reasoning and learning via SVM techniques

Thus, one might provide, in the learning/adaptation layer, five or more separate learning mechanisms to cover all possible technical implementations of the functional blocks of the AS. This sort of ensemble learning is used very successfully in other works (Dieterich 2000). Alternatively, as proposed by Domingos (2015), one could use a single hybridized and structured "master algorithm"—which builds on specific learning methods associated with the five reasoning approaches cited above—suitable for learning across all the functional blocks. Currently, this is a conjecture and an active area of research. We believe, however, that other, more novel approaches of generating and sharing knowledge will be required.

Learning will be critical for the AS to exhibit task, peer, and cognitive flexibility. It will need to learn about its task and how it relates to other tasks so that it can maintain task flexibility. It will need to learn how to interact with each individual to maximize performance and ensure trust. And it will need to modify its internal representation so that it can exhibit cognitive flexibility. Moreover, an AS must be able to learn beyond the knowledge it gains from its own sensors and experience: it must learn via shared knowledge (cultural learning) in an environment where humans and ASs enter and depart at will. To accomplish this calls for an agency constructed of humans and ASs: an Agile System of Systems (ASoS). Because strict communications protocols limit the advancement of cooperative AS, learning will need to be inherent in communications between ASs to enable knowledge sharing. This last statement cannot be understated.

Task flexibility provides a ready example of the importance of shared knowledge. Consider an AS platform that is physically capable of performing several tasks but may not have the knowledge to do so. For example, the AS could possess the knowledge on how to complete a series of combat search and rescue (CSAR) tasks, which necessitates the AS to have its own on-board fire support and sensors to locate and support the recovery operation. The AS is therefore equipped with the sensors and the processing, including algorithms, to support locating individuals within its region of concern and thus much of what is necessary to do manhunting—but not possess the necessary tracking models to support the manhunting mission. In this case, transfer learning could address the training requirement for developing tracking models (Pan 2010). Transfer learning could be a way that knowledge is shared from the ASoS *cultural* perspective, and some level of *mental simulation* (internal to the AS) could be used for the AS to generate an understanding of task completion without external stimuli. A broad range of technologies could be exploited to support machine-generated knowledge creation; we cover some of them here. Natural language processing (NLP) as used in the IBM Watson question-answer architecture is also useful for task flexibility (Ferrucci et al. 2010). Here, the AS might change its peer flexibility in a question-answer interaction with humans or other ASs to gain insight into the current battlefield conditions and then change its task based on that interaction. Graphical models, in general, could be used to enable task flexibility by maintaining tasks mapped to agents while maintaining task dependencies and metadata that could be used to determine when task flexibility may be enabled. For example, a high-priority task may not be switched to if it is dependent on the completion of another high-priority task that is not already complete. The fact that graphical models are used in decision support systems is indicative of their utility in task-change decision-making.

Peer flexibility emphasizes the necessity for ASs to learn the necessary communications approach to share their knowledge among other ASs in the ASoS so they can negotiate changes to their peer relationships. A broad range of technologies could be exploited to support peer flexibility; we note a few from a machine-learning perspective here. For example, peer flexibility could be enabled by way of the question-answering architecture employed in IBM Watson. For example, the AS could use this architecture to interact with the AS or human to improve system performance or acquire information from the operations community that the AS may not already possess. This interaction changes the peer relationship between the human and the AS. Furthermore, it could use a series of questions and answers to determine if, for example, a human operator is too fatigued to carry out his or her duties safely and assume control of the mission until the operator recovered or a relief operator is provided. Graphical models, in general, could serve an important role in peer flexibility, where the model maintains relationship status, as well as other metadata (e.g., peer status cannot be changed, etc.), between subordinates, peers, and supervisors so that the AS is more informed about how or when to change peer relationships.

Cognitive flexibility is currently significantly hindered by the ASs knowledge representation not being accessible by other ASs. Knowledge in current cognitive solutions is encoded as a model, model parameters, and the algorithms used to manipulate the model and is *internal* to that AS. A common knowledge substrate accessible by all agents in the ASoS is key to cognitive flexibility. Although the best approach to developing this knowledge sharing is still a matter of research, some existing technologies could support development in this area. Transfer learning could be used to transfer knowledge from agents in an ASoS (cultural knowledge creation between ASs). Reinforcement learning can contribute greatly to cognitive flexibility by allowing the system to learn by playing out simulations internal to the AS. This approach was used by Google DeepMind to train its system to play Atari 2600 games against humans with good success (Mnih et al. 201) and by the Google DeepMind AlphaGo system to play the game of Go against humans (Silver et al. 2016). Google DeepMind uses DL in its solution with a large degree of success. DL has demonstrated its ability to solve a wide variety of problems and should be considered in cognitively flexible solutions. IBM Watson uses a question-answering architecture (Ferrucci et al. 2010) that would be helpful for an AS to determine, from an operator, if its current performance is acceptable and, if not, make changes to the underlying cognitive approach until the operator is satisfied with the results the system is producing. Graphical models, in general, fit well in a system that requires cognitive flexibility. Not only are graphical models good for various machine learning tasks, but they also are used in decision support systems and could be used to make a decision regarding a change in cognitive approach in completion of a task. Finally, ensemble learning (Dietterich 2000) and mixed neural architectures (Merényi et al. 2014) could be integral in a system needing cognitive flexibility.

The technologies described above and elsewhere (e.g., NLP, question-answer architectures, DL, transfer learning, reinforcement learning, ensemble learning, and mixed neural architectures) are important to AS learning and adaptation, whether singly or as a member of an ASoS construct. However, knowledge acquisition, adaptation, and dissemination are still areas in which concepts and technologies are not sufficiently mature and are deserving of additional basic and applied research to advance the state of the art. This is clearly a fertile area for investment.

5.3.8 Domain-Specific Knowledge Base

The **domain-specific knowledge base** layer supports storage of and access to domain-specific knowledge that different modules need to perform their functions. There are several options a designer has for the knowledge base design, and the approach to a knowledge base depends on the domain and the problem. By segmenting the knowledge base layer from the other functions, it is possible to separate out as much domain-specific knowledge as possible and leave domain-independent "engines" in other modules of other layers of the AS.

Such an approach was pioneered in ESs in the early 1980s: separate "inference engines" applied rules (from a domain-specific ruleset) to a domain-specific "fact base" to generate additional facts (i.e., knowledge) and eventually conclusions sought by the user of the ES (Barr and Feigenbaum 1981; Hayes-Roth et al. 1983). This supports the reapplication of the ES framework, its functions, and their component engines in different domains, with most of the application-specific changes localized to the knowledge base layer.

More contemporary AI approaches continue in this tradition. Currently, statistical relational models provide a mechanism to overcome the fragile nature of the traditional rule-based ES (Getoor and Taskar 2007; Pfeffer 2016), but they also allow for two primary types of domain knowledge: structural and parametric. The most common statistical relational model is the Bayesian Belief Network, which links domain states together based on their dependencies (directly causal or simply associated); the probabilities associated with those dependencies are stored as conditional probability tables (CPT; Russell and Norvig 2010).

To illustrate, consider an SA function implemented using a BBN representation of the situation, where each node of the network represents a specific variable defining the situation or a component of it, the links connecting the nodes represent the primary causal linkages between nodes, and the CPTs within a node represent the conditional dependencies between variables within a given node. In this case, the *structural knowledge* consists of the variables or nodes, the links between nodes, and the dimensions of the CPTs (rows and columns); the *parametric knowledge* consists of the conditional probabilities comprising the CPTs. All of these BBN attributes, both structural and parametric, can be stored in the knowledge base to represent the domain-specific situations of interest and their interdependencies. The actual SA inferencing conducted by the BBN-based SA module would be done by a domain-independent BBN "engine" (or data-independent algorithm) implementing Bayes's law across the BBN node array in some prespecified fashion. Parameters associated with the engine's operation or algorithmic rules could be stored with the engine itself or within the knowledge base partition associated with the BBN SA function.

A similar parsing of a function's engine/algorithm from its data is envisioned for other functions in the AS, such as an expert system's partitioning of inference engine from its ruleset and fact set; a case-based reasoning function partitioning of its assessment of "case similarity" from its full set of cases; or even a lower-level estimation and control function used for traditional closed-loop or robotic control of, say, key AS mobility functions.

A simple engineering example of this last function is a model-based FCS, where the FCS *structure* is fixed across multiple aerodynamically similar platforms, all having architecturally similar dynamic equations of motion, and where the FCS *parameters* (the flight control "gains") are chosen on the basis of the parameters of the equations of motion of the specific platform for which the FCS is being developed. The FCS design effort then becomes one in which most of the effort is focused on developing a sufficiently accurate structured and parametrized model of the platform and then simply calculating FCS gains based on the model parameters (Zacharias 1974). Redesign of the FCS for a new aerodynamically similar platform is then simply a matter of specifying the new set of platform parameters.

This separation into structure and parameters also affords us the potential for more focused "learning" of the structure and parameters associated with the task and the environment, providing us with a means of conducting systems identification and model development of the outside world in which the autonomous system is to operate. For example, one could envision an initial learning phase focused on systems identification to define the basic structure of the task/environment, including major and minor subsystems and their interlinkages. Subsequent learning phases could refine the components and parameters to extend the knowledge of the outside world to gain greater knowledge of the operating environment and the goals of the AS. This same approach could be focused internally, on the AS itself, to support health maintenance monitoring, growth of expertise, and similar self-assessment functions.

The ideas of *task*, *peer*, and *cognitive flexibility* are best addressed in the functional components described earlier, because of the specificity of this particular knowledge base. However, the idea that code can modify itself, and that data and code are referential, means that there are research ideas that could result in the domain-specific knowledge base layer manifesting these flexibilities. The primary mechanism that could introduce all three forms of flexibility into the domain-specific knowledge base layer is self-modifying code, where the code modifies its own instructions while it is executing (Ortiz 2008). This is often implemented in assembly language and instruction sets but is also a core idea in s-expressions in the list processing (LISP) functional programming language (McCarthy 1960). Since everything in LISP (code and data) is a list, LISP macros can restructure the lists and, in essence, rewrite the executing program. Functional programming in general has experienced an upsurge with Javascript becoming the most in-demand programming language for developers.[27] This can lead to a better understanding of how to develop self-modifying programs.

At a higher abstraction level, researchers have been leveraging Bayesian probabilistic languages (Ellis et al. 2016) and abduction in inductive logic systems (Cropper et al. 2016). These are more often presented as statistical machine learning algorithms but are unique in that during learning, they are learning and revising a program or rulesets and not just modifying the data representation (Schmidhuber 2007).

---

27. See, for example, Stack Overflow, https://insights.stackoverflow.com/survey/2016.

5.3.9 Toolsets and Technologies

The **toolsets and technologies** substrate of our AS framework provides a general repository for toolsets (e.g., MATLAB at https://www.mathworks .com/products/matlab.html), technologies (e.g., optimization algorithms), and data repositories (e.g., those available at http://deeplearning.net/datasets/) to be used by one or more component functions illustrated in figure 5.1. Although it is not feasible to provide an up-to-date and exhaustive list of toolsets/ technologies/databases applicable to the design and implementation all of the functions shown in figure 5.1, there are several that currently exist that will keep researchers from starting from scratch. In some cases, a toolset or data repository may be useful for more than one function. As an example, the Stanford Natural Language Processing toolbox (https://nlp.stanford.edu/soft ware/) is useful for the *Learning and Adaption* layer as well as the *Human Computer Interfaces and Collaboration Environments* layer.

Table 5.1 provides a sample of the toolsets and datasets that might be considered for our AS framework; a more complete table is provided in appendix E. We have done our best to note the diversity in the applications but only consider primary and secondary applications based on our judgment of toolset/dataset relevance; it does not mean that the toolset or dataset does not apply to other functions. Furthermore, for some functional areas, such as the *Learning and Adaptation* layer, there are several available toolsets and datasets; yet in others, such as *Planning and Replanning*, there are very limited resources available.

5.4 Functions not Represented in the Example Framework

We close this section on the example AS framework with a brief enumeration of some of the functions that are not explicitly represented and that may be rich areas for further research.

5.4.1 Goal Generation and Prioritization

Critical to an AS's autonomous behavior is the generation, prioritization, and maintenance of multiple goals needed for mission success.

AS we discussed earlier in section 3.1, the *generation* of goals in the AI community is typically dealt with via the generation of subgoals or tasks to be achieved, as in a multistage planning problem (Russell and Norvig 2010). This is fine for implementing a "divide and conquer" approach to problem solving but fails to address the innovation of a truly autonomous agent that comes with a grasp of the "big picture" and the given goal(s) in that context and how the goal space might be reconfigured to achieve better/faster/cheaper solutions to a different goal than that originally specified. Reasoning by analogy (Salmon 2013), which humans are particularly good at, may be one approach to the problem, but it is clear, based on the limitations of CBR (Kolodner 1993), that the AI community has a long way to go in this area.

The *prioritization* of multiple goals in the face of changing situations (current and future) is another area that needs attention, since it is rare that any one goal will maintain top priority of an AS operating in a "real world" situation over long periods of time (unlike academic problem sets). Prioritization is compounded by the fact that goal values or utilities need to be "normalized" in some (not necessarily one-dimensional) space for an AS to make comparisons and assign priorities. Again, humans seem to deal well with noncommensurate goal metrics/utilities and can handle trade-offs in priorities over time; progress clearly needs to be made in this area in the AI community if we are to achieve the desired *task flexibility* we outlined earlier.

Finally, the *maintenance* of multiple goals—more correctly, the engagement of behaviors to achieve those goals—requires an executive controller that enables multitasking, in which the goal to be currently attended to is brought to the forefront *and* the set of behaviors needed to achieve that goal are retrieved ("goal shifting and "rule activation," respectively; Rubenstein et al. 2001), something also not explicitly represented in the example framework. This representation of an "executive controller" does not appear to be insurmountable, given the theory developed in the cognitive psychology community (Miyake and Shah 1999; Miyake et al. 2000) and instantiations developed in the agent research community, such as the EPIC model (Meyer and Kieras 1997), or Soar (Laird 2012).

5.4.2 High-level Meaning Making

Critical to survival in the real world is *meaning making* (also called *sense making*). Many studies have shown that human cognition strives to make meaning of objects, events, and situations in the world (Klein 1998). For the human, meaning making is an endogenous, mental event.

As discussed earlier in section 2.4, meaning making can be conceptualized as *sign interpretation* or semiotics (Hoopes 1991; Peirce 1960). The meaning of an object, event, or situation is its interpretation by an individual as a sign denoting some other (determining) object, event, or situation. For example, the meaning of a traffic jam during a morning commute would be its interpretation as a sign denoting that the person will be late for work. Meaning making

Table 5.1. Example list of toolsets and datasets for the toolsets and technologies substrate of the example framework of figure 5.1

Section	Primary Function	Secondary Function	Toolset or Dataset Name	Toolset Dataset Type	Type	Environment	License	URL
4.2.2.1	Databases		MongoDB	Toolset	No SQL Database	Windows/Linux/Unix	Open Source	https://www.mongodb.com/
4.2.2.2	Sensors		SUPPRESSOR	Toolset	**Electronic Warfare** Mission modeling	Ubuntu	GOTS	
4.2.2.2	Sensors		**Vigilant Hammer**	Toolset	**Electronic warfare**		GOTS	
4.2.2.2	Sensors		Mental Ray	Toolset	Optical modelling			
4.2.2.3.1	Sensor Processing		OPPORTUNITY Activity Recognition Data Set	Dataset	Multisensor data	Universal	Open Source	https://archive.ics.uci.edu/ml/datasets/opportunity+activity+recog.
4.2.2.3.1 & Fusion	Sensor Processing & Fusion		Vector Informatik, Advanced Driver Assistance Systems (ADAS) development	Toolset	vADAS Developer	Windows (Dev in Visual Studio)	Commercial	https://vector.com/vi_vadasdeveloper_en.html
4.2.2.3.2 Fusion	Data Mining & Fusion	Sensor Management & Data Mining	International Business Machine SPSS (CRISP-DM)	Toolset	Data Mining Process		COTS	http://www.ibm.com/analytics/us/en/technology/spss
4.2.2.3.2 Fusion	Data Mining & Fusion	Sensor Management & Data Mining	AdvancedMiner	Toolset				
4.2.2.4.1	Perception & Event Detection	Learning & Adaption	OpenCV	Toolset	Computer Vision		BSD	http://opencv.org/
4.2.2.4.2	**Situation Assessment**		NetSA	Toolset	**Network Situational Awareness**	Linux, Solaris, OpenBSD, Mac OS X, and Cygwin	GOTS	http://www.cert.org/netsa/
4.2.2.4.3	Decision Making		1000Minds	Toolset	PAPRIKA method Application	Online Web Application	COTS	https://www.1000minds.com/
4.2.2.4.4	Planning & Replanning		ICAPS Competitions	Dataset	Planning and Scheduling	N/A	Open Source	http://www.icaps-conference.org/index.php/Main/Competitions
4.2.2.4.5	Execution Management		DAM(Decentralized Asset Manager)	Toolset			COTS	
4.2.2.6	Human Computer Interface & Collaboration		OpenVibe	Toolset	Brain-Computer Interface	Windows	GNU Affero General Purpose License	http://openvibe.inria.fr/
4.2.2.6	Learning & Adaption Environments		H2O	Toolset	General	Java, R, Scala, Python	Open Source	http://www.h2o.ai/#
4.2.2.7	Learning & Adaption		ImageNet	Dataset	Various images	Universal	Open Source	www.image-net.org
4.2.2.8	Domain Specific Knowledge Base	Learning & Adaption	The Graphical Models Toolkit (GMTk)	Toolset	Dynamic Graphical Models & Dynamic Bayesian Networks Code	Homebrew, MackPorts, Linux Remote Package Manager, Source	Open Source	http://melodi.ee.washington.edu/gmtk/

can also be conceptualized as *frames* (Minsky 1975), which are remembered data structures representing stereotypical situations adapted to a given instance of reality. Higher levels of a frame can represent context, while lower levels correspond to specific elements or data, and the relation between the two may provide a way of representing meaning. During meaning making, frames may help define relevant data, while data may drive changes to existing frames (Klein et al. 2006).

Sign interpretation, at which humans are very adept (Patterson and Eggleston 2017), and frames represent complementary approaches to meaning making. Assigning meaning to an object, event, or situation by an individual in terms of a *sign* denoting some outcome would depend on context or frame. Alternatively, the context or frame helps determine which objects, events, or situations are interpreted as signs and what those signs denote.

For the AS to work effectively and autonomously with the human, the actions of the AS and human would need to reflect a common frame (i.e., recognize the same context) to interpret the same objects, events, or situations as the same signs. In other words, the AS and human would need to make the same meaning of the same objects, events, or situations; to do so, the actions of the AS and human would need to occur within the same frame or context. The challenge here is to define at the appropriate level of abstraction and with the appropriate elements exactly what frame or context is, how it could be common to both AS and human, and how it would affect meaning making.

In section 5.3.6 earlier, we noted the importance of having natural language interaction via an NLP capability on the part of the AS. This would not only provide for more natural and efficient interactions but also could support deeper understanding of the AS goals, the task constraints, the subsystem status, etc., all of which contribute to better and more effective achievement of AS goals. But a deeper NLP capability—for example, via argumentation theory (Toulmin 1959)—would provide opportunities for mutual human/AS formulation of primary mission goals, reprioritization, mutual problem solving, human/AS negotiation, and the like, all of which would contribute to more foundational teammate interactions and trust. Again, this capability for high-quality natural communications (beyond surface-level phrase parsing) is not explicitly represented in the example framework, and it is clearly highly dependent on the existence of common frames of reference and commonly understood signs, for human and AS alike.

5.4.3 Dual-channel Cognitive Processing

There is much research indicating that human decision making is governed by two systems or sets of cognitive processes, as we noted in section 2.2 earlier. As Evans and Stanovich (2013) point out, the Type 1 process, one that is *intuitive/ unconscious,* involves unconscious situational pattern recognition unconstrained by working memory limitations and is independent of conscious "executive" control, large in capacity, and fast. This type of cognition likely involves *implicit* learning, procedural memory, and knowledge, which cannot be consciously recollected. When intuitive cognition renders an unconscious decision, it posts the result to consciousness as a feeling or an "aha" experience. The Type 2 process, one that is *analytical-reflective/conscious,* entails conscious deliberation that draws on limited working memory resources and can be effortful, rule-based, symbolic, limited in capacity, and slow. This type of cognition involves declarative memory and explicit knowledge (i.e., knowledge that can be consciously recollected). These two systems for human decision making can be dissociated experimentally and neurologically (Patterson 2017).

Intuitive cognition presents challenges for the concept of situation awareness— the perception, comprehension, and future projection of elements in the environment (Endsley 1995a, 1995b), described at length in section 5.3.4. SA has been traditionally defined as involving conscious awareness (Endsley 1995a), and its measurement frequently relies on individuals' working memory and verbalization (e.g., Durso et al. 2007; Endsley 1995b). Accordingly, in many cases, conceptualizations and measurements of SA may not reflect intuitive cognition. SA appears to be representing only a portion (analytical component) of the cognitive processing humans use to interact with the world. New objective behavioral techniques need to be developed for defining and assessing human SA when intuitive cognition is active, so that when humans and ASs are teamed together they can maintain a common situation understanding and decision-making framework.

In addition, for an AS to work effectively and autonomously with a human, we need to understand how the human relies on one cognitive process or another (analytical or intuitive), given a specific task or work environment. It may be that the task environment can be structured to elicit primarily analytical or intuitive cognition (Hammond et al. 1987). The relative weight given to processing in one or the other system may depend on task factors such as the number of cues (intuitive: many; analytical: few) and the type of cues (intuitive: perceptual; analytical: symbolic, rule-based). Intuitive cognition seems to dominate responding in humans and likely is the default system whenever analytical cognition cannot solve the task (Patterson and Eggleston

2017). Thus, for an AS-human intuitive-cognition team, the task or work environment may involve speeded judgments about perceptual material with multiple cues and no symbolic calculation. For an AS-human analytical-cognition team, the task or work environment may entail deliberative judgments involving symbols and rules. The challenge here is to determine exactly how certain task or work environments drive analytical or intuitive processing and how that knowledge plays into the design of AS-human teaming.

As we noted earlier, it is also important for the human-AS team to develop and maintain a common frame (i.e., recognize the same context) to interpret the same objects, events, or situations with the same meaning. One method for developing this common context or frame is joint human-AS training for joint expertise development under simulated operational conditions, as we noted in section 3.4. However, in humans, the development of expertise involves a shift in cognitive processing from an initial emphasis on analytical processing towards one that emphasizes intuitive processing as expertise is gained (Reyna and Lloyd 2006; Reyna et al. 2014); human cognitive processing becomes more unconscious as the human becomes more skilled in the task.

The challenge here is to determine how the AS and human will communicate and develop teaming if the human's knowledge of his or her own cognitive processing grows largely unconscious with skill acquisition. Although there are techniques (verbal report) designed to investigate the cognitive nature of work (Hoffman and Militello 2009), there is much evidence suggesting that verbal reports can be reactive and invalid (Johansson et al. 2005; Nisbett and Wilson 1977; Russo et al. 1989; Schooler et al. 1993). Accordingly, new objective behavioral techniques will need to be developed for exploring the intuitive cognition nature of work so that a common context or frame can be developed with the AS, as training drives the human's reliance on intuitive cognition.

5.4.4 Affective Representation

Finally, the model framework does not explicitly provide for an affective computing capability (Picard 1997, 2003; Hudlicka 2003)—that is, the ability to recognize and adapt to emotions in humans or to instantiate emotions within the AS itself. The former capability requires sensors and perception units focused on emotional signals generated by humans (e.g., facial expressions, voice intonations, verbal content, etc.), elements that could be accommodated within the current framework but not called out in the current exposition. The latter capability requires a separate computational model of emotion—a current area of active research—and a means of expressing it to other agent teammates, human or machine, via simulated facial or verbal signals. Significant research is ongoing at several university-affiliated research labs, notably the Affective Computing Research Group at MIT's Media Lab (http://affect.media.mit.edu/); the Emotions and Cognition Lab at USC (http://emotions.ict.usc.edu/); and the Emotion, Cognition, and Social Research Lab at Jacobs University (http://akappas.user.jacobs-university.de/), to name a few.

Chapter 6

Recommendations

The previous chapters have addressed general principles, sample frameworks, and promising technologies for the development of ASs. In this chapter, we provide specific recommendations that can facilitate this development and provide guidance for a long-term research and development plan for successful prototyping of and experimentation with these novel systems.

Our general approach to developing these recommendations does not follow the conventional process of: (a) making observations on the current situation; (b) generating findings based on those observations; and (c) making recommendations to deal with the findings. Instead, we have focused on the basic properties needed to ensure proficiency of these systems (in chapter 2), the tenets of trust needed to ensure human-system compatibility in high-risk situations (in chapter 3), and how they combine to give us the principles of behavioral flexibility that truly define autonomous behavior (as introduced in chapter 1). In addition, we have provided background on several different communities working different aspects of the problem (in chapter 4) that could serve as the foundations of one or more unifying frameworks for AS development. We presented one example (of potentially several) to guide the engineering development of ASs, along with promising technology that can address particular functional needs (in chapter 5). This path has led us to a set of recommendations that not only focus on AS *behavioral objectives* (including proficiency, trustworthiness, and flexibility) and the *architectural/technical* issues underlying all three but also on the nontechnical issues of the *process of development* and the *structure of the organization* needed to support technical successes in this domain.

We present here six categories of recommendations in the remainder of this chapter, five of them in line with the structuring we have done with our review of past studies presented in appendix A and one serving as an integration platform:

- *Behavioral Objectives*: These are basically generalized design requirements specifying how we want an AS to behave, in terms of proficiency, trustworthiness, and flexibility.

- *Architectures and Technologies*: This covers unifying frameworks and architectures that will support cross-disciplinary research and development, along with the technology investments needed to support desired functionalities within an architecture.

- *Challenge Problems*: Addressed here are both domain-independent (or functional) problems, like dynamic replanning, and domain-dependent (or mission-oriented) problems, like multidomain fusion.

- *Development Processes*: This includes processes—in contrast to our traditional waterfall process of requirements specification, milestone satisfaction, and end-state test and evaluation (T&E)—that support innovation, rapid prototyping, and iterative requirements development to support rapid AS development and fielding.

- *Organizational Structures*: This includes organizing around a project (or outcome) focus, rather than, say, along traditional technical specialty domains.

- *Knowledge Platform*: This provides us with a holistic means of integrating across AS behavioral principles, architectures/technologies, challenge problems, developmental processes, and organizational structures.

To provide some context, it is appropriate to summarize the findings of previous autonomy studies outlined in Appendix A, organized in these same five categories.

Past recommendations in the category of *behavioral objectives* can be broken down into two subcategories. The first is fairly general and deals with the performance of the AS itself, such as ensuring that behaviors are directable and predictable and that the AS can accomplish tasks with adequate flexibility and adaptivity. We would claim that all of these requirements are well covered by our flexibility principles described earlier and our performance properties described in chapter 2. The second focuses on human-system teaming, including the desirability of being able to set mutual goals, to maintain adequate mutual shared awareness of the team and the adversary, and to communicate and coordinate effectively. Again, these general attributes have been well described in our trust tenets put forth in chapter 3.

Past recommendations in the category of *architectures and technologies* were put forth in only one of the studies we reviewed (Potember 2017), and they support and complement our discussion of frameworks and enabling technologies of chapter 4. Several hybrid architectures were discussed—that is, those that combine neural networks with more traditional approaches like game theory—with a recommendation to consider biomimetic cognitive systems development as well. Specific technologies are broken into "classical" algorithmic approaches and the more contemporary DL network approaches enabled by access to "big data," much like our earlier discussion in chapter 4.

Past recommendations in the category of *challenge problems* influenced our distinction between domain-independent (or functional) problems, like dynamic replanning, and domain-dependent (or mission-oriented) problems, like multidomain fusion across air, space, and cyber. The domain-independent problems cover a range of general functional areas: collection/sensing/fusion of information, generic decision-aiding (with a human) and decision-making (autonomous) subsystems, fractionated autonomous platforms, and operation in adversarial environments that demand improvisation. The domain-dependent problems range from generic concepts (autonomous swarms) to specific operations in air (including ISR, air operations planning, electronic warfare, and logistics), space (including fractionated platforms and embedded health diagnostics), and cyber (defensive operations, offensive operations, and network resiliency).

Past recommendations in the category of *development processes* cover six major areas:

- The need to actively track adversarial capabilities and usage of ASs

- The importance of human capital management, including the attraction and retention of experts in AI and software engineering, the introduction of AS capabilities into professional military education (PME) and wargaming, and the development of centralized AI/AS resources for the DOD's research community

- Continued support of basic and applied research in a broad area of underlying technologies (not just DL), coordinated across research communities and informed by operational experience and evolving mission requirements

- Support of advanced systems development, which separates the development of platforms from the autonomy software that governs them, augmented by the development of a discipline of AI engineering to accelerate progress

- Establishing processes for upgrading legacy systems with new AS capabilities

- Recognizing the difficulty of conducting T&E of these systems and establishing a research program for dealing with nontransparent, nondeterministic, and time-varying (learning) systems[1]

---

1. At the outset of this study, we realized the difficulty of dealing with this aspect of AS development, especially given that these systems are "moving targets" in terms of their expected operating environments, their overt behavioral characteristics, and their internalized software and hardware constructs. We have therefore chosen not to make recommendations at this early stage—except to advocate research and technology development aimed at advancing the state of the art of T&E for these systems—and instead have simply identified, in appendix I, some aspects of the problem we face and potential directions for the future.

In our survey of past studies summarized in appendix A, we found no recommendations dealing with the introduction of new *organizational structures* for AS R&D, nor for the concept of a unifying Knowledge Platform.

The remainder of this chapter presents our recommendations organized by the five categories just summarized (sections 6.1 through 6.5). We then discuss how we can address these individual recommendations in a holistic fashion, via the introduction of a Knowledge Platform (section 6.6), summarize our recommendations (section 6.7), and conclude with brief closing comments.

6.1 Behavioral Objectives

In our earlier discussions we have not attempted to explicitly define what autonomy is or what an AS does but rather have focused on the behavioral characteristics of an AS; in other words, the general behavioral characteristics of a to-be-designed/-developed AS. We proposed in chapter 1 that the key characteristic was flexibility, in terms of dealing with tasks, peers, and cognitive approaches. In chapters 2 and 3, we augmented these characteristics with expectations in terms of performance and trust and showed how they combine to give us the principles of behavioral flexibility that truly define autonomous behavior. In this section, we restate these desired attributes in terms of system design goals, to guide future AS design efforts.

Recommendation 1a: ASs should be designed to ensure *proficiency* in the given environment, tasks, and teammates envisioned during operations. Desired *properties for proficiency* include:

- *Situated Agency*. Provide for situated agency within the environment, which includes component abilities to sense or measure the environment, assess the situation, reason about it, make decisions to reach a goal, and then act on the environment, to form a closed loop of "seeing/thinking/doing," iteratively and interactively.

- *Adaptive Cognition*. Provide for a capability to use several different modes of "thinking" about the problem (i.e., assessing, reasoning, and decision making), from low-level rules to high-level reasoning and planning, depending on the difficulty of the problem, and the need for flexibility in dealing with unexpected situations.

- *Multiagent Emergence*. Enable an ability to interact with other ASs via communications and distributed function allocations (e.g., sensing, assessing, decision making, etc.), either directly or through a C2 network, in a manner that can give rise to emergent behavior of the group, in a fashion not necessarily contemplated in the original AS agent design.

- *Experiential Learning.* Provide for a capability to "learn" new behaviors over time and experience, by modifying internal structures of the AS or parameters within those structures, based on an ability to self-assess performance via one or more performance metrics (e.g., task optimality, error robustness, etc.) and an ability to optimize that performance via appropriate structural/parametric adjustments over time.

We also highlighted the importance of trust in an AS, even if it is deemed proficient, particularly when teamed with one or more human teammates. In addition to the conventional contributors to mistrust, there are particular challenges when dealing with ASs, including:

- *Lack of analogical "thinking" by the AS.* When the AS approaches and/or solves a problem in a fashion that is not at all like a human would attack the problem, trust can become an issue because of human concern that the approach may be faulty or unvalidated.

- *Low transparency and traceability in the AS solution.* Lacking an ability to "explain" itself, in terms of assumptions held, data under consideration, reasoning methods used, and so forth, the AS finds it difficult to justify its solution set and thus engender human trust.

- *Lack of self-awareness or environmental awareness by the system.* In the former, this might include AS health and component failure modes, while in the latter, this might include environmental stressors or adversary attacks. Either may unknowingly affect performance and proficiency and overstate the confidence in an AS-based solution made outside of its nominal "operating envelope."

- *Low mutual understanding of common goals.* When a human and AS are working together on a common task, a lack of understanding of the common goals, task constraints, roles, and more can lead to a lack of trust on the part of the human in terms of the system's anticipated proficiency over the course of task execution.

- *Non-natural communications interfaces.* The lack of conventional bidirectional, multichannel communications between human and system (e.g., verbal/semantic, verbal/tonal, facial expressions, body language) not only reduces communications data rates but also reduces the opportunity to convey nuances associated with operations by well-practiced and trusting human-only teams.

- *Lack of applicable training and exercises.* Lack of common training and practice together reduce the opportunities for the human to better understand the system's capabilities and limitations, as well as how it goes about "problem solving," and thus opportunities for understanding a system's "trust envelope"—that is, where it can be trusted and where it cannot.

To overcome these sources of distrust, we proposed several design recommendations.

Recommendation 1b: ASs should be designed to ensure trust when operated by or teamed with their human counterparts. Desired *tenets of trust* include:

- *Cognitive congruence and transparency.* If possible, build the system at the high level to be congruent with the way humans parse the problem, so that the system approaches and resolves a problem in a manner analogous to the way a proficient human does. Whether or not this is accomplished, provide some means for transparency or traceability in the system's solution, so that the human can understand the rationale for a given system decision or action.

- *Situation awareness.* Provide sensory and reasoning mechanisms supporting SA of both the system's internal health and component status and of the system's external environment, including the ambient situation, friendly teammates, adversarial actors, and so on. Provide a means for using this awareness for anticipating proficiency increments/decrements within a nominal system's "operating envelope" to support confidence estimates of future decisions and actions.

- *Human-systems integration.* Follow guidelines of good human-systems interaction design to provide natural (to the human) interfaces that support high bandwidth communications if needed, subtleties in qualifications of those communications, and ranges of queries/interactions to support not just tactical task performance but more operational issues dealing with goal management and role allocation (in teams).

- *Human-system teaming and training.* Before human-system teams are brought into operations, adapt or morph training programs and curricula to account for the special capabilities (and associated limitations) of humans teaming with ASs. Conduct extensive training so that the team members can develop mutual mental models of each other, for nominal and compromised behavior, across a range of missions, threats, environments, and users.

We also described the importance of achieving overall AS behavioral flexibility under different tasking, peer arrangements, and problem-solving approaches, summarized in terms of the following three principles of flexibility:

- *Task Flexibility*. An AS should be able to change its task or goal depending on the requirements of the overall mission and the situation it faces. Humans are not *optimizers* designed for only accomplishing one task, even if they are experts in one (e.g., Olympic athletes or world-class chess players); rather, they are *sufficers* in many tasks, flexibly changing from one to another as the need arises. Humans can accomplish multiple tasks, serially and in parallel, dynamically changing priorities over time, shedding tasks and taking on new ones, depending on the situation and motivation (rewards). We believe the same task flexibility needs to be embodied in ASs and that this capability is enabled by situated agency: sensing the environment, assessing the situation, deciding on a course of action to accomplish its tasking, and acting on that course of action, all the while closing the loop by monitoring the outcome and communicating with the other agents in its team.

- *Peer Flexibility*. An AS should be able to take on a subordinate, peer, or supervisory role, depending on the situation and the other agents, human or machine, populating the environment. Humans accomplish this type of relational flexibility as they move through different roles throughout the day, dynamically changing their relationships depending on the situation and the peers they are interacting with. We believe the same peer flexibility needs to be embodied in an AS, changing its relationship role with humans or other ASs within the organization, as the task or environment demands. An AS should participate in the negotiation that results in the accepted peer relation change, requiring the autonomous system to "understand" the meaning of the new peer relationship to respond acceptably. This capability is enabled by situated agency providing environmental and task awareness, an understanding of its peer population (humans and machines), and learning over time to develop proficiency.

- *Cognitive Flexibility*. An AS should be able to change how it carries out a task, both in the short term in response to a changing situation and over the long term with experience and learning. Humans accomplish tasks in multiple different ways, using visualization, verbalization, rote memory, solutions from first principles, and so on. They also change their approaches as they become more expert in tasks, with learning and skill acquisition over time. Finally, they may employ parallel approaches to problem solutions, and they may also act consciously or unconsciously. We believe that an AS should embrace this type of cognitive flexibility in addressing a problem by bringing to bear a variety of techniques to assess and then decide, selecting those techniques based on the current situation, past experience with given methods, the need to trade optimality vs timeliness, and so on. In the long term, the AS can also learn new "solution methods" over time, assessing and readjusting a technique's contribution to task performance for a given situation and mission tasking.

We then described how the proficiency properties and trust tenets could work in tandem to drive these flexibility principles, expressed as the following recommendation set.

Recommendation 1c: ASs should be designed to achieve *proficiency* and *trust* in a fashion that drives behavioral *flexibility*. This is illustrated in figure 6.1 and summarized as follows:

- *Task Flexibility*. An AS should be able to change its task or goal depending on the requirements of the overall mission and the situation it faces. This is enabled by situated agency: sensing the environment, assessing the situation, deciding on a course of action to accomplish its tasking, and acting on that course of action, all the while closing the loop by monitoring the outcome and communicating with the other agents in its team. This is also supported by providing for a capability for experiential learning to improve task proficiency and flexibility over time.

- *Peer Flexibility*. An AS should be able to take on a subordinate, peer, or supervisory role and change that role with humans or other ASs within the organization, as the task or environment demands. This capability is enabled by situated agency providing task/peer/environmental awareness, an understanding of the multiagent emergent behaviors of its peer population (humans and machines), and experiential learning to develop role-switching proficiency.

- *Cognitive Flexibility*. An AS should be able to change how it carries out a task, both in the short term in response to a changing situation and over the long term with experience and learning. In the short term, adaptive cognition can bring to bear a variety of techniques to assess and then decide, selecting those techniques based on the current situation, past experience with given methods, the need to trade optimality versus timeliness, and so forth. In the long term, experiential learning can improve proficiency and enable the acquisition of new behaviors over time, assessing and readjusting a technique's contribution to task performance for a given situation and mission tasking.

These *principles of flexibility* ensure that we focus on the development of systems that provide the USAF its newest asymmetric advantage, that of *improving every decision*. ASs that exhibit these principles will result in solutions that are interoperable, composable, and adaptable, allowing them to create and impose complexity through numbers, heterogeneity, spatial reach, speed, and deception—for example, the ability of flexible ASs to combine into an adaptive kill web composed of legacy/new, manned/unmanned, disaggregated, and distributed systems. These flexibilities will enable the development of systems than can self-organize into a system of systems (SoS) to work together (interoperability through peer flexibility), to solve problems that were not envisioned when the systems were created (composable systems achieved through task flexibility) and that will be adaptable to changes in the environment, tasking, and the adversary (because of their cognitive flexibility). This results in a USAF more able to overwhelm adversaries with complexity and speed, by creating simultaneous dilemmas across all domains.

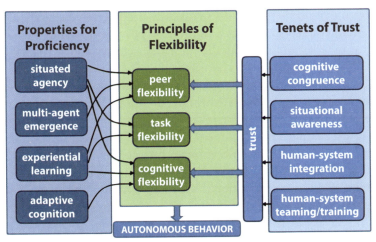

Figure 6.1. Relationship between autonomous system proficiency, trust, and flexibility

6.2 Architectures and Technologies

We have noted earlier that the time is right to begin to develop a common framework for describing, developing, and assessing ASs, based on our review of the research activities of a number of seemingly disparate communities, including:

- The robotics and cybernetics communities

- The cognitive psychology and neurosciences communities

- The "hard" AI and "soft" AI communities

In our view, these communities appear to be converging toward very broad common frameworks that describe the problem of agency and autonomous behavior from radically different viewpoints that, in their own way, may have significant contributions to make to the development of *proficient* and *trustworthy* ASs, in the sense that we have discussed those terms earlier in chapters 2 and 3. Bringing them together will necessitate common frameworks to bridge the gap across communities, to support the development of a common language to describe similar and related concepts in the different fields, and to provide the foundation for more rapid and efficient development (via cross-appropriation of validated concepts and modules) of conceptually well-founded future autonomous systems.

Recommendation 2a: Develop one or more *common AS architectures* that can subsume multiple frameworks currently used across disparate communities. Architectures should, at a minimum, provide for "end-to-end" functionality, in terms of providing the AS with a sensory ability to pick up key aspects of its environment; a cognitive ability to make assessments, plans, and decisions to achieve desired goals; and a motor ability to act on its environment if called upon. The architecture should be functionally structured to enable extensibility and reuse, make no commitment on symbolic vs subsymbolic processing for component functions,[2] incorporate memory and learning, and support human-teammate interaction as needed. An architecture can be deliberately engineering-focused (as presented earlier); or it could take a looser, nonfunctional approach provided by, say, (symbolic) rulebased systems[3] or (nonsymbolic) ANN structures;[4] or it could take a

---

2. And may include both, in a "hybrid" architecture; see, for example, Russell (2015), Davis and Marcus (2015), Booch (2016).

3. For example, where declarative knowledge is not even partitioned from procedural knowledge.

4. For example, where a layer's "function" in a multilayered network is not even defined until after learning stops.

strong biomimetic approach, assuming the underlying science is adequate to support engineering development. Whatever the form, an architecture should be extensible to tasks assigned, peer relationships engaged in, and cognitive approaches used. A key metric of an architecture's utility will be its capability of bridging the conceptual and functional gaps across disparate communities working autonomy issues.

If we are successful, the development of one or more common AS architectures could help us accomplish the following:

- Identifying the fundamental structure common to most or all ASs, in terms of the internal component functions, their relationship to each other and the environment, the principles governing their design, and overall control-flow and data-flow

- Finding a place in the autonomy "universe" for those working subsets of the general problem (e.g., data fusion, image classification, path planning, motor control, etc.) and providing connectivity to others working complementary subsets of the problem

- Helping develop a unifying "science of autonomy" underpinning the thousands of "one-offs" we now have in the engineering community

- Separating functionality from enabling technologies so architecture design can go on in parallel with technology development

- Pointing to where the S&T community needs to invest to develop "missing" functionalities and/or improve technology capabilities

- Dealing with the issue of meaning making and the need for a common frame or context

- And, in the longer term,

  o Serving as the foundation of a common Open Systems Architecture (OSA) to encourage reuse of developed software modules across applications and domains

  o Supporting interoperability across DOD (e.g., USAF ISR UAVs cooperatively teaming with Navy attack unmanned undersea vehicles)

In our discussion of a sample AS architecture in chapter 4, we identified a number of component functions and described for each a broad variety of technologies that could be used to implement those functions. It is not our objective here to single out one or another technology to pursue; that is the responsibility of the research and development community. Instead, we wish to encourage the parallel development of multiple technologies that might be used to implement a given function needed by an AS to achieve proficiency and trustworthiness in a broad range of operational environments.

Recommendation 2b: In parallel with the development of one or more AS architectures and the definition of component functions underlying those architectures, pursue the *development of enabling technologies* that provide the needed functionality at the component level. This includes technologies that support not only the basic "see/think/do" functions but also those that enable effective HCIs, learning/adaptation, and knowledge-base management, both of a general purpose and of domain-specific nature. The nature of technology development should range from basic research to exploratory development to early prototyping, depending on the maturity of the specific technology and its envisioned application.

Parallel development of technologies in this fashion can provide us with the following benefits:

- Faster and more efficient development of needed functionalities demanded by one or more AS architectures. Technology developers can focus on a narrower set of AS behaviors and rely on others to provide more general AS solutions via a common architecture and functional components developed elsewhere.

- A "best of breed" evolution of components and architectures over time, serving the entire community through common usage of proficient and validated components

- A natural approach to developing a capability for adaptive cognition and multiple approaches to dealing with different situations and problem sets, thereby supporting greater cognitive flexibility in AS behavior and resilience in the face of unanticipated conditions.

- An opportunity to focus on narrow but conceptually deep problems that are still on the leading edge of AI and AS development, such as explainable AI, context-adapting AI, and automated training of AI systems

- An opportunity to address similarly conceptually deep issues that drive cognitive science, including fast/slow thinking mechanisms, affective representations and behavior moderators, goal formation, free will, and consciousness.

Most of our discussion has focused on the architectural, functional, and algorithmic aspects of ASs, but effective and efficient development, validation, and modification will call upon additional engineering-focused considerations, notably "architectural patterns" afforded by multitier hardware and multilayer software architectures.[5] An example of such a pattern is illustrated in figure 6.2, composed of four hardware/software tiers/layers:

- *Human Machine Interface.* As described in sections 3.3 and 3.4, a well-designed HMI can afford effective human-systems integration and close human-AS teaming (or, equivalently, HCI), in a manner that engenders communications, task sharing, and trust. The HMI can be thought of both as a physical tier in that it may be implemented on a separate computer with hardware providing dedicated displays and controls across multiple sensory/motor modalities and as a logical layer, commonly referred to by the software community as a presentation layer or user interface.

- *Autonomous System Architecture.* This is primarily a software layer designed to provide the modularity and functionality of a selected AS architecture, such as that described in section 4.2 earlier. Ideally, it is a reusable domain-independent (or domain-insensitive) plug-and-play architecture that can be used across different domains with expandable/contractable functionality. The software community typically refers to this as an application service or business logic layer. In terms of a hardware tier, this service would likely be hosted on one or more embedded computers associated with a host platform, for example.

- *Computational Methods/Algorithms.* This is primarily a software layer—although special-purpose processors could be put to use here as additional hardware tiers—providing multiple common computational approaches to implementing a given function in the AS architecture (in the fashion described in section 4.2 earlier), for several functions. The software community typically refers to this as a business services layer, or low-level business layer, supporting one or more higher level functions.

- *Hardware/Software Platforms.* At the bottom of the figure we show hardware/software tiers and layers providing any needed software services (e.g., operating systems), computational power, and memory to instantiate the overall AS architecture, its layers and services, as well as sensors and effectors needed to support situated agency (as described in section 2.1), and networking protocols and communication channels to

5. We use the term tier to refer to physical hardware segmentation of some of the AS functions and layer to refer to the logical software segmentation of AS functions (Fowler 2002).

support multi-AS and/or human-AS communications. It is anticipated that this tier/layer will evolve rapidly with explosive growth in cloud computing and data storage, improvements in communications bandwidth, the push toward fog computing for special-purpose applications, and the inevitable rise in commercial Internet of Things applications. In addition to these technical enablers, operational verification and validation (V&V) and cybersecurity will have a strong influence on how these "platforms" evolve as well.

Several benefits accrue with the definition of a well-designed architectural pattern for AS development. As a result, we encourage an effort devoted to thoughtful hardware and software engineering design.

Recommendation 2c: Develop and promulgate a *multitiered hardware and multilayered software architecture* to support AS development, validation, operation, and modification, where each tier provides for physical structuring across distinct hardware implementations/hosts for given high- and low-level functions and each layer provides distinct software implementations of similar functions. Figure 6.2 provides a simple example of tiers/layers, but more complex architectural patterns may be needed to take full advantage of emerging technology trends, particularly in the commercial sector.

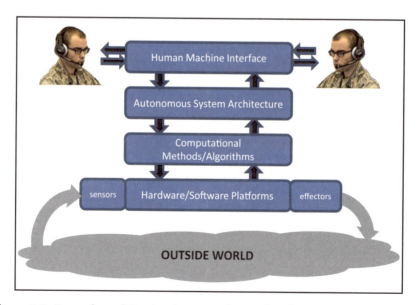

Figure 6.2. Example architectural pattern for AS development

A well-designed tiered/layered architecture can afford us the following benefits:

- By decoupling (or loosely coupling) different functions into different tiers/layers, the scope of development and testing for any one layer is less than for the full application, simplifying design and testing.

- Decoupling also allows for parallel development by separate teams, speeding up overall system development; in many situations, the development team can take advantage of already developed code, inside and outside of DOD.

- Multiple tiers/layers also support rapid modification of the application for new uses/domains, because changes can often be localized to a single layer without reworking the entire application.

- Different components of the application can be independently deployed, maintained, and updated, on different time schedules.

- Multiple applications can reuse the components.

- Tiers/layers make it possible to configure different levels of security for different components deployed on different hosts.

We conclude this section by noting that these three sets of recommendations will not be achievable without the context set by general and specific behavioral AS objectives, as expressed in "challenge problems." We address this in the next section.

6.3 Challenge Problems

Development of architectures and technologies for autonomous systems will not happen in a vacuum. What is needed is an active pursuit of specific challenge problems that drive different aspects of AS development, ranging from basic and exploratory S&T that focuses on architectures and functions (domain-independent) to more operationally focused system development applications to solve "real-world" operational problems (domain-dependent). We discuss this at greater length here.

6.3.1 General Considerations

Before discussing specific problem sets, we wish to emphasize three points. First, both domain-independent and domain-dependent problem sets should be guided by, and stay consistent with, the AS architectures and functionalities chosen to frame the R&D in this area (as we have emphasized in the previous two chapters). This provides a basis for a common reference within the community, a means for communicating "bottom up" and "top down," and possible concept/code reuse and offers a potential bulwark against the development of "one offs" that contribute little or nothing beyond improving the expertise of the teams that developed them. Second, we should be thinking in terms of challenge problems that truly push the bounds of the basic science and the engineering technologies; these are supposed to be challenges, after all. As noted by the Office of Net Assessment (ONA) in a recent study:

> *A great deal of current DOD thinking focuses on AI applications in terms of extending our current activities and capabilities. The "Loyal Wingman" concept is a good example: using AI applications to do better and more of a thing we already do, in this case, flying F-16s. We could distinguish this line of thinking from creating different, previously-impossible concepts of operations—using technologies to do things that weren't possible before these technologies were used. [We] might focus more thinking on the latter category.* (DOD ONA 2016)

Third, we wish to make clear that these challenge problems should not be fixed in stone and should be dynamically updated over time, when they are found to be, for example, ill-stated, "solved" by other technology advances (inside or outside the AF), or irrelevant because of adversary advances or countermeasures. These challenge problems need to be continually updated, based on our failures, successes, and changes in both the technology base and the operational world.

6.3.2 Domain-Independent or Functionally Oriented Challenge Problems

Much of the discussion of chapters 2, 3, and 4 has focused on the basic behaviors, architectures, functions, and technologies needed to support autonomous system operations.[6] Driving the development of these foundational AS structures and attributes—in a fashion that is domain- and mission-independent—calls for an appropriately scoped, scaled, and abstracted set of challenge problems that will allow the S&T community to focus down on different functions that subserve autonomous system behavior, as well as the technologies needed to create these functionalities. As we note in appendix A (and earlier), these functional areas include collection/sensing/fusion of information; generic decision-aiding (with a human) and autonomous decision-making (without a human); planning, replanning, and scheduling; operation

---

6. As well as concepts, theories, and algorithms, and the underlying "platforms" afforded by software, hardware, datasets, and communications channels.

in adversarial environments that demand improvisation; teaming with humans and other agents; and learning over time and experience. Table A.4 of appendix A provides greater detail on several domain-independent challenge problems nominated by earlier studies; figure 6.3 below provides a comparable set defined by DOD's Autonomy COI, several of which have already been discussed here in earlier chapters.

On reviewing these and other recommendations for domain-independent functional challenge problems, it quickly becomes evident that these lists: (1) do not provide the basic rationale for selection and prioritization of one challenge problem over the other; and (2) do not provide any indication that they are an appropriate "spanning set" of the functional space of AS behaviors, since no indication is given that they are (even approximately) exhaustive and exclusive. As a result, we are left with a number of proposed challenge problems, with no guidance as to how to select, prioritize, or integrate into any kind of holistic R&D roadmap.

Machine Perception, Reasoning and Intelligence (MPRI):
- Common Representations and Architectures
- Learning and Reasoning
- Understanding the Situation/Environment
- Robust Capabilities

Human/Autonomous System Interaction and Collaboration (HASIC):
- Calibrated Trust
- Common Understanding of Shared Perceptions
- Human-Agent Interaction

Scalable Teaming of Autonomous Systems (STAS):
- Decentralized mission-level task allocation/assignment
- Robust self-organization adaptation, and collaboration
- Space management operations
- Sensing/synthetic perception

Test, Evaluation, Validation, and Verification (TEW):
- Methods & Tools Assisting in Requirements Development and Analysis
- Evidence based Design and Implementation
- Cumulative Evidence through Research, Development, Test, & Evaluation (RDT&E), Developmental Testing (OT), and Operational Testing (OT)
- Run time behavior prediction and recovery
- Assurance Arguments for Autonomous Systems

Figure 6.3. DOD Autonomy Community of Interest: challenge areas (Bornstein 2015)

Recommendation 3a: Drive basic behavior, architecture, and function development of ASs with an appropriately scoped, scaled, and abstracted set of functionally oriented challenge problems that allow different members of the S&T community to focus down on different contributors to AS behavior. Select the set of challenge problems based on an initially nominated architecture and function set, in a fashion that spans the full set of functionalities represented in the architecture (exhaustiveness) and that minimizes the overlap in functionalities needed to address any two challenge problems (exclusivity). Iterate on the architecture and function set as results from addressing the challenge problems become available and, likewise, iterate on the challenge problem design based on the updated architecture/functions, following a model-based experimentation protocol (Mislevy et al. 2017). Finally, any challenge problem portfolio should be done with an awareness of what the rest of the R&D community is engaging in: a lead/leverage/watch strategy is clearly called for here.

Clearly, this will require the initial selection of one or more architectures and function sets to kick off the nomination of an initial set of domain-independent or functional challenge problems, and it will also require considerable collaboration and communication across the S&T community to address these problems in a coordinated and efficient manner. It will also require periodic updating of the selected architectures/functions, a reassessment of the adequacy of the challenge problem set, and updating this problem set as needed. This approach of model-driven inquiry will come at a significant cost of coordination across the community but should support:

- Evolution of one or more architectures to better enable and represent the desired set of AS proficiencies, trust relations, and flexibilities

- Better coverage of the full scope of AS functions, with less duplication of effort across the S&T community

- Faster development because of distributed tasking across the S&T community and the potential for sharing between labs architectures, functions, algorithms, and empirically derived datasets

- Enhanced V&V opportunities because of the co-evolution of architectures/functions with challenge problems

The key takeaway here is the dynamic iterative nature of the AS architecture, function set, technology enablers, and the challenge problems themselves, driven by the interaction of the architecture/function developers and the challenge problem empiricists.[7]

As these challenge problems become "solved" at the 6.2 and 6.3 level and the associated capabilities are transitioned to more operationally relevant de-

---

7. In a fashion analogous to that which occurs in the physics community: theoreticians, with their models, drive the experimentalists in their searches for, say, new particles, and experimentalists, with their new discoveries, drive theoreticians in finding new models to explain the data.

velopment efforts (see next section), the challenge problem portfolio will need to be updated to take on increasingly difficult behavioral, architectural, and functional challenges. Most of these have been commented on earlier, and we have indicated where; in addition, for three of the problems, we have pointers to sections in appendix F, where we provide additional details. We recommend taking on the domain-independent challenge problems given in table 6.1, presented as an unprioritized but functionally organized list; the research community clearly needs to address prioritization.

Table 6.1. Domain-independent challenge problems for autonomous systems development (with pointers to relevant sections)

Function	Domain-independent challenge problem	Section
Perception and awareness	Attention management and "active sensing"	5.3.2
	Nonlinear dynamic state estimation and event detection from heterogeneous sensors and databases	5.3.3, 5.3.4.1
	High-level meaning making and situational awareness	3.1, 3.2, 5.3.4.2, 5.4.2
Cognition and decision-making	Goal formation and prioritization	2.2, 5.3.4.3, 5.4.1
	Dual-channel "fast/slow" cognition to support adaptive cognition, human-system integration, and effective human-AS teaming	2.2, 3.3, 3.4, 5.3.4.3, 5.4.3
	Reasoning by analogy and cases	5.4.1
	Flexible planning/replanning in a dynamic uncertain environment	5.3.4.4
	Context adapting ASs to support task and cognitive flexibility	1.3, 2.2, 5.3.4.4
Execution management	Generic effector management to transform high-level plans to actions	5.3.4.5
	Plan execution monitoring methods and algorithms	5.3.4.5
Internal representation	Representation of AS world models to make them consistent, stable, and useful to a broader set of problems	5.3.8, App F.1.1
	Transparent and ready access to enabling toolsets/technologies	5.3.9

Table 6.1. Domain-independent challenge problems for autonomous systems development (with pointers to relevant sections) *(continued)*

Function	Domain-independent challenge problem	Section
Learning/ Adapting	Supervised/unsupervised learning paradigms needing minimal datasets	2.4, 4.3, 5.3.7
	Cultural or "fleet learning" across a heterogeneous agent populatior	2.4
	Generic cross-paradigm learning methods*	5.3.7
HCI	Explainable AI to support AS transparency and operator trust	3.1
	HSI taxonomies/protocols to support human-AS teaming	3.3, 3.4.1
	Multimodal communications for mixed-initiative teaming	3.4.1
	Theory and protocols for human-AS team training	3.4.2
Other	Dynamic agent communications, enabling agents to learn a communications protocol in a decentralized way	5.3.2, 5.3.5, App F.1.2
	Multiagent systems design methods for specifying and developing desired emergent behavior in multi-AS systems	2.3
	Affective computing by ASs, for better human-system teaming and to support dual-channel cognition	3.4, 5.4.3
	Consciousness and "free will"; ability to trade off directability/predictability vs autonomy/independence	2.2, 2.3, App F.1.3

Other functionally oriented and mission-independent challenge problems will naturally arise over time, either from the "bottom up" as we learn more from the basic sciences of, say, human cognition, or from the "top down" as we reach to apply AS technology to increasingly difficult operational problems. The S&T community will need to stay aware of these opportunities and be ready to change its basic and applied S&T problem portfolio appropriately.

6.3.3 Domain-Dependent or Mission-Oriented Challenge Problems

To complement the domain-independent challenge problems, we see a need for a set of domain-specific or mission-oriented challenge problems that focus on current and future operations where autonomous systems could contribute to

---

* In the spirit of *The Master Algorithm* (Domingos 2015).

close a "gap" or enable new capabilities. As we note in appendix A, problems range from generic concepts (for example, autonomous swarms), to specific operations in air (including ISR, air operations planning, electronic warfare, and logistics), space (including fractionated platforms and embedded health diagnostics), and cyber (defensive operations, offensive operations, and network resiliency). Such challenge problems would have two major purposes. First, they would serve to drive exploratory development and prototyping of relatively mature operational and technical concepts (e.g., a "Pilot's Associate" that could provide simple decision-aiding and subsystem management for a pilot during high workload situations[8]) for risk reduction in downstream acquisition decisions. Second, they would serve as "stretch" goals to drive longer term development of less mature systems that are both operationally relevant with a high payoff for success and that challenge the S&T community to make significant advances in the science and engineering of AS functionality (e.g., a Pilot's Associate that acted like a true human member of a two-person crew, in a manner that was proficient, trust-worthy, and flexible). Table A.5 of appendix A provides greater detail on several challenge problems nominated by earlier studies. There are clearly others that could be candidates as well.

As with the domain-independent challenge problems, selection and prioritization are an issue. Even if we chose to be entirely "operationally gap driven," it is unclear how we would prioritize the set of challenge problems that address those gaps. More critically, by constraining ourselves to fill current and envisioned gaps, we may be bypassing important and emergent AS capabilities that could fundamentally change, for the better, the nature of our current strategy, operations, and concepts of operations.

Recommendation 3b: Select mission-oriented challenge problems with the two objectives of: (1) addressing current or future operational gaps that may be well-suited for AS application; and (2) challenging the S&T community to make significant advances in the science and engineering of AS functionality. Ensure that the challenge problems can be addressed within the context set by the architectures and functions selected earlier, to ensure consistent efforts between the domain-independent and domain-dependent efforts,[9] and to avoid "one off" application efforts that end up having little to contribute to other mission-oriented problem sets, in the way of concepts, algorithms, or reusable software modules. Consider both "partial" mission-focused challenge problems (e.g., real-time pop-up threat detection/identification) as well as "end-to-end" chal-

---

8. A knowledge-based system sponsored by DARPA dating back to the 1990s designed to support the pilot in the management of onboard aircraft subsystems (Banks and Lizza 1991).

9. If consistency cannot be achieved with one or more important operationally relevant challenge problems, then reconsideration of the architecture and or functionalities is called for.

lenge problems (e.g., multi-INT ISR [MI-ISR] that drives multidomain C2 [MDC2]). Finally, *do not* allocate S&T resources to solving operational problems that have close analogs in other sectors, unless the USAF–specific attributes make the problem so unique that it can't be solved in an analogous fashion with simple modifications.[10] As with the domain-independent challenge problems, a lead/leverage/watch strategy is called for in shaping the operational challenge problem portfolio.

When we recommend *choosing challenge problems that are well-suited to AS application*, we mean that problem sets should be fundamentally focused on the pickup of information, its processing, and its dissemination; problem sets should not be focused on sensing and locomotion because even though AS behaviors may depend on these capabilities, that is not where their contributions lie. Their contributions lie in adding intelligence to the information processing and adding value via their proficient, trustworthy, and flexible behaviors in an operational environment. As a consequence, we would recommend, in the short term, minimizing any *in motion* investments in the development of autonomous platforms and focus on *at rest* investments that not only leverage the rapid advances currently being made in all aspects of information systems and AI but also avoid the high costs associated with assuring platform flightworthiness. This recommended de-emphasis on *in motion* investment is also consistent with our recommendation to leverage developments that may exist outside the Air Force: for example, it is quite likely that the technology being developed for autonomous commercial transports[11] may be directly applicable to Air Force transport and tanker operations as well as the next generation of UAVs, a family of platforms that are currently very far from being considered autonomous systems.

In recommending both "partial" and "end-to-end" challenge problems, we see a need for balancing digestible problems (that can be addressed by a small, close-knit team in a reasonable time horizon) with "stretch goal" problems (that not only drive the S&T community but also, if successfully solved, can add significant capability to the Air Force). In addition, an end-to-end challenge can provide a stress test for one or more of the selected AS architectures, iden-

---

10. For example, an intelligent human resources decision aiding system aimed at matching personnel with job openings might be readily brought over from the commercial sector and modified to support Air Force personnel career management and assignments, whereas a commercial airlines operations center autonomous planner/scheduler might not, because of the greater complexity/constraints of Air Force operations. The former is an acquisition community responsibility, the latter an S&T community responsibility.

11. See, for example, CNN Business, http://money.cnn.com/2017/10/05/news/companies/boeing-acquires-aurora-autonomous-797-air-taxi/index.html.

tifying shortcomings in the architecture not apparent before the end-to-end problem was fully analyzed. Both types of challenge problems are called for and should be related to one another, since partial solutions may be aggregated at some point to provide end-to-end solutions. As noted earlier in this chapter, other studies have put forth both classes of operational challenge problems, and we have summarized them in table A.5 of appendix A.[12] We have also folded into the set presented here problems of both classes described in appendix F. Table 6.2a presents a set of *partial* challenge problems along with an associated "stretch challenge" end-to-end problem. Table 6.2b (next page) presents a larger set of end-to-end challenge problems, based on our considerations during the course of this study and on those presented by several other studies and summarized in appendix A.

Table 6.2a. Domain-dependent partial and stretch challenge problems

Partial challenge	Stretch challenge
Workflow improvement allocator/scheduler decision aids	Autonomous workflow managers for allocation and scheduling
Single-intelligence fusion processors and event detectors	Multi-intelligence fusion processors and event detectors (see table 6.2b)
Planning and scheduling assistants	Planners and schedulers for Battle Management and Command and Control (see table 6.2b)
Realistic red entity emulators for constructive simulations	Constructive multiagent engagement simulations

Naturally, the eventual selection of an appropriate set of operational challenge problems, from these lists or others, will depend on other factors beyond the S&T challenges they afford or even the operational gaps they fulfill; they must, at the least, be selected in the context of the larger USAF vision, mission, and strategy, simply because autonomous systems have a tremendous opportunity to transform not only our systems but also the fashion in which they are used. As a result, simultaneously working the larger Air Force strategy is strongly recommended.

---

12. It is worth noting that the great majority of domain-specific challenge problems identified by the previous studies are platform focused, in air, space, or cyberspace; our goal here was to balance these with problems that are more platform agnostic and more closely associated with cross-domain ISR and C2 issues.

Table 6.2b. Domain-dependent end-to-end challenge problems

Challenge category	Domain-dependent challenge problem	Section
Multi-INT cross-domain data fusion*	Real-time pop-up threat detection/identification	App A.2.3
	Multidomain situation awareness (MDSA)	App F.2.1
	MDSA operational framework	App F.2.2
	ISR and PED for narrative generation	App F.2.3
	Multidomain situated consciousness (MDSC)	App F.2.4
Multidomain BMC2	Data-to-decisions (D2D) air-to-air (A2A) mission effect chain (MEC)	App F.2.5
	Targeting, resource allocation, planning, and scheduling, for cross-domain operations	
	Execution management, replanning, and BDA for cross-domain operations	
Logistics	Predictive logistics and adaptive planning	App A.2.3
	Adaptive logistics for rapid deployment	App A.2.3
Air platforms/on-board	Sensor fusion aids, decision aids ("Pilot's Associate")	
	Embedded platform health diagnostics	App A.2.3
Air platforms/off-board	Autonomous flight formation member ("Loyal Wingman")	
	Functionally fractionated platforms	App A.2.3
	Low-cost, autonomous flight systems for hyperprecision aerial delivery in difficult environments	App A.2.3

---

* Note that this subsumes the call for single-domain solutions proposed as challenge problems in appendix A.

Table 6.2b. Domain-dependent end-to-end challenge problems *(continued)*

Challenge category	Domain-dependent challenge problem	Section
Air platforms/off-board *(continued)*	Autonomous swarms that exploit large quantities of low-cost assets	App A.2.3
	Agile attritable UAVs capable of complex decision making	App A.2.3
Air platforms/on-board weapons/EW	High-precision low collateral damage munitions	App A.2.3
	Dynamic spectrum management for electronic attack/EW	App A.2.3
Space platforms	On-board satellite autonomy for defensive and offensive counterspace, incorporating sensors, event detection, SA, DM, resource allocation, execution management, and BDA	
	Embedded platform health diagnostics	App A.2.3
	Functionally fractionated platforms	App A.2.3
Cyber platforms/defensive	Agents to improve cyberattack indications and warnings (I&W)	App A.2.3
Cyber platforms/defensive *(continued)*	Active defensive cyber agents for intrusion-resilient cyber networks that achieve continued mission effectiveness under large-scale, diverse network attacks	App A.2.3
	Embedded system resilience: autonomous cyber resilience for platforms	App A.2.3
Cyber platforms/offensive	Active offensive cyber agents incorporating sensors, event detection, SA, decision-making, resource allocation, execution management, and BDA	

6.4 Development Processes

This section makes several recommendations focusing on the processes needed to develop innovative autonomous systems, delivering them in a timely and responsive manner. These recommendations are based on our own knowledge of Air Force processes inside and outside of AFRL, processes used in other R&D communities, and process recommendations made by others, notably those summarized in appendix A. Our recommendations cover four areas: developing the people with the needed skillsets (section 6.4.1); developing the architecture(s) and applications for the kinds of challenge problems just described (6.4.2); developing, curating, and distributing the data that is critical to the development of this class of systems (6.4.3); and developing the computational infrastructure that will enable the operation of the envisioned systems and the maintenance of their associated datasets and knowledge bases (6.4.4).

6.4.1 Workforce Development

Chapter 4 discussed six distinct communities we believe to be converging onto a common understanding of human behavior *and* autonomous system behavior, via the development of common computational models of cognition. To recap, these are: the robotics and cybernetics communities, which have driven a better understanding of machine-based autonomy; the cognitive psychology and neurosciences communities, which bring us closer to understanding human cognition; and the AI communities, both "hard" and "soft," which continue to provide us with nontraditional computational approaches to difficult perceptual/cognitive problems. We believe that the skillsets in all these communities are important to developing autonomous systems, but if we were to prioritize them for Air Force applications, it would probably be in an order reverse to which we have just (re)introduced them. Specifically, we believe that AI should be the primary domain and skillset the Air Force needs to emphasize in its workforce, since this will serve as the means for design and implementation of these systems, and that the cognitive sciences and the neurosciences should be a secondary domain of emphasis, since these will serve as a guide to understanding the behavior and processes of an already existent system, the human. Because we have deemphasized "in motion" applications challenge problems in the previous section, we would consider robotics (and its cousin, cybernetics) a tertiary skillset emphasis area. Finally, we have emphasized the need for human factors engineering and human-systems integration throughout the design and development process, and this still holds in the workforce development needs.[13]

---

13. Although AFRL has strong credentials in this area, the broader Air Force Materiel Command (AFMC) community is clearly lacking.

Unfortunately, AFRL does not have a strong AI contingent in the workforce: in a total workforce of approximately 5,000, we estimate those familiar with the technology to number in the low hundreds, and of those, perhaps fewer than 10 might be considered AI practitioners in the league of university researchers or commercial developers currently driving the field. We therefore propose a multipronged workforce development process to grow Air Force people in this critical area.

First, concentrate on employee education and application of that education to the types of challenge problems discussed in the last section, and do so in a deliberate and planned way. We propose that organizations such as AFRL send a group of researcher engineers to the Air Force Institute of Technology (AFIT), the USAF's graduate school, to attend its "Introduction to Autonomy" survey course to get a broad perspective of modern approaches to AI and autonomy. Following the survey course, this group would be embedded in an AI-focused special operations activity: an autonomy capabilities team (ACT)[14] to continue their education (in the same vein as an internship) while applying what they learn to solve one or more relevant AI-related AS problem.

Second, members of the ACT can be embedded at nongovernment organizations dealing with common sets of technical objectives. This approach has two distinct benefits: (1) it exposes those selected USAF individuals to the problems of scale and agility typically facing commercial efforts (and often academic efforts as well) and not usually encountered in government laboratories; this has been done successfully in a limited fashion with Google, IBM, Facebook, and a few academic institutions schools; and (2) it creates new collaborations with industry and academic partners that may not have existed in the past, which supports the kind of architecture and application development described in the next section.

Third, the ACT could support summer interns from universities as a way to expose them to a wide range of our challenges and support the recruiting of potential future employees. The AFRL directorates at both Wright-Patterson AFB and Rome, New York, have had great success with this process and have attracted premier talent in AI-related fields of study.

If an ACT is commissioned as part of our organizational recommendations in section 6.5, the first year should see an increase of about 20 staff and serve to initiate an on-ramp for an additional 20 through the AFIT Autonomy course later in the year. During the second year, a new set of 40 residents could participate with the AI special operations unit, and 40 more staff could on-ramp into the AFIT Autonomy course. The overall goal would be to grow

---

14. See additional description in section 6.5.

AFRL's organic capability in AI by an order of magnitude in four years, from approximately 20 in 2018 to 200 by 2022.

This educational and experiential effort needs to be supplemented with a strong retention program, since competition for these skillsets is fierce, especially in the commercial sector.[15] This group should be closely monitored regarding compensation, advancement, and quality of research/acquisition postings and afforded the same considerations we currently afford our pilot specialties.

Finally, we should "track (via a knowledgeable cadre) and invest in (via a 6.1 research portfolio) the most dynamic and rapidly advancing areas of AI" (Potember 2017), as well as provide long-term support of the extramural leaders in the area.

Recommendation 4a: Create an educational and intern-like personnel pipeline to send selected staff to AFIT for an introductory autonomy short course, focusing on AI enablers. Individual members would then be embedded into an AI-focused special operations activity: an ACT to learn how to apply the skillsets they acquired in addressing USAF autonomy needs. Support this effort over the course of four years to grow AI manpower by an order of magnitude over today's level. Assure retention via several special incentive programs. Supplement this cadre with appropriate and long-term support of key extramural researchers.

6.4.2 Architecture and Application Development

In discussing appropriate challenge problems in the previous section, we alluded to processes for generating candidate challenge problems, downselecting, and prioritizing how they are to be addressed via autonomous systems development. In this section, we discuss this process in more detail.

Figure 6.4 illustrates a three-phase framework for iterative selection of challenge problems, for modeling the impact of potential solutions, and for solution development, prototyping, and assessment.

The first phase is composed of *wargaming* and is illustrated in figure 6.5a. It is conceptual in nature, looking at future threats and capabilities, with the goal of developing conceptual autonomous systems to address those threats or take advantage of potential opportunities. The sophistication of the wargaming can range from tabletop exercise with BOGSAT[16] assessments of capabilities and

---

15. The fact that business schools (such as Harvard Business School, Stanford's Graduate School of Business, and MIT's Sloan School of Management) have initiated courses on how to manage AI algorithms and applications for more informed business decisions indicates that the demand will only grow for AI practitioners across the commercial ecosystem and not just in the high-technology sector (Simons 2016).

16. Bunch of Guys/Gals Sitting Around the Table.

"moves" to more sophisticated simulation-generated outcomes and metric-based assessments. The major activity is exploratory, aimed at exploring the full span of options, to avoid fixating on too narrow a solution that misses a game-changing opportunity. The major product of this phase is the identification of challenge problems that drive AS-based conceptual solutions that have a high potential for dramatically affecting future operational outcomes. This has been accomplished in a limited number of recent wargames (for example, the Air Force strategic wargames of 2013 and 2015 [USAF 2013, 2016]), but more such exercises need to be conducted to fully explore the potential of deployed autonomous systems.

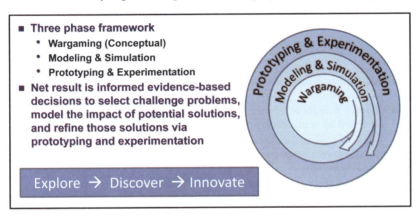

Figure 6.4. Framework for AS development process

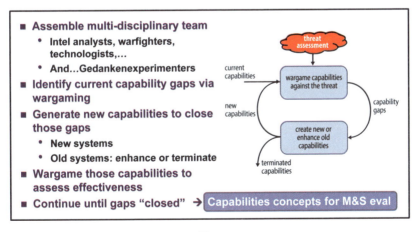

Figure 6.5a. Wargaming of concepts[17]

---

17. A "Gedankenexperiment," a term invented by Einstein, is a thought experiment, using a conceptual approach that relies on thinking through an idea rather than doing an actual experiment. See more at https://www.britannica.com/science/Gedankenexperiment.

The second phase is composed of *modeling and simulation* and is illustrated in figure 6.5b. The goal here is to provide a deeper assessment of promising candidates identified in the first phase, via formalization of those concepts with quantitative models, simulations, and parameters of performance. Again, the level of sophistication of the models and the breadth and depth of the simulations can run the gamut from purely constructive to virtual or some combination of the two. The major activity is discovery of the advantages and disadvantages of a particular proposed AS solution and its potential impact on overall wargame outcomes, at a "higher" level of fidelity than that done in the first phase. The major product of this phase is the identification of promising capabilities that need to be looked at more closely via prototyping and experimentation.

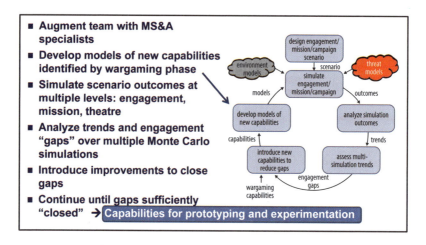

Figure 6.5b. Modeling, Simulation, and Analysis[18]

The third phase focuses on the *design, prototyping, and experimental assessment* of one or more (primarily software) engineering prototypes of promising AS capabilities/functions identified in the previous phase; it is illustrated in figure 6.5c. Desired maturity would be at a proof-of-concept level or breadboard level (a Technology Readiness Level or TRL of 3 or 4 [NASA 2012]), and structured experiments—beyond simple demos—would

---

18. A Monte Carlo simulation is one in which a model is "run" multiple times with parameters or inputs varied randomly (hence the name "Monte Carlo") so that trends in outputs can be analyzed statistically over a range of conditions of interest. See "Monte Carlo Method" entry in Wikipedia, accessed 12 February 2019, https://en.wikipedia.org/wiki/Monte_Carlo_method.

be carried out to assess performance and potential operational impact. The major activities include AS innovation, with rapid prototyping, structured experimentation, data-driven assessment, and design iteration.[19] The major product of this phase is a prototype AS that can effectively address the key challenge problem(s) identified in the earlier phases and that can serve as: (1) a design prototype for acquisition; and (2) a design driver for additional needed S&T.

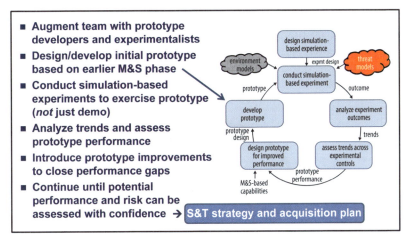

Figure 6.5c. Prototyping and experimentation

Recommendation 4b: Use a three-phase framework for iterative selection of challenge problems, for modeling the impact of potential solutions, and for solution development, prototyping, and assessment. Conduct an initial phase of wargaming-based assessment, looking at future threats and capabilities, with the goal of identifying key challenge problems and AS-based solutions that can address those threats or take advantage of potential opportunities. Provide a deeper assessment of promising AS candidates, via formalization of those concepts with quantitative models and simulations (M&S) and parameters of performance. Finally, focus on the design of one or more (primarily software) engineering prototypes of promising AS candidates identified in the M&S studies. Develop and experimentally evaluate a prototype AS that can serve as: (1) a design prototype for acquisition; and (2) a design driver for additional needed S&T.

---

19. These are the kinds of activities currently undertaken by the Strategic Development Prototyping and Experimentation (SDPE) Office under AFMC. We are proposing a similar effort, focused purely on autonomous systems development.

As part of this prototyping and experimentation process, we envision conducting a number of technical integration experiments (TIE), illustrated schematically in figure 6.6. The diagram shows that there is a minimal amount of development prior to a demonstrable product, as indicated by the blue dashed circle and the blue triangle after TIE3. Each TIE thereafter may or may not include a demonstration, depending on the goals. The key issue then is to define the TIEs in a meaningful way so as to deliver operationally useful capabilities as different challenge problems are addressed while extending the underlying knowledge base and functionality of the autonomous system framework. Appendix G presents a specific example of a TIE approach for the spiral development of a variety of AS applications that iteratively build on one another.

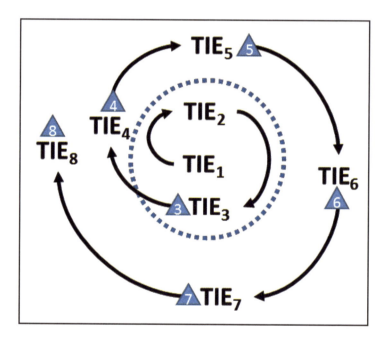

Figure 6.6. A series of technical integration experiments with demonstration events (blue triangles) for developing a variety of AS applications

6.4.3 Data Development

Data development, storage, and use have tended to be aligned with the particular functionals and major commands in the Air Force that depend on that data, leading to stovepiping and lack of interoperability—and even lack of awareness of existence. A broader view at the enterprise level needs to be taken, recognizing that data can be a valued asset beyond the original "owners" of the data, especially if stovepipes can be broken down and disparate datasets can be combined in new ways to generate new insights. Corporately, the USAF should seek to formalize the process to gathering, curating, and indexing the data with a central or distributed repository with appropriate data descriptions/labels and make it available to the global Air Force community (both intramural and extramural) working on data-intensive processes and processing. Wherever possible, operational data should be saved for future research use in support of AI for DOD-unique missions (Potember 2017). Fortunately, the Air Force is now starting down this path with the recent creation of a chief data officer position, serving at USAF Headquarters, to ensure interoperability of data and applications as well as effective access to non-USAF data sources.

The Defense Information Systems Agency owns the Defense Enterprise Computing Centers (DECC). Establishing a data warehouse for USAF data in a DISA DECC could be the first step in collecting and managing data across the Air Force enterprise. Establishing directives and a process by which data-producing organizations upload their data to a DISA DECC is warranted. Such data-producing organizations should have a data curator to manage the data from acquisition to maintenance. Links to common open source datasets should also be included.[20] An efficient process to control access to data should also be pursued, to improve upon the current process of using the DD2875 form, which was designed for system authorization access, when there is a need to have access to a DOD computer system. For example, role-based and clearance-based access using the common access card would reduce the requirement for DD2875 forms, thus reducing the management complexity associated with granting access to data and total turnaround time to get access to the data.

Recommendation 4c: Through the USAF Chief Data Office (CDO), acquire space to store Air Force air, space, and cyber data so that AI professionals can use it to create autonomy solutions to challenge problems. Establish data curator roles in relevant organizations to manage the data and to create streamlined access and retrieval approaches for data producers and consumers.

---

20. See appendix E for a list of open source datasets, along with relevant toolsets.

6.4.4 Computational Platform Development

Building an autonomous system requires a significant amount of processing capability, whether we are considering embedded processing for unmanned systems or large high-performance computing (HPC) centers for ISR tasks. In some cases, the processing requirements during the early stages of prototyping are unknown, and often, an evaluation of alternative computational approaches is needed. Processes to gain access to USAF HPC are well established. But more research is needed for leveraging quantum computing in a general computing paradigm, since it has the potential to drastically alter existing processes.

Recommendation 4d: Support the movement to cloud-based computing while also leveraging quantum computing as a general computational paradigm that can be exploited to meet embedded and HPC processing demands.

6.5 Organizational Structures

Developing and deploying autonomous systems is a daunting challenge, as we have described earlier. But it is not just a technical challenge—it is also an organizational challenge, one requiring the contributions of many different disciplines acting together in a project-focused manner. This section focuses on the organizational structures needed to address the challenge problems proposed in section 6.3 while implementing the development processes just described in section 6.4. We first describe the Air Force organizational framework we propose for autonomous system R&D (6.5.1), outline an approach to technology employment (6.5.2), and close with a summary recommendation for an USAF organizational structure focused on autonomous system R&D (6.5.3).

6.5.1 Air Force Organizational Structure

In chapter 5 we presented a possible generic architecture for the development of a broad variety of mission-specific ASs. Realizing this architecture requires a diverse set of skills. Currently, much of this expertise exists in various AFRL Technology Directorates (TD). Bringing this capability together into a project- or program-focused team calls for a cross-directorate exercise. But experience shows that cross-directorate initiatives or programs see only a small portion of their realizable potential, because there is no clear line of authority to execute. Anecdotally, much of the success of successful cross-directorate efforts can be attributed to a "cooperation of the willing" and individual leadership, an approach that is not sustainable. But such a challenge is not unique to AFRL. Johnson formed Lockheed Martin's Skunk Works during World War II to deliver the Air Force's first jet fighter,[21] which was developed by a highly skilled, highly motivated, cross-functional team. Furthermore, the perception exists that the AFRL TDs can be focused on process versus product, and this needs to change if the Lab is to deliver autonomous capabilities to the warfighter. How do we get there?

6.5.1.1 Organizational Model within AFRL. A new organizational model is a way to start the transformation from process to product. A nice summary of five common organizational models is provided in Morgan (2014, 2015):

- Traditional hierarchical (e.g., the DOD)

- Flatter organizations with limited hierarchy where communications lines are open; many commercial sector companies use this model

- Flat organizations with no hierarchy, and everyone works as a peer (e.g., Valve [2012])

- Flatarchy, a more traditional hierarchy with small flat organizational spinoffs via internal incubator/innovation offices; for example: IBM's Emerging Business Opportunities incubator and DuPont's Market Innovation program (Wunker 2012)

- Holocratic (e.g., Zappos [Pontecraft 2015])

The autonomous systems R&D effort will need to be product focused, with all R&D geared toward building the product. This will require the integration of 6.1–6.4 experts under one roof. To accomplish this and allow the proposed ACT to be successful, we recommend a flatarchy (Morgan 2015b): an organization formed within the existing hierarchy of the Lab but with a clear line of authority, with internal AFRL talent matrixed in from the TDs and answering to the team lead. Augmenting this core team, and where technical gaps exist, the ACT will be supplemented by contractor and academic partners, forming an augmentation team.

We recommend striving for a 90 percent production ratio, that is, having 90 percent of the personnel costs being product-focused and 10 percent allocated to management and administrative overhead. We recognize that this is not achievable in the traditional hierarchical organizational structure and

---

21. See "Missions Impossible: The Skunk Works Story," Lockheed Martin, accessed 21 February 2019, https://www.lockheedmartin.com/us/100years/stories/skunk-works.html.

have thus recommended a flatarchy for the ACT, an organizational structure that has the potential to achieve a 90 percent production ratio.

Figure 6.7 shows an organization that sits outside of the TDs, operating as a peer of the TDs. It has a small cadre with a single management authority, a Chief (say at a COL/O6 level) that reports directly to the AFRL commander. It has a single technical authority, an Autonomy ST that maintains consistency in technical direction by establishing a vision, has hire and fire authority for all technical contributors, and ultimately determines any organizational changes to meet the technical mission. The Autonomy ST has a small technical staff that helps establish capability delivery methods, establishes a customer base for capability off-ramps, directs the construction of the technology solution, establishes a technical road map along with a strategy-to-task layout as appropriate, and establishes the daily battle rhythm. The Chief Scientist establishes the strategic and tactical science vision and direction on frameworks and technologies. The Chief Technology Officer establishes the strategic and tactical direction on tools, software, facilities, and computational environments to enable implementation of the framework and development of specified applications. The Lead Integrator is responsible for overseeing all development activities, to include managing development and delivery schedules and providing direction on product deployment.

A larger core technology team focuses on science and engineering. The core team consists of major theoretical and applied engineering disciplines—discussed at length in chapters 4 and 5—that should cover most of the technology needs for an AS development effort: experts in synthetic agency to create the sensory, cognitive, and effector capabilities needed for agent-based interaction with the real world; experts in AI to support knowledge acquisition and learning within and across autonomous systems; cognitive scientists to support the development/use of cognitive architectures that support human-system teaming and trust; mathematicians/statisticians to concentrate on the theoretical and applied work supporting autonomy, including validation and verification; human factors engineers for development of effective human-computer interfaces; and software architects and developers to ensure a supportable and deployable systems are produced.

ACT is intended to support both long-term AS foundational needs (e.g., AS frameworks and domain-independent capabilities) as well as short-term AS application needs (e.g., domain- and mission-dependent problem sets), spiraling capabilities on in the fashion described earlier in section 6.4 under the TIES discussion. For the short-term efforts, ACT will be flexible, and the team composition will be dynamically shaped to be responsive to customer needs and product delivery times. Customer representatives will act as customer advocates during production development, and problem domain experts will provide insights into customer needs workflows. The support team will include a program manager to manage budget and contract needs; developers to support production and basic and applied exploration; a data curator to manage all aspects of the data; and a hardware and information technology support person to keep the infrastructure up and running.

Although figure 6.7 identifies the number of functions desired, the goal is to have people that are "pi-shaped." That is, they have expertise in more than one area while having a broad perspective and set of experiences.

Populating the ACT in an organization like AFRL is effectively accomplished by the matrix approach. The matrix model is effective when there is a focused effort, like AS development, but the human capital needed to address the corporate need is fragmented in different sub-organizations (like AFRL and its TDs) (Stuckenbruck 1979). Given the approach AFRL uses for evaluating employee contributions to the sub-organization, the matrixed organization results in two chains of command, one for the ACT and one for the TDs. Care must be taken to ensure that employees' supervisory chain within the TD is aware of their contributions to the ACT's mission, so that their performance evaluations reflect their contributions accordingly. While assigned to the ACT, members should not be accessible to their home TDs unless arrangements are made with the ACT leadership ahead of time.

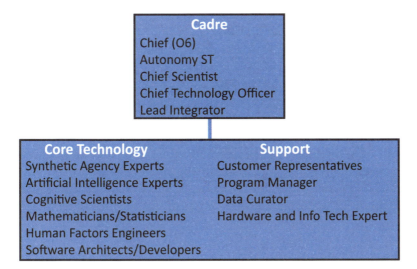

Figure 6.7. Autonomy Capabilities Team with a flatarchy organizational structure

6.5.1.2 Organizational Support outside of AFRL. The DOD has an autonomy community of interest that has as members the services and combat support agencies, providing an opportunity for the services and agencies to share information on their ongoing autonomy efforts. The USAF needs to stay engaged with this COI and contribute thought leadership to integrated challenge problems, relevant core AI technologies, and opportunities for cost savings through joint and complementary R&D. From the S&T perspective, this is a role that AFRL is best equipped to manage, and it can serve as an intermediary between the ACT and the broader DOD community.

AFWERX is a *catalyst for agile Air Force engagement across industry, academia, and nontraditional contributors to create transformative opportunities and foster an Air Force culture of innovation.*[22] If an organization has a capability or potential capability that is of interest to the Air Force, AFWERX will help make connections between the organization and the USAF customer base. This is very much an adoption of the platform business model, where AFWERX focuses on matching vendors with consumers.[23] However, there is also a need to ensure that the technology supporting the AFWERX-identified capability is managed in a supportable way. A logical scenario is to embed organizations of interest within the ACT organizational structure on a contractual basis. From an acquisitions perspective, the USAF should no longer buy an end-to-end AI-based system. They should buy "plug-ins" that fit one or more autonomous systems architectures, such as the one described earlier in chapter 5, that can be specialized for particular domains or mission sets. ACT could therefore provide AFWERX guidance on technology needs, while AFWERX could serve as a discriminating broker for innovations appearing in the commercial sector.

AFRL has a basic research office, the USAF Office of Scientific Research (AFOSR) funding domestic and international basic research (6.1 funded), across academia, industry, and within the other AFRL TDs as well. A pathfinder to grow people, exchange science and engineering ideas, and fill critical gaps in technology areas of interest to the USAF is illustrated in figure 6.8. Such a model enables significant intellectual transfer between the ACT, the AFRL TDs, and AFOSR (and their academic and industry partners) and would support the dynamic staffing needs of the ACT as different autonomous system development efforts come and go. This model also provides consistency across the USAF S&T community in terms of AS behavioral

---

22. More can be found at afwerxdc.org.
23. We discuss this platform business model at greater length in the next section.

requirements, architectural frameworks, component functionality and enabling technologies, and lessons learned as well as "best process" for development. Finally, this model supports exposure and growth of personnel in the TDs and AFOSR in the technology of autonomous systems, in both theory and application areas, as well as end-user USAF mission-focused needs.

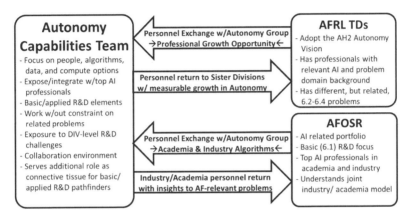

Figure 6.8. Pathfinder between ACT, AFRL TDs, and AFOSR

6.5.1.3 Business Processes: The ACT. We have already described, in the previous section, a set of recommended autonomous systems development processes. Here, we wish to focus on "cultural" processes to be followed by those staffing the ACT. A natural model is Johnson's Skunk Works model,[24] with 14 guiding rules or principles applying to management, team, or product categories. We present those rules in table 6.3 (next page), suitably modified to accommodate the ACT organization operating within the AFRL organizational umbrella.

---

24. "Skunk Works," Lockheed Martin.com, n.d., https://www.lockheedmartin.com/en-us/who-we-are/business-areas/aeronautics/skunkworks.html.

Table 6.3. Skunk Works 14 rules adapted to ACT and AFRL

Rule	Category	Description
#1	Team	The ACT Autonomy ST must be delegated practically complete control of his/her program in all aspects, to include the hiring and firing of personnel, establishment of the science, and the particulars of the experiments. He/she should work with a high performing Division Chief (O6) that reports directly to the AFRL/CC or his/her designated representative.
#2	Management	Strong but small project offices must be provided both by the government and industry. The project offices need to be light, yet well equipped to handle the dynamic needs of the initiative. This also includes reach back into the AFRL TDs to gain access to specialized equipment or spaces.
#3	Team	The number of people having any connection with the project must be restricted in an almost vicious manner. A small number of high caliber people (approximately 30, say) with the right expertise should participate. The number will fluctuate based on project needs.
#4	Team	A very simple documentation and documentation release system with great flexibility for making changes must be provided.
#5	Management	There must be a minimum number of reports required, but important work must be recorded thoroughly. A quarterly reporting requirement is reasonable.
#6	Management	There must be a quarterly cost review covering not only what has been spent and committed but also projected costs to the conclusion of the program. This should be kept at a minimum and handled by a Program Manager.
#7	Management	The team must be delegated and must assume more than normal responsibility to get good vendor bids for any contract work needed on the project. The Program Manager must have flexible funding options to fully support the ACT.
#8	Management	The inspection system currently employed by AFRL is adequate, although reducing the overhead associated with inspection should occur. Do not add inspection requirements unless there is a concern (see Rule #12).
#9	Product	The team must be delegated the authority to test the final product in some form of operational test and evaluation.

Table 6.3. Skunk Works 14 rules adapted to ACT and AFRL *(continued)*

Rule	Category	Description
#10	Product	The specifications applying to the hardware and software must be specified by and agreed to by the ACT.
#11	Funding	Program funding must be dedicated, timely, and not subject to budget drills; otherwise, progress will be limited and delayed, and that will put the initiative at risk.
#12	Management	There must be mutual trust between the team and the delegated office to which the team reports. This cuts down misunderstanding and correspondence to an absolute minimum.
#13	Management	Access by outsiders to the project and its personnel must be strictly controlled by appropriate security measures and at the discretion of the Autonomy ST.
#14	Management	Because only a few people will be used in engineering and most other areas, ways must be provided to reward good performance and not be left to the internal directorate appraisal process. Any ilitary involved should abide by their respective Officer or Enlisted Performance Appraisal Systems.

6.5.2 Organizational Model for Technology Employment

Having a team in place to create the technology is of great value. However, if the USAF is to employ ASs within the operational community, a different integration model needs to be considered. The approach to doing integration in the air domain is based on the connection of sensor assets to shooter assets via command and control assets, while receiving assistance via support assets (recall the platform-centric view of today's Air Force shown in fig. 1.9). Connections are hardwired that have limited flexibility, and that often leave out other assets that may need connectivity;[25] if a new configuration is needed, it requires a new set of connections (and likely interfaces, etc.). The current way of conducting intelligence and operations is built around this *physical* platform model.

We need a new way to employ technology and have it integrate with existing business practices (i.e., operations). The platform *business model* is such an approach. To retouch on our vision for the future described in section 1.5, information-intensive industries can most benefit from the platform revolu-

---

25. The nonconnectivity between the F-22 and F-35 is a canonical example (Everstine 2018).

tion (Sarkar 2016).[26] And we *are* an information-intensive organization, given our activities in collecting and processing ISR data, identifying threats and opportunities for strikes, conducting resource allocation and planning activities, making operational and tactical decisions, executing and monitoring of our actions, and conducting battle damage assessment for the next activities in the cycle across multiple domains of air, space, and cyber. The platform approach is used to solve these types of large-scale information intensive problems by connecting vendors and consumers, and, in our case, information vendors with information consumers. The integration is not about data or technology; it is about making connections across the network of these vendors and consumers. The role the USAF leadership plays then is to facilitate a move to this business model—concentrating on building the network and enforcing a common language across network nodes—and the role the S&T community plays is to provide the technology to make the exchange more useful. This is a transformation in the way we think but one that can be beneficial. We discuss it at greater length in the next section.

6.5.3 Organizational Structure Recommendation

Recommendation 5: Establish the ACT within AFRL, incorporating a flatarchy business model to bring 6.1–6.4 experts into a single product-focused organization to develop the science of autonomous systems while delivering capabilities to the warfighter. Collaborate with AFOSR and other key AFRL TDs and coordinate with USAF organizations outside AFRL, including the DOD ACOI, AFWERX, and other offices that can facilitate technology transition to the warfighter. Within the ACT, incorporate product-focused business processes based on a Skunk Works–like set of "guiding rules" and facilitate the move toward an information-centric business platform model for the future Air Force.

6.6 Knowledge Platform

Combining the IT platform approach with a platform business model is an industry best practice used in Silicon Valley and in some combat support agencies. It has revolutionized the connection of consumers to vendors. For example, Uber connects people needing transport services to people capable of providing transport services in the needed timeframe. Amazon originally connected people to books but had the *IT platform* well established, making

---

26. And can be most vulnerable from competition if they do not make the move to benefit from such advances.

the connection of people to any number of products a seamless process. Amazon web services connect organizations to big data solutions (*tool development*) through their scalable, reliable, big-data platform. The National Security Agency (NSA) and DISA similarly have a Big Data Platform that enables efficient tool development for cyber operations, which has been made available to the entire DOD. Facebook has revolutionized connecting people-to-people through ts social media platform. TaskRabbit connects homeowners to safe and reliable services for residences. Google has taken the platform idea into the AI arena, providing a platform to make machine learning easier for a range of users.[27] These are all modern platform business models that are enabled by IT.

Knowledge is what an autonomous system uses to create the meaning from its observations. An AS is a system that creates the knowledge necessary to remain flexible in the tasks it undertakes, its relationships with humans and machines, and its ability to entertain multiple cognitive approaches to problem solving. This concept of knowledge creation is echoed in Domingos's book *The Master Algorithm* (Domingos 2015) and provides foundational understanding of knowledge in support Air Force situation assessment and decision-making activities. Tool-based solutions have provided much value over the years, but the solutions only work for a limited set of problems, under certain circumstances, and, typically, only for a single domain. In short, they do not scale. However, a knowledge focus toward problem solving provides us with a means of breaking these constraints, transitioning us from the traditional tools-based approach that solve a small number of problems to a knowledge platform approach applicable to a far greater set of problems.

A KP designed for the multidomain-operating Air Force should monopolize the connection of observation agents with knowledge creation agents and warfighting effects agents, any and all of which can be either human or machine-based agents (AS), based in air, space, or cyber. Collaborative ASs would be able to dynamically and opportunistically team and separate as necessary and as dictated by the unfolding battlespace events. Each AS would need to be able to communicate with ASs to achieve some particular desired multidomain effect through a dynamic and variable vocabulary. This move from a specialized representation of knowledge to a more flexible creation, representation, and consumption of knowledge is cited as what facilitated animal species in nature expanding their purview (Newell 1990). ASs need to be focused on creating knowledge and appropriately applying that knowledge to maximize their contributions to a range of missions.

---

27. Google Cloud presents an array of services and products; see more at https://cloud .google.com/products/machine-learning.

A KP overcomes the limitation of current AI machine-learning solutions where knowledge is tied in a linear manner, from the observables to the creation of acceptable meaning for those observables, by that AI agent, and for a small number of specialized consumers of that meaning (e.g., a cyber access protection agent "consuming" the product of a face recognition agent). In contrast, a KP takes in a range of observables (many of which cannot be predefined) and creates a variety of acceptable meanings for a broad population of agents (humans and ASs), to enable a range of effects. The span of knowledge necessary for enabling this scalable KP can be partially programmed, but the majority will be learned both through experiential learning and "cultural" learning (from other ASs), as discussed earlier in section 2.4, as well as through "cognition" or some facsimile of it, by generating its own experiences using its ability to *simulate* interactions with the world.

The KP provides the ecosystem necessary to create capabilities, and those capabilities are used to create combat effects. This ecosystem will come to fruition by exploiting the three behavioral principles of autonomy (section 6.1), the architectures and technologies that enable those behaviors (6.2), the domain-independent and domain-dependent challenge problems (6.3), the developmental processes across people, architectures/applications, data, and computational infrastructures (6.4), and, finally, the organizational structures that need to be in place to advance the technology, exploit it, and deliver

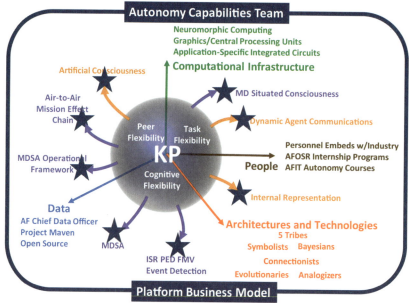

Figure 6.9. Knowledge Platform components

capability (6.5). This is conceptualized in figure 6.9 where the AS architectures and technologies are represented by the blue sphere (section 6.2); the resultant behaviors represented as white text in the blue sphere, illustrating task, peer, and cognitive flexibility (6.1); the spin-offs are technology advancements (associated with the domain-independent challenge problems of 6.3) and capabilities (associated with the domain-dependent challenge problems of 6.3); the foundational components and processes represented as axes in the diagram covering people, architectures/technologies, data, and computational infrastructures (6.4); and the ACT (6.5) and the platform business model (6.6) represented by the outer box, which jointly serve to create and employ the KP.

This chapter provided recommendations related to each of the elements of the above figure. They assimilate into a KP core, which is a technology solution illustrated in figure 6.10. The KP (shown in the purple-blue box) is the assimilation of operationally relevant data (blue arrow), architectures and technologies (red box), computational infrastructures (green outline), domain-independent technologies (orange boxes), providing domain-dependent applications (purple ovals) with the needed flexibilities (maroon boxes) to support the warfighter (brown outline).

A key to developing a KP is to understand the left and right limits of each of the three flexibilities. The property for the left limit is a single hard-coded

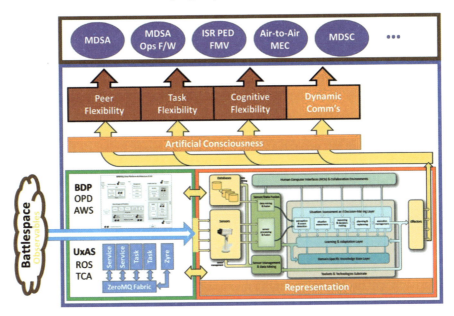

Figure 6.10. Knowledge Platform core

view of the world; no flexibility. In this left limit, systems have a single task (e.g., vacuuming the floor), a single relationship (e.g., always a subordinated peer), and have a single way to solve that task (e.g., a sensor-driven rule-based system). The right limit is a conscious simulation over the AS's total knowledge base to determine which subtasks need to be accomplished to complete its main task or which tasks need to be accomplished to meet the mission intent; determine the peer relationships to accomplish that task and have multiple cognitive approaches to solving tasks and making decisions on peer relationships. There needs to be a well-defined path that demonstrates increasing levels of task, peer, and cognitive flexibility from the left limit to the right limit; a description of its knowledge complexity; the source of knowledge that is represented to capture the complexity of the solution space; concrete examples of systems that may or may not exist; and description of their internal representation. The intent is not to be prescriptive, but for clarity an example is provided in appendix G where we describe a series of TIEs that could be taken to build a knowledge platform.

A path on how to get to the envisioned KP is to leverage the current architecture and technologies we have available and have reviewed in chapter 5 and tackle the proposed domain-independent and domain-dependent challenge problems outlined in section 6.3, in a way that focuses development of platform components that demonstrate the desired behaviors outlined in chapters 2 and 3. This sort of development will require novel processes, as outlined in section 6.4, and a deviation from the current organizational structure to create the platform and use it for operations, as discussed in section 6.5.

Recommendation 6: Develop a Knowledge Platform (KP) centered on combining an IT platform approach, with a platform business model. A KP designed for the multidomain operating Air Force should monopolize the connection of observation agents, with knowledge creation agents and with warfighting effects agents, which can be either human or machine-based agents (ASs). The KP provides the ecosystem necessary to create capabilities, and those capabilities are used to create combat effects. This ecosystem will come to fruition by exploiting the three behavioral principles of autonomy (section 6.1), the architectures and technologies that enable those behaviors (6.2), the driving challenge problems (6.3), the developmental processes across people, architectures/applications, data, and computational infrastructures (6.4), and, finally, the organizational structures that need to be in place to advance the technology, exploit it, and deliver capability (6.5). This approach will provide us with the means of transitioning the USAF from the traditional tools-based approach that solves a small number of problems to a knowledge platform approach applicable to a far greater set of problems.

6.7 Summary of Recommendations

We now summarize our recommendations made earlier, grouped by category.

6.7.1 Behavioral Objectives

These are basically generalized design requirements specifying how we want an AS to behave, in terms of proficiency, trustworthiness, and flexibility.

- *Recommendation 1a*: ASs should be designed to ensure *proficiency* in the given environment, tasks, and teammates envisioned during operations. Desired *properties of proficiency* include situated agency, a capacity for adaptive cognition, an allowance for multiagent emergence, and an ability to learn from experience.

- *Recommendation 1b*: ASs should be designed to ensure trust when operated by or teamed with their human counterparts. Desired *tenets of trust* include cognitive congruence and/or transparency of decision-making, situation awareness, design that enables natural human-system interaction, and a capability for effective human system teaming and training.

- *Recommendation 1c*: ASs should be designed to achieve *proficiency* and trust in a fashion that drives behavioral *flexibility*, across tasks, peers, and cognitive approaches. Desired *principles of flexibility* for an AS include an ability to change its task or goal depending on the requirements of the overall mission and the situation it faces. It should be able to *take on a subordinate, peer, or supervisory role* and change that role with humans or other autonomous systems within the organization. And it should be able to *change how it carries out a task*, both in the short term in response to a changing situation and over the long term with experience and learning.

6.7.2 Architectures and Technologies

This covers unifying frameworks and architectures that will support cross-disciplinary research and development, along with the technology investments needed to support desired functionalities within an architecture.

- *Recommendation 2a*: Develop one or more *common AS architectures* that can subsume multiple frameworks currently used across disparate communities. Architectures should, at a minimum, provide for "end-to-end" functionality, in terms of providing the AS with a sensory ability to pick up key aspects of its environment, a cognitive ability to make assessments, plans, and decisions to achieve desired goals, and a motor ability to act on its environment, if called upon. The architecture should be functionally structured to enable extensibility and reuse, make no commitment on symbolic vs subsymbolic processing for component functions, incorporate memory and learning, and support human-teammate interaction as needed. Whatever the form, an architecture should be extensible to tasks assigned, peer relationships engaged in, and cognitive approaches used. A key metric of an architecture's utility will be its capability of bridging the conceptual and functional gaps across disparate communities working autonomy issues.

- *Recommendation 2b*: Pursue the *development of enabling technologies* that provide the needed functionality at the component level. This includes technologies that support not only the basic "see/think/do" functions but also those that enable effective HCIs, learning/adaptation, and knowledge-base management, both of a general purpose and of domain-specific nature. The nature of technology development should range from basic research to exploratory development to early prototyping, depending on the maturity of the specific technology and its envisioned application.

- *Recommendation 2c*: Develop and promulgate a *multitiered hardware and multilayered software architecture* to support AS development, validation, operation, and modification, where each tier provides for physical structuring across distinct hardware implementations/hosts for given high- and low-level functions and each layer provides distinct software implementations of similar functions. A variety of complex architectural patterns may be needed to take full advantage of emerging technology trends, particularly in the commercial sector.

6.7.3 Challenge Problems

Addressed here are both domain-independent (or functional) problems, like dynamic replanning, and domain-dependent (or mission-oriented) problems, like multidomain fusion.

- *Recommendation 3a*: Drive basic behavior, architecture, and function development of ASs with an appropriately scoped, scaled, and abstracted set of functionally oriented challenge problems that allow different members of the S&T community to focus down on different contributors to AS behavior. Select the set of challenge problems based on an initially nominated architecture and function set, in a fashion that spans the full set of functionalities represented in the architecture (exhaustiveness) and that minimizes the overlap in functionalities needed to address any two challenge problems (exclusivity). A broad range of domain-independent challenge problems is presented in table 6.1 and appendix F.

- *Recommendation 3b*: Select mission-oriented challenge problems with the two objectives of: (1) addressing current or future operational gaps that may be well suited for AS application; and (2) challenging the S&T community to make significant advances in the science and engineering of AS functionality. Ensure that the challenge problems can be addressed within the context set by the architectures and functions selected earlier, to ensure consistent efforts between the domain-independent and domain-dependent efforts and to avoid "one off" application efforts that end up having little to contribute to other mission-oriented problem sets. Consider both "partial" mission-focused challenge problems as well as "end-to-end" challenge problems. Finally, *do not* allocate S&T resources to solving operational problems that have close analogs in other sectors, unless the AF-specific attributes make the problem so unique that it can't be solved in an analogous fashion. A broad range of domain-dependent challenge problems is presented in table 6.2 and appendix F.

6.7.4 Development Processes

This includes processes—in contrast to our traditional waterfall process of requirements specification, milestone satisfaction, and end-state T&E—that support innovation, rapid prototyping, and iterative requirements development to support rapid AS development and fielding.

- *Recommendation 4a*: Create an educational and intern-like personnel pipeline to send selected staff to AFIT for an introductory autonomy short course, focusing on AI enablers. Individual members would then be embedded into an AI-focused special operations activity: an ACT to learn how to apply the skillsets they acquired in addressing USAF autonomy needs. Support this effort over the course of four years to grow AI manpower by an order of magnitude over today's level. Assure retention via a number of special incentive programs. Supplement this cadre with appropriate and long-term support of key extramural researchers.

- *Recommendation 4b*: Use a three-phase framework for iterative selection of challenge problems, for modeling the impact of potential solutions, and for solution development, prototyping, and assessment. Conduct an initial phase of wargaming-based assessment, with the goal of identifying key challenge problems and AS-based solutions that can address those threats or take advantage of potential opportunities. Provide a deeper assessment of promising AS candidates, via formalization of those concepts with quantitative M&S and parameters of performance. Finally, focus on the design of one or more engineering prototypes of promising AS candidates identified in the M&S studies. Develop and experimentally evaluate a prototype AS that can serve as: (1) a design prototype for acquisition; and (2) a design driver for additional needed S&T.

- *Recommendation 4c*: Through the USAF CDO, acquire space to store USAF air, space, and cyber data so that AI professionals can use it to create autonomy solutions to challenge problems. Establish data curator roles in relevant organizations to manage the data and to create streamlined access and retrieval approaches for data producers and consumers.

- *Recommendation 4d*: Support the movement to cloud-based computing while also leveraging quantum computing as a general computational paradigm that can be exploited to meet embedded and HPC processing demands.

6.7.5 Organizational Structures

This includes organizing around a project (or outcome) focus, rather than along traditional technical specialty domains.

- *Recommendation 5*: Establish the ACT within AFRL, incorporating a flatarchy business model to bring 6.1–6.4 experts into a single product-focused organization to develop the science of autonomous systems while delivering capabilities to the warfighter. Collaborate with AFOSR and other key AFRL technical directorates and coordinate with USAF organizations outside AFRL, including the DOD COI, AFWERX, and other offices that can facilitate technology transition to the warfighter. Within the ACT, incorporate product-focused business processes based on a Skunk Works–like set of "guiding rules" and facilitate the move toward an information-centric business platform model for the future Air Force.

6.7.6 Knowledge Platform

This provides us with a holistic means of integrating across AS behavioral principles, architectures/technologies, challenge problems, developmental processes, and organizational structures.

Recommendation 6: Develop a KP centered on combining an IT platform approach with a platform business model. A KP designed for the multido-main operating Air Force should monopolize the connection of observation agents with knowledge creation agents and with warfighting effects agents, which can be either human or machine-based agents (ASs). The KP provides the ecosystem necessary to create capabilities, and those capabilities are used to create combat effects. This ecosystem will come to fruition by exploiting the three behavioral principles of autonomy (section 6.1); the architectures and technologies that enable those behaviors (6.2); the driving challenge problems (6.3); the developmental processes across people, architectures/applications, data, and computational infrastructures (6.4); and, finally, the organizational structures that need to be in place to advance the technology, exploit it, and deliver capability (6.5). This approach will provide us with the means of transitioning the USAF from the traditional tools-based approach that solves a small number of problems to a knowledge platform approach applicable to a far greater set of problems.

6.8 Closing Comments

Our goal with this document has been twofold: to provide a vision for USAF senior leaders of the potential of autonomous systems and how they can be transformative to warfighting at all levels and to provide for the science and technology community a general framework and roadmap for advancing the state of the art *while* supporting its transition to existing and to-be-acquired systems. We believe that we need to be more aggressive in maturing this technology and have therefore broadened our set of recommendation beyond the usual technological ones to include such issues as processes and organizations. Specifically, we have made recommendations in the areas of:

- The behaviors these systems must have if they are to be proficient at what they do, trusted by their human counterparts, and flexible in dealing with the unexpected

- The unifying frameworks, architectures, and technologies we need to bridge across not only insular S&T communities, but also operational stovepipes and domains

- The focused challenge problems, both foundational and operational, needed to challenge the S&T community while providing operational advantages that go far beyond our traditional platform-centric approach to modernization

- New processes for dealing with people, systems, data, and computational infrastructures that will accelerate innovation, rapid prototyping, experimentation, and fielding

- A new organizational structure, the autonomy capabilities team ACT, that brings together technical specialties into a single organization focused on innovative product development, with outreach to other organizations and communities as needed

- A Knowledge Platform for holistically integrating across AS behavioral principles, architectures/ technologies, challenge problems, developmental processes, and organizational structures

The AFRL, and specifically the ACT, cannot simply limit its attention to the *research space* of autonomous systems. Nor can it simply perpetuate the model of applying modern AI and AS technology to provide incremental mission capability improvement in one-off demonstrations. Challenge problems, like those described earlier in this chapter, must be chosen to advance the Knowledge Platform's ability to provide, in an agile fashion, ASs that exhibit proficient, trustworthy, and flexible behaviors, in transformational applications. In addition to project-focused efforts, the ACT can serve to prioritize and coordinate AFRL's entire autonomy S&T portfolio—synchronizing efforts to maximize investment impact—bringing autonomous systems capabilities to mission challenges at scale, and in a timely fashion, all while "sharing the wealth" of new architectures, technologies, and processes across the S&T directorates. Finally, when successful, the ACT can serve as an "existence proof" of how the Lab can transform itself from its legacy of a discipline-focused organization to one that is more cross-disciplinary and project-oriented, solving transformative USAF enterprise-wide problems.

We have a unique opportunity to transform the USAF away from an air platform–centric service, where space and cyber often take a back seat, to a truly multidomain and knowledge-centric organization. The USAF needs to invest in the platform business model as the way to generate combat effects. By tightly coupling the platform business model with the autonomous systems that can be delivered to the warfighter by way of the Knowledge Platform, every mission in air, space, and cyber will be improved—and not just incrementally, but multiplicatively. We will become an enterprise that is service-oriented, ubiquitously networked, and information-intensive. Our vision is to become:

An agile, information-centric enterprise making timely decisions executed via friction-free access to exquisitely effective peripherals.

Appendix A

Review of Past Studies

In this appendix, we review seven past studies of autonomous systems for military applications and summarize and categorize their major findings and recommendations. In section A.1, we provide individual summaries of each study and list the major issues and recommendations identified by the study. In section A.2, we combine the across-study issues identified and recommendations made into three major categories:

- Areas of functionality needed in autonomous systems and areas for future science and technology (S&T) investment

- Potential mission-relevant "challenge problems" to drive autonomous systems development

- Processes and organizational structures for developing the functionality and addressing the challenge problems

This then serves as background to motivate, in chapter 1, our identification of three broad considerations for successful autonomous system development and deployment: principles of autonomous behavior, properties for proficiency, and tenets for trust. This also serves to inform the resulting recommendations in chapter 6, covering recommendations for S&T investments and challenge problems to be addressed, development processes to accelerate that development, and organizational structures to ensure a continued focus in this emerging area.

A.1 Summaries of Individual Studies

In this section, we summarize—on an individual study basis—the major findings and recommendations of seven prior studies dealing with autonomous systems, six of which focus on specific DOD applications. Findings and recommendations have sometimes been grouped into more general categories to avoid a "laundry list" presentation style and to support more general across-study groupings later on. The studies are as follows:

- G. Klein, D. Woods, J. Bradshaw, R. Hoffmann, and P. Feltovich, "Ten Challenges for Making Automation a 'Team Player' in Joint Human-Agent Activity," *IEEE Journals & Magazine* 19, no. 6 (2004): 91–95

- W. Dahm, *Technology Horizons: A Vision for Air Force Science and Technology 2010-2030* (Maxwell AFB, AL: Air University Press, 2011)

- Defense Science Board (DSB), "DSB Autonomy Study: The Role of Autonomy in DoD Systems" (Washington, DC: Office of the Under Secretary of Defense for Acquisition, Technology, and Logistics, 2012)

- DSB, "Report of the Defense Science Board Summer Study on Autonomy" (Washington, DC: Office of the Under Secretary of Defense for Acquisition, Technology, and Logistics, 2016)

- R. Potember. "Perspectives on Research in Artificial Intelligence and Artificial General Intelligence Relevant to DOD" (JSR-16-Task-003) (McLean, VA: The MITRE Corp., 2017)

- A. Hill and G. Thompson, "Five Giant Leaps for Robotkind: Expanding the Possible in Autonomous Weapons," War on the Rocks, December 28, 2016, https://warontherocks.com/2016/12/five-giant-leaps-for-robotkind-expanding-the-possible-in-autonomous-weapons/

- M. L. Hinman, "Some Computational Approaches for Situation Assessment and Impact Assessment," *Proceedings of the Fifth International Conference on Information Fusion 2002*, Vol. 1, 687–93

(A) G. Klein, D. Woods, J. Bradshaw, R. Hoffmann, and P. Feltovich, "Ten Challenges for Making Automation a 'Team Player' in Joint Human-Agent Activity," *IEEE Journals & Magazine* 19, no. 6 (2004): 91–95

Recommendations are in the category of basic research and primarily focus on human-system teaming with automation but clearly hold for dealing with autonomous systems (AS) as well. One extremely important contribution of the work is the "Basic Compact" idea: when humans team with autonomous agents or autonomous agents team with each other, the agent team members must enter into a compact expressing their intent to work cooperatively. They must work to maintain a common situation awareness and to stay mutually predictable and directable. All of the study's conclusions are directly applicable to achieving "peer flexibility" as we define it in the main body of this report. Recommendations cover the following general categories of behavioral expectations:

- Teaming & coordination
  - Setting mutual goals

- Engaging in goal negotiation

- Forming and maintaining the basic contract

○ Communicating and coordinating

- Observing and interpreting signals of status and intentions

- Forming and maintaining adequate models of others' intentions and actions

- Effective signaling of pertinent aspects of [self] status and intentions

- Controlling the costs of coordinated activity

• Expectation management

○ Maintaining adequate directability

○ Maintaining predictability without hobbling adaptivity

• Attention management

• Incorporate [new] technologies that are incremental and collaborative

(B) W. Dahm, *Technology Horizons: A Vision for Air Force Science and Technology 2010–2030* (Maxwell AFB, AL: Air University Press, 2011)

Several general issues are raised that can be interpreted in terms of performance needs or requirements to be met for successful autonomous system operation, including, as noted in *Autonomous Horizons* volume I, the need for flexible and resilient autonomy (Endsley 2015c). Also pointed out is the need for effective verification and validation (V&V). But few direct recommendations are made in terms of underlying functions, behaviors, or technology enablers. Generic system/mission application areas are identified for the use of ASs, and mission-oriented "grand challenge" problems are also identified. Recommendations fall into the following three categories:

• General issues and needs

○ Need for "flexible autonomy" at different levels

○ Need for "resilient autonomy"

○ Need for trust in autonomy

○ Recognize and address difficulty of doing V&V

- ○ Adversarial use of autonomous systems
- System/mission application areas
  - ○ Platforms
    - ▪ Agile attributable unmanned aerial vehicles (UAV) capable of complex decision making
    - ▪ JPADS: Joint Precision Air Drop System
    - ▪ Embedded platform health diagnostics
    - ▪ Long-Range Strike/Bomber (LRSB)
    - ▪ (Functionally) fractionated platforms
  - ○ Weapons
    - ▪ High-precision, low collateral damage munitions
  - ○ Cyber
    - ▪ Cyber defenses
  - ○ Intelligence, surveillance, and reconnaissance (ISR) and command and control (C2)
    - ▪ Cross-domain ISR
    - ▪ Cross-domain C2
- Mission-oriented "grand challenges"
  - ○ Inherently intrusion-resilient cyber networks
    - ▪ Explore, develop, and demonstrate autonomous and scalable technologies that enable large, non-secure networks to be made inherently and substantially more resilient to attacks entering through network or application layers and to attacks that pass through these layers.
    - ▪ Emphasis is on advancing technologies that enable network-intrusion tolerance rather than traditional network defense, with the goal to achieve continued mission effectiveness under large-scale, diverse network attacks.
  - ○ Trusted, highly autonomous decision-making systems
    - ▪ Explore, develop, and demonstrate technologies that enable current human-intensive functions to be replaced, in whole or in part, by

more highly autonomous decision-making systems and technologies that permit reliable V&V to establish the needed trust in them.

- Emphasis is on decision-making systems requiring limited or no human intervention for current applications, where substantial reductions in specialized manpower may be possible and for future applications involving inherent decision time scales far exceeding human capacity.

o Fractionated, composable, survivable, autonomous systems

- Explore, develop, and demonstrate technologies that can enable future autonomous aircraft or spacecraft systems achieving greater multi-role capability across a broader range of missions at moderate cost, including increased survivability in contested environments.

- Emphasis is on composability via system architectures based on fractionation and redundancy. This involves advancing methods for collaborative control and adaptive autonomous mission planning, as well as V&V of highly adaptable, autonomous control systems.

o Hyperprecision aerial delivery in difficult environments

- Explore, develop, and demonstrate technologies that enable single-pass, extremely precise, autonomously guided aerial delivery of equipment and supplies under Global Positioning System (GPS)-denied conditions from altitudes representative of operations in mountainous and contested environments and winds representative of steep, mountainous terrain.

- Emphasis is on low-cost, autonomous flight systems with control authority capable of reaching target point within specified impact limits under effects of large stochastic disturbances.

(C) Defense Science Board (DSB), "DSB Autonomy Study: The Role of Autonomy in DoD Systems" (Washington, DC: Office of the Under Secretary of Defense for Acquisition, Technology, and Logistics, 2012)

Recommendations fall into two broad categories: desired functionality and development process recommendations. The study also provides strong support for the use of "challenge problems" to drive autonomous system development. Recommendations below have been paraphrased and categorized for clarity and brevity:

- Desired AS functionality/performance attributes

- Natural user interfaces (UI) and trusted human-system collaboration
  - Predictable and understandable behaviors [by the AS]
  - Effective human-system dialog
- Perception and situation awareness to operate in a complex battle space
  - Airspace deconfliction for manned/unmanned system ops
  - Real-time pop-up threat detection/identification
  - High-speed obstacle detection in complex terrain-
  - Multisensor integration
- Large-scale manned/unmanned teaming
  - Collaborative, mixed-initiative dynamic planning and task execution
  - Shared synchronized common operating picture
  - Anticipation of future teammate response
- Test and evaluation
  - Dealing with complex, nontransparent software-intensive systems
  - Dealing with nondeterministic systems
  - Challenges of simulating a complex mission space
- Process-oriented recommendations
  - Research
    - Abandon the debate over definitions of levels of autonomy and embrace a three-facet (cognitive echelon, mission timelines, human-machine system trade spaces) autonomous systems framework
    - Attract and retain artificial intelligence (AI) and software engineering experts
    - Stimulate the DOD's S&T program with realistic "challenge problems"
    - Establish a coordinated S&T program guided by feedback from operational experience and evolving mission requirements
  - Acquisition and testing and evaluation (T&E)
    - Structure autonomous systems acquisition programs to separate the autonomy software from the vehicle platform

- Establish a research program to create the technologies needed for developmental and operational T&E that address the unique challenges of autonomy

  ○ Training

  - Include the lessons learned from using autonomous systems in the recent conflicts into professional military education, war games, exercises, and operational training

  ○ Intel

  - Track adversarial capabilities with autonomous systems

(D) DSB, "Report of the Defense Science Board Summer Study on Autonomy" (Washington, DC: Office of the Under Secretary of Defense for Acquisition, Technology, and Logistics, 2016)

Recommendations fall into two broad categories: operations oriented and development-process oriented. In the operations-oriented recommendations, there are interesting ideas in autonomous sensing/swarms, cyberspace indications and warnings (I&W), autonomous logistics, and dynamic spectrum management. Process-oriented recommendations include challenges in mitigating increasing cyber vulnerabilities by adopting autonomous systems and countering adversary use of autonomy. Recommendations given below have been grouped into common categories for greater clarity; note that some are not USAF relevant:

- Operations-oriented recommendations (current and future)
  ○ Platforms
    - Onboard autonomy for sensing
    - Unmanned undersea/underwater vehicle (UUV) for mine countermeasures (defensive)
    - UUVs for maritime mining (offensive)
    - Organic UAVs for ground forces
    - Autonomous swarms that exploit large quantities of low-cost assets
  ○ Cyberspace
    - Agents to improve cyberattack I&W
    - Automated cyber response to attacks

- Intrusion detection on the Internet of Things
- Autonomous cyber resilience for military vehicle systems
  - Logistics
    - Predictive logistics and adaptive planning
    - Adaptive logistics for rapid deployment
    - Autonomous air operations planning
  - Miscellaneous
    - Time-critical intelligence from seized media
    - Dynamic spectrum management
    - Early warning system for understanding global social movements
- Process-oriented recommendations
  - Tackling the engineering, design, and acquisition challenges
  - Mitigating cyber issues introduced by increasingly autonomous and networked systems
  - Creating new test and evaluation and modeling and simulation paradigms
  - Integrating technology insertion, doctrine, and concepts of operations
  - Developing an autonomy-literate workforce
  - Improving technology discovery
  - Improving DOD governance for autonomous systems
  - Countering adversary use of autonomy

(E) R. Potember. "Perspectives on Research in Artificial Intelligence and Artificial General Intelligence Relevant to DOD" (JSR-16-Task-003) (McLean, VA: The MITRE Corp., 2017)

This study, more than the others, focuses on underlying S&T enablers for autonomous system development. It contrasts "traditional" AI that can accomplish specific narrow tasks with artificial general intelligence (AGI) that "can successfully perform any task that a human might do" and have "general cognitive capabilities." Much of the focus is on enabling technologies, specifically artificial neural networks (ANN) and especially on the "deep learning revolution" starting around 2010 with deep learning (DL) networks and big data (BD), and the possible shortcomings that can be encountered. Recommendations cover potential S&T investment areas and high-level process-oriented recommendations:

- Potential S&T investment areas
  - Technologies
    - Classical algorithmic approaches
      - Capturing expert judgment (other than for labeling BD sets)
      - Physical and deterministic modeling
      - Direct computation on complex symbolic data representations
      - Purpose-designed and rule-based systems designed for error-free response
    - Learning networks
      - Combined DL ANNs and BD
      - Graphical stochastic models, such as Bayesian networks
      - Probabilistic Programming Languages
      - Hardware and algorithms for implementing deep neural networks (DNN)
  - Hybrid architectures
    - AlphaGo's use of a DNN for rapid evaluation of a given game state or configuration, combined with "clever" pruning of a game tree to explore potential evolution of the game states
    - Generative Adversarial Networks (GANs), which plays off one DNN against the other, executed in the framework of a min-max two-player game
    - Autoencoders using back to back DNNs in an encoder-decoder configuration
    - Wide but shallow Boolean logic search trees combined with narrow but deep DNNs
  - Biomimetic cognitive systems
- Process-oriented recommendations
  - Track (via a knowledgeable cadre) and invest in (via a 6.1 research portfolio) the most dynamic and rapidly advancing areas of AI, including but by no means limited to DL.

○ Support the development of a discipline of AI engineering, accelerating the progress of the field through Shaw's "craft" and (empirical) "commercial" stages. A particular focus should be advancing the "ilities" in support of DOD missions.

○ DOD's portfolio in AGI should be modest and recognize that it is not currently a rapidly advancing area of AI. The field of human augmentation via AI is much more promising and deserves significant DOD support.

○ Support the curation and labeling, for research, of DOD's unique mission-related large data sets. Wherever possible, operational data should be saved for future research use in support of AI for DOD-unique missions.

○ Create and provide centralized resources for DOD's intramural and extramural researchers, like the Marine Operating and Support Information System, including labeled data sets and access to large-scale GPU training platforms.

○ Survey the mission space of embedded devices for potential breakthrough applications of AI, and consider investing in special-purpose accelerators to support AI inference in embedded devices for DOD missions, if such applications are identified.

(F) A. Hill and G. Thompson, "Five Giant Leaps for Robotkind: Expanding the Possible in Autonomous Weapons," War on the Rocks, December 28, 2016, https://warontherocks.com/2016/12/five-giant-leaps-for-robotkind-expanding-the-possible-in-autonomous-weapons/

This study focuses on five operationally oriented challenge problems:

- Hostage rescue using discriminating lethality
- Deployment under disrupted/degraded comms
- On the spot improvisation of materiel solutions
- Adaptation for discovery of new tactics, techiniques, and procedures (TTP)
- Evoking "disciplined initiative" when original plan is failing/illegal/immoral

(G) M. L. Hinman, "Some Computational Approaches for Situation Assessment and Impact Assessment," Proceedings of the Fifth International Conference on Information Fusion 2002, Vol. 1, 687–93

These recommendations focus on AS mission-oriented applications:

- Adversary inferencing
  - Predicting enemy goals
  - Predicting enemy courses of action (COA) and associated probabilities
  - Identifying appropriate indicators (and associated metrics) that the enemy is pursuing a specific COA
- ISR tasking
  - Determining specific additional information requirements (to enable the above tasks) for input to the ISR tasking system
- Conveying recommendations and insights to appropriate decision makers in an effective manner
- Battle damage assessment (BDA)
  - Assessing whether desired effects have been achieved

A.2 Categorized Cross-Study Recommendations

This section brings together all of the above findings and recommendations and aggregates them into five common categories: behavioral objectives, technologies and architectures, mission-oriented "challenge problems," development processes, and organizational structures. The same nomenclature is used as in the previous section (with some editing for brevity), with traceback to the source study, making the particular recommendation indicated by the letter (A through G) associated with the specific study above.

A.2.1 Behavioral Objectives

Under behavioral objectives, we have identified two main categories of recommendations. The first is fairly general and deals with behavioral objectives that are desired in ASs (shown in table A.1), such as ensuring that behaviors are directable and predictable, and that the AS can accomplish tasks with adequate flexibility and adaptivity. The second focuses on human-system teaming, and specific behaviors desired when those ASs are interacting with humans[1] (shown in table A.2), including the desirability of being able to set mutual goals, to maintain adequate mutual shared awareness of the team and the adversary, and to communicate and coordinate effectively.

---

1. Including humans that are both friendlies and adversaries.

Table A.1. General behavioral objectives

General behavior objectives	Specific behavior objectives	Ref
Limiting scope/range of behaviors	Maintaining adequate directability	A
	Predictable and understandable behaviors	C
	Maintaining predictability without hobbling adaptivity	A
Flexibility under different situations	Need for "flexible autonomy" at different levels	B
	Need for "resilient autonomy"	B

Table A.2. Human-system teaming needs

General teaming needs	Specific teaming needs	Ref
Setting mutual goals	Engaging in goal negotiation	A
	Forming and maintaining the basic contract	A
Maintaining shared situation awareness (SA) of environment, intent, and teammate actions	Attention management	A
	Shared synchronized common operating picture	C
	Forming and maintaining adequate models of others' intentions and action	A
	Effective signaling of pertinent aspects of [self] status and intentions	A
	Collaborative, mixed-initiative dynamic planning and task execution	C
	Anticipation of teammate's future response	C
	Conveying recommendations and insights [for an AS-based decision aid] to appropriate decision-makers in an effective manner	G
Maintaining SA and intent inferencing of adversary	Predicting enemy goals	G
	Predicting enemy COA and associated probabilities	G
	Identifying appropriate indicators (and associated metrics) that the enemy is pursuing a specific COA	G
Communicating and coordinating	Observing and interpreting signals of status and intentions	A

Table A.2. Human-system teaming needs *(continued)*

General teaming needs	Specific teaming needs	Ref
Communicating and coordinating *(continued)*	Controlling the costs of coordinated activity	A
	Assuring effective human-system dialog	C
Need for trust in autonomy		B

A.2.2 Architectures and Technologies

Only one study (Potember 2017) made recommendations in the area of architectures and technologies (see table A.3) that could enable the development of ASs, but we have included them here to support and complement our discussion of enabling technologies and frameworks in chapter 4. It also provides an indication of how many of the studies focus on desired AS functionality or development processes rather than the actual enablers that might be used to bring forth that functionality. A number of hybrid architectures are discussed, with a recommendation to consider biomimetic cognitive systems development as well. Specific technologies are broken into "classical" algorithmic approaches, and the more contemporary deep learning network approaches enabled by access to big data.

Table A.3. Architectures and technologies

Architecture/ technology families	Architecture/technology specifics	Ref
Hybrid Architectures	Use of a DNN for rapid game state evaluation, combined with "clever" pruning of game tree to explore potential evolution of the game states (used in AlphaGo)	E
	Generative Adversarial Networks (GAN) playing one DNN against the other, executed in the framework of a min-max two-player game	E
	Autoencoders using back-to-back DNNs in an encoder-decoder configuration	E
	Wide but shallow Boolean logic search trees combined with narrow but deep DNNs	E
Classical algorithmic approaches	Capturing expert judgment (other than for labeling big data sets)	E
	Physical and deterministic modeling (for example, physics-based computer vision)	E

Table A.3. Architectures and technologies *(continued)*

	Direct computation on complex symbolic data representations	E
Classical algorithmic approaches *(continued)*	Purpose-designed and rule-based systems designed for error-free response (for example, the "classical" side of control theory)	E
Learning networks	Combined DL ANNs and BD	E
	Graphical stochastic models, such as Bayesian networks	E
	Probabilistic Programming Languages	E
	Hardware and algorithms for implementing DNNs	E
Biomimetic cognitive systems		E

A.2.3 "Challenge Problems"

Most of the study recommendations focused in this area of challenge problems. We have separated them into two main categories—one being domain-independent (or functional), as shown in table A.4, and the other being domain-dependent or mission-focused (air, space, cyberspace), as shown in all three versions of table A.5. The domain-independent problems cover a range of general functional areas, such as collection/sensing/fusion of information, generic decision-aiding (with a human) and decision-making (autonomous) subsystems, fractionated autonomous platforms, and operation in adversarial environments that demand improvisation. The domain-dependent problems range from generic concepts like autonomous swarms, to specific operations in air (including ISR, air operations planning, electronic warfare, and logistics), space (including fractionated platforms and embedded health diagnostics), and cyber (defensive operations, offensive operations, and network resiliency).

Table A.4. Domain-independent or functional challenge problems

Function	Domain-independent challenge problem	Ref
Collection, sensing, fusion, and BDA	Early warning system for understanding global social movements	D
	Determining specific additional information requirements for input to the ISR tasking system	G

Table A.4. Domain-independent or functional challenge problems (*continued*)

Function	Domain-independent challenge problem	Ref
Collection, sensing, fusion, and BDA *(continued)*	Extracting time-critical intelligence from seized media	D
	Onboard autonomy for sensing	D
	Multi-sensor integration	C
	Cross-domain ISR	B
	Autonomous BDA for effects assessment	G
Trusted, highly autonomous decision-aiding/-making systems	Technologies that can enable replacement of human-intensive functions by more highly autonomous and trustworthy decision-making systems	B
	Decision-making systems requiring limited or no human intervention for current/future applications, to support reductions in manpower and decision times	B
Fractionated, composable, survivable, autonomous systems [platforms]	Technologies to enable future autonomous aircraft or spacecraft [or cybercraft] achieve greater multi-role capability across a broader range of missions at moderate cost, including increased survivability in contested environments.	B
	Composability via system architectures based on fractionation and redundancy	B
	Methods for collaborative control and adaptive autonomous mission planning, as well as V&V of highly adaptable, autonomous control systems	B
Cross-domain C2		B
Operations in denied or degraded environments	AS Deployment under disrupted/degraded communications	F
	Evoking "disciplined initiative" when the original plan is failing/illegal/immoral	F
	On the spot improvisation of materiel solutions	F
Adaptation for discovery of new tactics/TTPs		F

Table A.5a. Domain-dependent or mission-focused challenge problems: Air

Function	Domain-dependent challenge problem	Ref
Concepts	Large autonomous swarms of low-cost assets	D
	[Functionally] fractionated platforms	B
	Agile attritable UAVs capable of complex decision making	B
	LRSB	B
Operations	Autonomous air operations planning	D
	Airspace deconfliction for manned/un-manned system ops	C
	Hostage rescue using discriminating lethality	F
ISR	Real-time pop-up threat detection/identification	C
Weapons and subsystems	High-precision low collateral damage munitions	B
	Dynamic spectrum management for electronic warfare (EW)	D
	Embedded platform health diagnostics	B
Logistics	JPADS: Joint Precision Air Drop System	B
	Single-pass, precise, autonomously guided aerial delivery of equipment and supplies in mountainous terrain under GPS-denied and contested environments	B
	Low-cost, autonomous platforms with control authority capable of reaching target point within specified impact limits under effects of large stochastic disturbances	B
	Predictive logistics and adaptive planning	D
	Adaptive logistics for rapid deployment	D

Table A.5b. Domain-dependent or mission-focused challenge problems: Space

Function	Domain-dependent challenge problem	Ref
Concepts	[Functionally] fractionated platforms	B
Weapons and subsystems	Embedded platform health diagnostics	B

Table A.5c. Domain-dependent or mission-focused challenge problems: Cyberspace

Function	Domain-dependent challenge problem	Ref
Defensive	Agents to improve cyberattack I&W	D
	Intrusion detection on the Internet of things	D
Offensive	Automated cyber response to attacks	D
Network resilience	Inherently intrusion-resilient cyber networks	B
	Autonomous and scalable technologies that make non-secure networks inherently resilient to attacks entering through network or application layers, and to attacks that pass through these layers	B
	Autonomous technologies that enable continued mission effectiveness under large-scale, diverse network attacks	B
Embedded systems	Autonomous cyber resilience for military vehicle systems	D

A.2.4 Development Processes

Many study recommendations focused on AS development processes, as shown in table A.6, covering six major areas: (1) the need to actively track adversarial capabilities and usage of autonomous systems; (2) the importance of human capital management, including the attraction and retention of experts in AI and software engineering, the introduction of AS capabilities into professional military education (PME) and wargaming, and the development of centralized AI/AS resources for DOD's research community; (3) continued support of basic and applied research in a broad area of underlying technologies (not just deep learning), coordinated across research communities and informed by operational experience and evolving mission requirements; (4) support of advanced systems development, which separates the development of platforms from the autonomy software that governs them, augmented by the development of a discipline of AI engineering to accelerate progress; (5) establishing processes for upgrading legacy systems with new autonomous system capabilities; and (6) recognizing the difficulty of conducting T&E of these systems, and establishing a research program for dealing with nontransparent, nondeterministic, and time-varying (learning) systems.

Table A.6. Development processes

Category	Process	Ref
Intelligence	Track adversarial capabilities/usage of autonomous systems	B, C
Human capital management	Attract and retain AI and software engineering experts	C
	Include lessons learned from using autonomous systems in recent conflicts, into professional military education, wargames, exercises, and operational training	C
	Create and provide centralized resources for DOD's intramural and extramural researchers, including labeled data sets and access to large-scale GPU training platforms	E
Basic and applied research	Establish a coordinated S&T program guided by feedback from operational experience and evolving mission requirements	C
	Track (via a knowledgeable cadre) and invest in (via a 6.1 research portfolio) the most dynamic and rapidly advancing areas of AI, including, but not limited to, deep learning	E
	Emphasize research into AI-enabled human augmentation over AGI, recognizing former is more promising and latter is not currently a rapidly advancing area of AI	E
	Stimulate the DOD's S&T program with realistic "challenge problems"	C
	Support the curation and labeling, for research, of DOD's unique mission-related large data sets. Wherever possible, operational data should be saved for future research use in support of AI for DOD-unique missions.	E
Systems development	Abandon debate over definitions of levels of autonomy; embrace a three-facet AS framework: cognitive echelon, mission timelines, human-machine system trade spaces	C
	Support the development of a discipline of AI engineering, accelerating the progress of the field through Shaw's "craft" and (empirical) "commercial" stages. Focus should be on advancing the "ilities" in support of DOD missions.	E
	Structure autonomous systems acquisition programs to separate the autonomy software from the vehicle platform	C

Table A.6. Development processes *(continued)*

Category	Process	Ref
Upgrading systems using AI	Incorporate [new] technologies that are incremental and collaborative	A
	Survey the mission space of embedded devices for potential breakthrough applications of AI. Consider investing in special-purpose accelerators to support AI inference in those devices/missions.	E
Test and evaluation	Recognize and address difficulty of doing both developmental and operational V&V	B
	Address complex, nontransparent software-intensive systems	C
	Address issues of nondeterministic systems	C
	Address challenges of simulating a complex mission space	C

A.2.5 Organizational Structures

No specific recommendations regarding new or modified organizational structures were identified in any of the studies.

Appendix B

Frequently Asked Questions

This appendix, developed by the Air Force Research Laboratory (AFRL), presents a number of frequently asked questions (FAQ) concerning autonomous systems. The goal is to provide a self-consistent position on the subject that can be used to facilitate discussions on some of the underlying concepts, the science and technology challenges, and the potential benefits for addressing capability gaps. It attempts to build upon the main body of work presented here but is not intended to exclude alternative definitions. Where there may be a conflict between what is presented in the main body of this report and what is presented in this appendix, the main body should take precedence.

B.1 What Is an Autonomous System (AS)?

We have defined an autonomous system in terms of its attributes across three dimensions—namely, proficiency, trustworthiness, and flexibility:

- An AS should be designed to ensure proficiency in the given environment, tasks, and teammates envisioned during operations. Desired *properties for proficiency* include situated agency, adaptive cognition, multiagent emergence, and experiential learning.

- An AS should be designed to ensure trust when operated by or teamed with its human counterparts. Desired *tenets of trust* include cognitive congruence and transparency, situation awareness, effective human-systems integration, and human-systems teaming/training.

- An AS should exhibit flexibility in its behavior, teaming, and decision-making. Desired *principles of flexibility* include flexibility in terms of being able to conduct different tasks, work under different peer-to-peer relationships, and take different cognitive approaches to problem-solving.

We have defined these characteristics in greater depth in chapters 2, 3, and 6. We believe that all of these dimensions need to be satisfied to some degree if we are to effectively field and use ASs in the Air Force. Stated another way, a failure to satisfy the design space across all three dimensions will lead to a failure of a fielded AS: low proficiency will lead to the use of other systems, low trust will lead to disuse, and low flexibility will lead to an AS that fails to exhibit true autonomy under changing circumstances that may not have been envisioned during the design phase.

A natural question is whether or not these dimensions have any meaning toward how much (or what level of) autonomy the system has. The answer is no. Purposefully, there is no intent to use the three dimensions to define or guide some notion of levels of autonomy. This is simply because it is not clear this is a useful construct, as has been discussed here and in earlier studies cited here.

We now address some autonomous system FAQs.

B.1 General Concepts

What is intelligence? What is artificial intelligence?

Intelligence is the ability to gather observations, create knowledge, and appropriately apply that knowledge to accomplish tasks. Artificial intelligence (AI) is a machine that possesses intelligence.

What is an AS's internal representation?

Current ASs are programmed to complete tasks using different procedures. The AS's internal representation is how the agent structures what it knows about the world, its knowledge (what the AS uses to take observations and generate meaning), how the agent structures its meaning and its understanding, for example, the programmed model used inside of the AS for its knowledge base. The knowledge base can change as the AS acquires more knowledge or as the AS further manipulates existing knowledge to create new knowledge.

What is meaning? Do machines generate meaning?

Meaning is what changes in a human's or AS's internal representation as a result of some stimuli. It is the meaning of the stimuli to that human or system. When you, a human, look at an American flag, the sequence of thoughts and emotions that it evokes in you is the meaning of that experience to you at that moment. When the image is shown to an AS, and if the pixel intensities evoked some programmed changes in that AS's software, then that is the meaning of that flag to that AS. Here we see that the AS generates meaning in a manner that is completely different than from how a human does it. The change in the AS's internal representation, as a result of how it is programmed, is the meaning to the AS. The meaning of a stimulus is the agent-specific representational changes evoked by that stimulus in that agent (human or AS). The update to the representation, evoked by the data, is the meaning of the stimulus to the agent. Meaning is not just the posting into the representation of the data; it is all the resulting changes to the representation. For example, the evoking of tacit knowledge, or a modification of the ongoing simulation (consciousness; see below), or even the updating of the agent's knowledge resulting from the stimuli, is included in the meaning of a stimulus to an agent. Meaning is not static and changes over time. The meaning of a stimulus is different for a given agent depending on when it is presented to the agent.

What is understanding? Do machines understand?

Understanding is an estimation of whether an AS's meaning will result in it acceptably accomplishing a task. Understanding occurs if it increases the belief of an evaluating human (or evaluating AS) that the performing AS will respond acceptably. Meaning is the change in an AS's internal representation resulting from a query (presentation of a stimulus). Understanding is the impact of the meaning resulting in the expectation of successful accomplishment of a particular task.

What is knowledge?

Knowledge is what is used to generate the meaning of stimuli for a given agent. Historically, knowledge comes from the species capturing and encoding via evolution in genetics, experience by an individual animal, or animals communicating knowledge (via culture) to other members of the same species. With advances in machine learning, it is a reasonable argument that most of the knowledge that will be generated in the world in the future will be done by machines.

What is thinking? Do machines think?

Thinking is the process used to manipulate an AS's internal representation; a generation of meaning, where meaning is the change in the internal representation resulting from stimuli. If an AS can change or manipulate its internal representation, then it can think.

What is reasoning? Do machines reason?

Reasoning is thinking in the context of a task. Reasoning is the ability to think about what is perceived and the actions to take to complete a task. If the system updates its internal representation, it generates meaning and is doing reasoning when that thinking is associated with accomplishing a task. If the system's approach is not generating the required "meaning" to acceptably accomplish the task, it is not reasoning appropriately.

What is cognition? What makes a system cognitive?

Cognition is the process of creating knowledge and understanding through thought, experience, and the senses. A system that can create knowledge and understanding through thinking and experience and sensing is cognitive. As an example, a Cognitive Electronic Warfare (CEW) system gathers data from its senses and creates knowledge. It uses relevant knowledge to accomplish its EW mission, which demonstrates a level of understanding the system has with respect to its task.

What is a situation?

A situation is the linkage of individual knowledge entries in the AS's internal representation that can be combined to make a new single-knowledge entry. This new single-knowledge entry becomes a situation due to its linkage to the individual entries it is composed of. Situations are the fundamental unit of cognition. Situations are defined by their relationship to, and how they can interact with, other situations. Situations are comprehended as a whole.

What is situated cognition?

Situated cognition is a theory that posits that knowing is inseparable from doing by arguing that all knowledge is situated in activity bound to social, cultural, and physical contexts. This is the so-called see/think/do paradigm.

What is learning? What is deep learning?

Learning is the cognitive process used to adapt knowledge, understanding, and skills, through experience, sensing, and thinking, to be able to adapt to changes. Depending upon the approach to cognition the agent is using (its choice of a representation, for example, symbolic, connectionist, etc.), learning is the ability of the agent to encode a model using that representation (the rules in a symbolic agent or the way artificial neurons are connected and their weights adjusted, for a connectionist approach). Once the model has been encoded, it can be used for inference. Deep learning is a subset of the connectionist approach incorporating many neuronal processing layers, with a learning paradigm that has overcome past limitations associated with the multilayer "credit assignment" problem (i.e., which weight should be adjusted to improve performance), has made use of big data and multiple instantiations for training, and has made advances in computational infrastructures. Deep learning has received much attention in recent years due to its ability to process image and speech data; it is largely made possible by the processing capabilities of current computers, the dramatic increase in available data, and modest modifications in learning approaches. Deep learning is basically a very successful big data analysis approach.

B.2 Examples

Is a garage door opener automated or automatic?

A garage door opener opens the door when it is signaled to do so and stops based on some preset condition (number of turns the motor makes, or by a switch). When it is closed, it opens when it is signaled to do so and stops based on the same sort of preset condition. A garage door opener is an automatic system since it performs a simple task based on some trigger mechanism and stops at the completion of its task, also based on some trigger mechanism.

Is an automatic target recognition system automated or automatic?

Current methods used for target recognition work under a set of assumed operating conditions, against known targets, and can reject target-like objects resulting in some level of robustness. As such, they are automated solutions.

Is a Ground-Collision Avoidance System (GCAS) autonomous?

A GCAS takes control of an aircraft if there is concern that the pilot will cause the aircraft to collide with the ground. Here, the system takes command and control (C2) away from the pilot to keep the aircraft from colliding with the ground, then the pilot can regain C2 (either explicitly relinquished control or the pilot takes it back). The GCAS demonstrates *peer flexibility* and is therefore addressing a key challenge for an AS noted earlier. Notice that in this description the system does not, however, exhibit *task* or *cognitive flexi-*

bility. Some argue that GCAS is merely an automated system, due to its lack of cognitive flexibility.

Is an autopilot system autonomous?

An autopilot system has the task of flying a particular trajectory, at a particular speed, and at a particular altitude (or altitude profile) set by the pilot. The autopilot does not change its task or its peer relationship with the pilot, nor does it change *how* it controls the aircraft. It therefore does not satisfy any of the three flexibility principles of autonomy. It does, however, *reflexively* adapt to changing conditions that impact its heading, speed, and altitude to maintain the parameters provided to it. It is therefore an automated system.

Is an adaptive cruise control system autonomous?

A cruise control system exists to maintain a constant speed. An adaptive cruise control system also maintains its speed but adapts to sensed changes in front of it by changing its speed, without permission from the driver, to maintain a safe distance behind the car in front of it. It may also brake in case the need arises. An adaptive cruise control is automated since it never changes its peer relationship, never changes its task (only the way it is accomplishing its task in a preprogrammed manner), and does so with no cognitive flexibility.

Is an air-to-air missile autonomous?

An air-to-air missile, or even a cruise missile, has a fixed peer relationship with the human that launched it. The missile is doing a predefined task and doing it in a preprogrammed way. None of the three principles of flexibility are demonstrated—and it is therefore not autonomous. The system is remarkable and able to complete a very complex task, but it is merely automated.

Are the Google Car and Tesla-with-autopilot autonomous?

The Google Car drives to a location as directed, but it can change its task from driving to making an emergency stop to avoid running into a pedestrian, for example. When the Tesla autopilot locks the driver out and won't allow engagement because the driver is taking his/her hands off the wheel, or when the VW autonomous car takes control and brakes to prevent a head-on collision, we are seeing instances of peer flexibility, as in GCAS described earlier. But it is not autonomous.

Is the Roomba autonomous?

The Roomba is a popular home commodity that serves as the homeowner's proxy to vacuum the carpet. It is quite capable, and new versions incorporate modern robotics, to include the ability to determine its current location while mapping out its environment. The Roomba has a sole task—to vacuum. It does not have an ability to change its peer relationship, and it does not change its model for completing its task. As such, the Roomba is not autonomous, but it is a very capable and useful automated system.

Is IBM's Watson intelligent? Is Watson an AI?

Watson has *knowledge* that is gathered and/or generated by a combination of human programming and the application of those programs to large stores of data. It is capable of efficiently storing and retrieving potentially relevant *knowledge* so that it may respond to queries. One could argue that when Watson is allowed to use its programming to search and appropriately index large repositories of data, it is gathering information that it later applies appropriately. In doing its search, it uses an ensemble learning approach, which means it changes its model so it can provide better results, an aspect of cognitive flexibility. As such, Watson is therefore addressing a key challenge for an AS, but we would not consider it autonomous. However, Watson is a combination of hardware and software that exhibits intelligence. As defined previously, it is therefore an AI.

Is Siri intelligent? Is Siri an AI?

The information gathered by Siri to respond to queries is done via its programming. It does gather data in response to a query and often appropriately uses that knowledge to provide value to answering it and is a good automation. Since Siri is a combination of hardware and software and exhibits intelligence as defined previously, we would label Siri an AI.

Does Siri understand what I'm asking?

One can say that Siri understood when we have a reasonable expectation that it will give an answer that can be used and say it did not understand when we have a reasonable expectation that the answer will not be acceptable. But with all AI systems, the user must realize the meaning generated by the system is not "human meaning" and thus must be used judiciously. As an example, an AI can call a school bus an ostrich with high confidence, yet any human looking at the image will not be able to understand how the AI could possibly make that error. The reason is that, to the AI, the meaning is a location in a vector space reached by processing pixel intensities and colors and an ostrich is an object category that does not possess the rich meaning humans associate/generate in our meaning.

Does AlphaGo understand the game of Go?

AlphaGo's understanding of the game Go can only be assessed from the perspective of another agent. As a non-Go player, one might be willing to say AlphaGo understands the game because, from a naïve perspective, it responds acceptably to the task of playing the game. Here, AlphaGo has generated "meaning" of any board states that facilitate an expected acceptable response. From the perspective of one who wants to define "understanding the game" as "the AS has generated internal to itself the meaning of what a game is," then one might conclude AlphaGo does not understand the game of Go. AlphaGo is an automated system, and because it uses knowledge to generate its meaning of those board states that facilitate its response, it is an AI. Table B.1 attempts to summarize some of the categorizations we have made above.

Table B.1. Mapping of several systems to the principles of flexibility, artificial intelligence, automated systems, and automatic systems

		Garage door opener	Automatic target rec.	GCAS	Autopilot	Adaptive cruise control	Air-to-air missile	Google Car	Roomba	IBM Watson	Siri	AlphaGo
Flexibility	Peer			X				X				
	Task											
	Cognitive									X		
Artificial intelligence										X	X	X
Automated system			X	X	X	X	X	X	X		X	X
Automatic system		X										

B.3 Practical Limitations of AI

Where do current AI systems fail?

Current approaches to AI rely on knowing all that is needed to know about the environment and programming in acceptable responses for all possibilities. These approaches are unable to respond correctly when they are unable to get all of the data they expect or if they encounter a stimulus they do not have a programmed response for. The problem is compounded when both conditions occur.

How can we be sure an AI will do what we want it to do versus something we absolutely do NOT want it to do?

A significant challenge faced when using AI to perform tasks is the problem of validation and verification of that AI's performance. Any AI is programmed to generate an internal representation (its meaning), and that representation has to capture all aspects of what delineates acceptable and unacceptable behavior to be confident it will only do what it is supposed to do. It is impossible to test an AI under all possible operating conditions; it will not be known when it will fail or perform unacceptably. The same issues are faced with human agents. The Air Force goes to great lengths to train its Airmen to do tasks and cannot possibly test them on all the operating conditions they will face. Training is continually refined based on feedback on performances of those Airmen doing their jobs. The same will have to be done with AIs. There is the additional challenge/advantage in that the meaning of a given stimulus can be programmed into the AI and therefore tested against.

Appendix C

Sensor Processing and Fusion

C.1 Stimuli and Sensing

This appendix expands on section 5.3.3.1, providing a more formal mathematical description of sensor processing and fusion. The discussion on fusion that follows is based on the premise of the following diagram:

$$E \xrightarrow{s} D^{raw} \xrightarrow{f} D \xrightarrow{p} F \xrightarrow{x} X$$

where E is the environment that contains the stimuli and s is the senor that senses that stimuli to create the raw data D^{raw}. In the diagram, f, p, and x are a series of processes on the intermediate results D^{raw}, D, and F respectively that ultimately produce a set X. Set X could be detected as changes between two different images as described in section 5.3.3.1 in the change detection scenario. Set X could also be a classification of all of the objects in the environment collected by the sensor. The details of the processes and intermediary sets are covered in more detail in subsequent sections. It is first important to understand how to get from *stimuli* in the environment to raw data as collected by a sensor that can be used for the fusion task.

The basis for various *stimuli* that can be transformed by sensors into signals, images, text, and so on can be physical (imaging sensors), electronic (signal sensors such as electronic intelligence sensors), or psychological (Twitter, email). Both imaging and signal sensors use wave propagation to remotely capture the energy reflected or emitted in the form of electromagnetic energy. Stimuli sensed by imaging sensors are governed by physical laws such as Maxwell's equations to determine wave reflection and propagation, Newton's laws that govern the dynamics of the world, and geometrical laws that govern shape and the relationships between objects. Different imaging wavelengths are sensitive to different phenomena. For example, infrared wavelengths are sensitive to differences in temperature both reflected and emitted. Hyperspectral stimuli are sensitive to material properties. Polarization stimuli provide orientation information or surface properties. Hence, imaging sensors sense the stimuli of the physical world. On the other hand, signal sensors capture signal modulations that are generated by electronic circuits and software-programmable waveform devices. These adversary signal generators may include radars, communication, navigation, or even jammers. The signal sensors can be used to perform a function or can be used to disrupt

Air Force mission objectives. These signal stimuli are not governed by physical laws, as with the stimuli sensed by imaging sensors, but rather are constrained by the adversary engineers' and scientists' skill and imagination coupled with the physical constraints of size, weight, and power provided by the emitting platform. An exploding source of stimuli being collected is based on cyber technology or open-source methods that continually mine the internet. Given that the source of these stimuli is the human mind, the scope of possible contributions to a fusion inference is enormous.

Each type of stimuli conveys different types of information, which is potentially very powerful, as the fusion of different and independent sources can provide high-confidence estimates and decisions. But the different types of information also introduce some technical challenges. These technical challenges include space/time alignment and an evidence combination of disparate stimuli (images, signals, text). The challenge is not only based on the difference in stimuli (1-D signals verses 2-D images) but also, predominantly, on the difference in the meaning of the different stimuli (e.g., images conveying shape and text conveying sentiment).

These challenges occur within a class of stimuli such as images and also between stimuli such as images and text. Within a class of stimuli, the challenges are significant, but common representations are more straightforward. For example, one could use a 3D world representation to combine various stimuli from imaging sensors. However, it is much more difficult to conceive of common representations to fuse psychological state of mind with a signal waveform.

Before the stimuli can be analyzed and fused, they must first be *sensed*. The sensing of the stimuli is a significant challenge. Images and signals have their own sensing considerations such as line of sight, sensitivity, and resolution, which restricts the range that the stimuli can be sensed and discriminated. Environmental conditions such as atmosphere or obscuration also affect the sensing range. Finally, the stimulus itself can also limit the sensor's effectiveness by its appearance variability (pose, configuration) or its signal variability (waveform agility, spoofing). Clearly, it is vitally important to understand the performance of the various sensors under the myriad of conditions that they must operate. This performance understanding is necessarily encoded in a performance model to facilitate the fusion and employment of the various sensors.

Cyber or open source, on the other hand, has the degrees of freedom of language to encode the thoughts and intents of humans. The challenge here is not with the sensing—since the information is already encoded in text or language—but is instead in the sifting and interpretation of the text or language. In addition, there are certainly cyber-sensing challenges, which may force one to focus on the content of the cyber and open source data rather than the sensing.

C.2 Defining Fusion

A general, yet simple, definition of fusion is: "Combining things (or objects) for better results." This definition often raises several questions.

First, what are the objects that are to be combined? The word "object" is used to denote a generic entity. For example, objects can be the raw sensor data, filters, exploiters, features, entire sensor-exploitation systems, or even situations.

Second, what does "combining" mean? Combining is the combining of objects using some "rule" that outputs an object from the category of objects of interest (DTIC 2005). The word "rule" is used generically to convey the function property (i.e., given an input for which the rule is defined, there exists a unique output). If the objects are specified, then the name used for the rule can be specific. For example, if the data from two different sensors are numerical time-varying data, for example from signals intelligence (SIGINT) collections, then a transformation (the rule) acting on the two SIGINT signals could output a new signal or some meaningful numerical data. Features from two different feature sets can be associated to produce a new single-feature set. A simple association is to form ordered pairs of features, thus concatenating the two feature sets in a special way. But to discover commonality in the two feature sets would prove fruitful.

Third, what does "better" mean? The designer has to choose a performance criterion—specifically, a real-valued function p is used to quantify the criteria of interest. Assume larger values mean better performance and that there are two SIGINT systems under consideration, $\mathbf{A}$ and $\mathbf{B}$. If $p(\mathbf{A}) > p(\mathbf{B})$, then system $\mathbf{A}$ is better than $\mathbf{B}$ with respect to the performance function p.

Fourth, what does "results" mean? Suppose system $\mathbf{C}$ uses a rule $\mathbf{r}$ that is designed to combine the two systems, $\mathbf{C} = \mathbf{r(A,B)}$. If $p(\mathbf{C}) > \max\{p(\mathbf{A}), p(\mathbf{B})\}$, then combining $\mathbf{A}$ and $\mathbf{B}$ with ruler $\mathbf{r}$ has yielded a new system with better results. If $p(\mathbf{C}) = p(\mathbf{A})$ or $p(\mathbf{C}) = p(\mathbf{B})$, then combining $\mathbf{A}$ with $\mathbf{B}$ using rule $\mathbf{r}$ has not produced better results.

Simply stated, the goal of sensor fusion is to combine sensor-exploitation systems for improved performance.

C.3 Various Forms of Fusion

Fusion occurs at different locations within the processing chain. It can be accomplished at the sensor data level (e.g., signal or image), at the feature level (e.g., physical dimension of an object), or at the decision level (e.g., radar says "tank" with 90 percent confidence and EO says "T72" with 95 percent confidence). Each approach has its advantages and challenges, and these are discussed in the following sections.

C.3.1 Sensor Data Level Fusion

Although fusion at the sensor data level can provide remarkable performance improvements, fusion at the signal/pixel level also puts the most stringent demands on spatial registration and timing. For example, multiple platforms that have phase coherence can perform exquisite geolocation of emitters. Imaging sensors that are fused at the pixel level can reason about parts at a very detailed level. For example, fused EO and IR imaging systems allow the higher resolution EO system to be enhanced by thermal information that could reveal the state of the target (e.g. engine hot, wheels hot—indicating recent movement). Fused low-frequency radar with 3D laser radar could penetrate camouflage and provide shape information not visible to EO systems. Fusion at the signal level is also motivated by the signal processing inequality, which states that any intermediate decision in a processing chain will lose information unless a very special, rare circumstance is met called the "sufficient statistic." Thus, combining information at the data level, while hard due to the above considerations, is the most powerful form of fusion between two sensors.

C.3.2 Feature Level Fusion

Fusion can also occur at the feature level. The simplest feature combination rule is to concatenate features from the multiple sources, but sometimes this leads to a reduction in performance (Mura et al. 2011). This reduction in performance can be due to: a) redundant information; b) lack of a clear representation or measurement of the common environment; or c) the lack of enough labelled data to accommodate the increase in dimensionality induced by the concatenation, and thus over-fitting. A particularly promising approach to fusion is using graph-based feature fusion (Liao et al. 2015), which includes the data mining necessary for appropriate processing of the streaming sensor data. The graph-based approach to feature level fusion introduces structure into the representation and, moreover, can represent the conditional dependencies vital to the proper combination of information to make accurate decisions. The concatenation of features puts the entire burden on the classifier and the data to discover structure and dependencies in the data. The graphical model can represent the conditional probabilities among sensors, and it can also encode the nuisance parameters or operating conditions that affect performance

C.3.3 Decision Level Fusion

The most prevalent form of fusion is decision (also called label) level fusion. This form of fusion has the most flexibility as it puts lower communication and knowledge constraints on the sensors and platforms; however, it is also the form of fusion that has the lowest performance—again, due to the signal processing inequality. To get the most out of decision level fusion, several challenges, many of which are shared with feature level fusion, must be addressed. A key challenge is to understand and include the dependencies among the various deciders in the fusion calculation. Today's decision level fusion approaches, in general, make a conditional independence assumption called the naïve Bayes approach. This assumption is rarely true and will give optimistic results that, in turn, create decision errors and nonrobust performance. To overcome these errors, the joint statistics and dependencies must be modeled, which is extremely difficult. In addition, these joint dependencies are, in turn, a function of operating conditions causing the size of the hypothesis space to be extremely large and difficult to model. Now, model-based approaches attempt to model these statistical dependencies explicitly with the aforementioned graph-based approaches being a significant direction. Learning-based approaches learn these dependencies implicitly via training data. This approach is tractable if the sensors are tightly coupled— like our touch, sight, sound, and smell sensors or like the video-text example given earlier in this section. More loosely coupled, distributed sensor fusion approaches may be problematic for pure learning-based approaches due to the difficulty in providing training data of all the possible situations in which they could interact. For these distributed type of fusion challenges, hybrid-model and learning-based approaches should be considered to leverage their respective advantages.

C.4 Mathematical Formulation of Fusion

The goal of this section is to present a mathematical formalism for fusion. A parallel modeling approach to sensor-data fusion, feature-level fusion, and decision level fusion are presented. Finally, a sequential fusion model is briefly described.

C.4.1 Fusion Foundations

Let E be a population set of stimuli from the same environment. Let **s** be a sensor that produces datum as its output, *i.e.*, $s : E \to D^{raw}$, where D^{raw} is the raw sensor data set. Assume that the sensor has the "function" property, that is, given an input stimulus (in its domain of definition), there exists a unique output datum in D^{raw}. Typically, there is "noise" in this raw datum due to the randomness of the stimuli in the environment and/or the sensor itself. As such, the noise parameter can be modeled as a random variable (also called a random mapping or random transformation in higher dimensional output). If the raw sensor datum needs to be preprocessed (*e.g.*, filtered, standardized, or refined, as is described in section 5.3.3.1), then there needs to be a processor $f:D^{raw} \to D$. The output in this data set D may be too difficult to quantify, so another processor, **p**, defined on D will extract an object called a feature F (F is a data set). Typically, a feature is a vector $F \subseteq R^N$ for some positive integer N. Thus, the phrase "feature vector" is often used. However, a feature could be some other object such as a string of symbols, text, or a graph.

From F, one may wish to make a decision concerning a feature or a subset of features or exploit this refined data set further. Let **x** denote an "exploiter" mapping a feature from F to an element in X, the exploited data set. Thus, an exploiter is a special processor. If **x** is a classifier, then X might be a label set corresponding to the "names" or the labels assigned to objects in the classes, e.g., $X=\{$tank, truck, jeep$\}$ or $X=\{$(tank, truck, jeep), (truck, tank, jeep), (jeep, truck, tank),...$\}$ (notice the "labels" are ordered to denote an order of choice). If each of these processors (functions) are designed such that they can be composed

$$x \circ p \circ f \circ s$$

where o denotes the composition symbol, then a graphical representation of these processors is given in the following diagram:

$$E \xrightarrow{s} D^{raw} \xrightarrow{f} D \xrightarrow{p} F \xrightarrow{x} X$$

where an arrow denotes that a processor originates at the input set and points to the output set.

Define **A=xopofos** to be the exploitation **system**, so for a stimulus $e \in E$ then $A(e)$ is an element defined in X. Objects can be sensor-exploitation systems, say, $\mathbf{A} = x_a \text{o}\mathbf{p}_a \text{o}f_a \text{o}s_a$ and $\mathbf{B} = x_b \text{o}\mathbf{p}_b \text{o}f_b \text{o}s_b$ both sensing from the same stimuli environment set E. Consider the two sensor-exploitation systems, $\mathbf{A}$ and $\mathbf{B}$, diagrammed below.

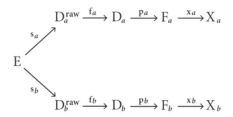

The sensors $\mathbf{s}_\alpha$ and $\mathbf{s}_b$ might be sensing a target of interest, yet have different modality. For example, $\mathbf{s}_\alpha$ might be an electro-optical (EO) sensor and $\mathbf{s}_b$ an infrared (IR) sensor. Suppose there is a "rule" $\mathbf{r}$ for combining the exploitation data sets X_α and X_b together such that $\mathbf{r}: X_\alpha \times X_b \to X_c$ into an exploitation set X_c. This yields the diagram

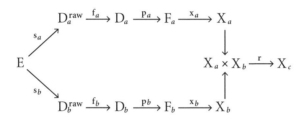

The rule $\mathbf{r}$ makes a new system $\mathbf{C}$ defined by the compositions

$$\begin{aligned} \mathbf{C} &= \mathbf{r} \circ (x_a \circ p_a \circ f_a \circ s_a, x_b \circ p_b \circ f_b \circ s_b) \\ &= \mathbf{r} \circ (\mathbf{A}, \mathbf{B}). \end{aligned}$$

The rule for combining the exploited data is a special processor.

To check the performance of the new system, it is necessary to see if $p(\mathbf{C})$ = $p(\mathbf{ro}(\mathbf{A},\mathbf{B})) > \max\{p(\mathbf{A}), p(\mathbf{B})\}$. If it is, then $\mathbf{r}$ is a fusion rule with respect to the performance function p, or simply, $\mathbf{r}$ is a **fusor**. It is possible that $p(\mathbf{A}) = p(\mathbf{B})$, so one system is not better than the other. But combining might yield a new, better system with respect to the performance function p. If there are several rules of interest, say the set R, then the fusion designer would search through this collection of rules to find the best fusor $\mathbf{r}^*$ from among all fusors. That is,

$$p\left(\mathbf{r}^* \circ (A,B)\right) = \max_{\mathbf{r} \in \mathcal{R}} p\left(\mathbf{r} \circ (A,B)\right) > \max \{p(A), p(B)\}.$$

This means that "sensor processing and performance criteria" will make up sensor fusion.

There may be several performance criteria a fusion designer seeks to satisfy. These criteria might compete with (or contradict) each other. As such, multi-objective performance fusion techniques are a class of fusers that need to be considered. If q is another performance function (assume larger is better), then does

$p(\mathbf{r} \circ (A,B))$	>	$\max\{p(A), p(B)\}$
$q(\mathbf{r} \circ (A,B))$	>	$\max\{q(A), \sigma(B)\}$

hold true? Or does

$p(\mathbf{r} \circ (A,B))$	>	$\max\{p(A), p(B)\}$
$q(\mathbf{r} \circ (A,B))$	<	$\max\{q(A), q(B)\}$

hold true? Suppose there is another rule t such that

$p(\mathbf{t} \circ (A,B))$	>	$\max\{p(A), p(B)\}$
$q(\mathbf{t} \circ (A,B))$	>	$\max\{q(A), q(B)\}$

and

$$p(\mathbf{t} \circ (A,B)) > p(\mathbf{r} \circ (A,B))$$

then rule $\mathbf{t}$ is the fusor with respect to the multi-objective performance functions p and q.

C.4.2 Diagraming Fusion Systems

Fusion systems can be diagramed to aid in understanding the processes and input/output relationships associated with those process. Specifically, in this section, diagrams for sensor-data fusion, feature-level fusion, decision level fusion, and a hybrid of feature- and decision level fusion are presented. In the diagrams, processes are denoted as: s is a sensor; $\mathbf{f}$, $\mathbf{p}$, and $\mathbf{u}$ are processors; $\mathbf{r}$ is a rule; and $\mathbf{x}$ is an exploiter. Sets are denoted as: E is the environment and serves as the starting point; D are data; F are features; X are exploitation sets; and L are label sets.

Sensor-data fusion (on the filtered data) is diagrammed as

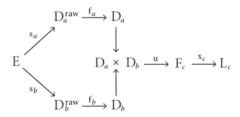

and a similar diagram exists for sensor data fusion on the raw sensed data (not shown here). Sensor data fusion can occur with a single sensor, diagramed as

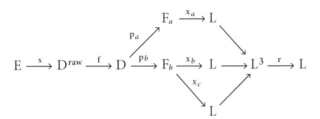

In the case where two sensor-exploitation systems are sensing two distinct environments (E_a and E_b at different locations and/or different stimuli) and there is a need to process the combined stimuli to discover a relationship between them, then combining the stimuli sets via concatenation yields

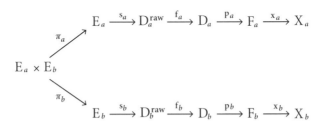

where π_a and π_b are projectors onto the respective environment / stimulus sets, E_a and E_b, that is, $\pi_a:E_a \times E_b \rightarrow E_a$ and $\pi_b:E_a \times E_b \rightarrow E_b$ are "identity" functions in the sense that $\pi_a(e,e')=e$ and $\pi_b(e,e')=e'$ for any stimulus pair $(e,e') \in E_a \times E_b$. It is tempting to call this "event" fusion or "stimulus" fusion, but it requires a sensor data rule, feature rule, decision rule, or any mixture to produce an exploited datum as its final output in order to discover any relationship between the stimuli in these two distinct environments.

Feature level fusion is diagrammed as

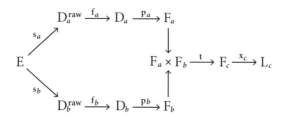

The following diagram shows decision (label) fusion, where L_a and L_b are label sets, and $\mathbf{x}_a$ and $\mathbf{x}_b$ are classifiers.

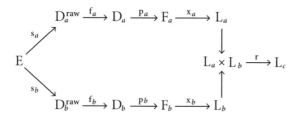

where label set L_c could equal $L_a, L_b, L_a \times L_b, L_a \cup L_b$, the union of both label set, or something completely different.

A mixture of fusion approaches can also be of use. As an example, feature-level fusion and decision level fusion can be combined to produce a fused result and is diagrammed as

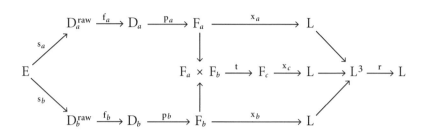

C.4.3 Series Fusion Architecture

So far, the fusion discussion has centered on parallel architectures. Another fusion architecture allows for "series" architectures call sequential fusion. This is related to "sequential analysis" as described in statistics. The medical community regularly performs "tests" sequentially to determine the correct diagnosis. Similarly, air warfighters will operate sequentially, especially if there the exploited data is not sufficient to determine between a "friend or foe" (the labels). More data might be needed to help determine the "unknown" label to "friend or foe." Or maybe the original sensor data needs to be processed in another way that would be faster than having to physically take "another look." Typically, many of these "processors" have parameters that can vary during real-time operations. For example, if exploiters (e.g., classifiers) depend on a (vector) parameter $\theta \in \Theta \subset R^M$, say $\mathbf{x}_a^{(\theta)}$ and $\mathbf{x}_b^{(\phi)}$, for (vector) parameter $\phi \in \Phi \subset R^N$, then the performance of the following diagram

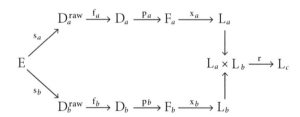

not only depends on the choice of rule $\mathbf{r}$ but also on the choice of the parameter pair (θ,ϕ). Hence, we should optimize over all parameters in $\Theta \times \Phi$. Define

$$
\begin{aligned}
C^{(\theta,\phi)} &= \mathbf{r} \circ \left(\mathbf{x}_a^{(\theta)} \circ p_a \circ f_a \circ s_a , \mathbf{x}_b^{(\phi)} \circ p_b \circ f_b \circ s_b \right) \\
&= \mathbf{r} \circ \left(A^{(\theta)} , B^{(\phi)} \right).
\end{aligned}
$$

then the goal is to find the best $(\theta^*, \phi^*) \in \Theta \times \Phi$ and $\mathbf{r}^* \in R$ such that

$$p(\mathbf{r}^* \circ (A^{(\theta^*)}, B^{(\phi^*)})) = \max_{\substack{\mathbf{r} \in R \\ \theta \in \Theta \\ \phi \in \Phi}} p(\mathbf{r} \circ (A^{(\theta)}, B^{(\phi)})) > \left\{ \max_{\theta \in \Theta} p(A^{(\theta)}), \max_{\phi \in \Phi} p(B^{(\phi)}) \right\}.$$

Appendix D

Human-Systems Integration Project Example

The Air Force Research Laboratory has engaged in a program focused on human–autonomous system (AS) teaming in the command and control (C2) of multiple heterogeneous unmanned vehicles (UXV)[1] in a base defense mission. The system has been designed to implement a "playbook" delegation-type architecture (Parasuraman et al. 2005; Miller and Parasuraman 2007) extended to enable more seamless transition between control levels (from manual to fully autonomous). At one extreme, the human can manually control a specific UXV. At another level, the operator makes numerous inputs to specify all the details of a "play" that defines the tasking that one or more UXVs autonomously perform once the play is initiated. At the other extreme, the human can quickly task UXVs by specifying only two essential details (play type and location), and then a "C2" AS determines all the other tasking details. For example, if the human operator calls a play to achieve air surveillance on a building, the C2 AS recommends which UXV to use (based on sensor payload, estimated time en route, fuel use, environmental conditions, etc.), a cooperative control algorithm—another limited-capability AS—provides the shortest route to get to the building (taking into account no-fly zones, etc.), and the C2 AS monitors the play's ongoing status (e.g., alerting if the vehicle will not arrive at the building on time) (Draper 2017). The human-machine interfaces (HMI) also support the human communicating any other play detail, and this additional information informs the AS on how to optimize the recommended plan for the play. For example, the human may have information that the AS does not like the target size and current visibility. With these HMIs, the human operator can, at any time, tailor the role of the AS depending on the task, vehicle, mission event, or the human's trust in the AS (or a unique combination of these dimensions).

Detailed descriptions of the HMIs and the rationale for their design are available elsewhere (Calhoun et al. 2017a; Calhoun et al. 2018). Here, only brief introductions are provided on HMIs for several steps of employing a play-based human-AS teaming approach. Note that the displays and controls all feature video gaming–type icons that represent unmanned vehicle type, play type, and/or a play detail specific to a base defense scenario. This approach presents information in a concise, integrated manner and also helps

---

1. X is used to denote Air, Ground, or Sea vehicle (A, G, S).

maintain the human's visual momentum (Woods 1984) as information is retrieved from the various HMIs. The icons also support the human's direct perception and manipulation, with the human able to act directly with the icon (Shneiderman 1992).

To call a play to define the actions of one or more unmanned vehicles, the human operator needs to specify at least two details to the AS:

1. Play location, via a speech command or by manually selecting a map location or one of many predefined locations on a pull-down menu.

2. Play type, with speech command or selecting corresponding icon on a play-calling HMI, as shown in figure D.1.

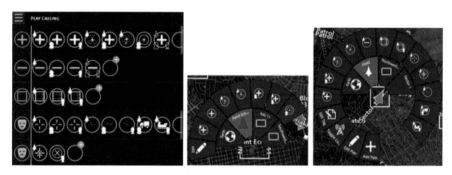

Figure D.1. Three HMIs to specify play and unmanned vehicle type with mouse or touch. *Left*: All base defense plays, including surveillance of a location, road, or area and others involving friendly or unfriendly entities (e.g., overwatch, escort, shadow, and cordon). *Center*: Radial menu pop-up includes only the play options relevant to the map location selected (i.e., no sea vehicle for land location). *Right*: Radial menu pop-up includes only the play options relevant to the vehicle selected on the map.

Once the human operator communicates desired play type and location, the AS reasons on available assets and other relevant details to come up with one or more recommended plans for the play. Figure D.2 illustrates two HMIs that provide the human operator with feedback on its processing and the rationale for its recommendation. In the polar coordinate plot on the left, several of the AS-identified feasible plans are summarized (Hansen 2016). The HMI to the right shows more details on the AS's rationale for the proposed play plan.

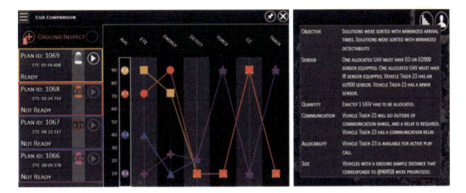

Figure D.2. AS-generated plans. *Left*: Plan Comparison HMI shows trade-off of several candidate AS-generated plans for a play. *Right*: AS's additional rationale for its recommended play plan.

As shown in figure D.3, the AS's recommended plan is also presented on a map with dashed symbology, as well as in a Play Workbook that provides details of the play, including the assumptions and constraints the plan is based on. For example, the Workbook's shaded icons (sun, clock, and "HI" circled icons on the Workbook's right page) indicate that the play plan assumes there is good visibility and that the play is high priority and is optimized to get an asset to the location as soon as possible.

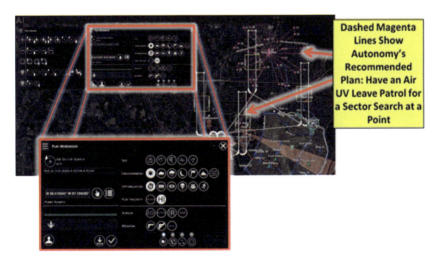

Figure D.3. Map showing Play Workbook and proposed vehicle and route on the map (AS is proposing to add an air vehicle to perform a sector search at Point Romeo, as indicated by the dashed magenta symbology.).

By issuing speech commands or changing selections on the Play Work-book, the human operator can prompt the AS to come up with new plans re-flecting the revised details. For instance, if the operator designated the size of the target on the first Workbook page row or changed the environmental icon to one indicating cloudy (second row), the AS will consider vehicle assets with a more appropriate sensor payload and change the proposed plan for the play. By interacting with the Play Workbook (or issuing speech commands), the operator can quickly specify/revise as many play details as desired, both when calling a play and after the play is under way (the latter likely in re-sponse to a change in mission requirements). Additionally, the HMIs serve as the basis for the human posing "what-if" queries to the AS. Changes in Work-book selections result in new AS-generated play plan(s). Another query method is for the human operator to verbalize a question (e.g., "When can a lethal weapon arrive at Gate Delta?"), with the answer presented aurally re-flecting the AS's deliberations. If the operator replies "Show me," the corre-sponding information (e.g., play plan) is shown on the map and in a Play Workbook.

Once the human operator consents to a particular play plan, the AS handles play execution by controlling UXV movements such that they complete the tasks specified for the play. The human operator can track the progress of the AS-supported play in several ways: watching the movement of UXV symbology on the map, retrieving information on a row dedicated to each play in a table (fig. D.4), and viewing a Play Quality Matrix display (fig. D.5).[2]

---

2. Note: the symbology associated with each ongoing play has a unique color that is em-ployed both on the map and in the Active Play Table. Besides aiding visual momentum, use of color coding by play helps the human operator maintain a global perspective in terms of which vehicles are coordinating on the same play.

Figure D.4. Active Play Table HMI. Each row provides additional information on plays under way as well as the ongoing patrol. For each play, the row shows (from *left* to *right*) the type of play, play location, assigned vehicles, functionality to cancel or pause the play, and a color indicator of the progress of the play. Selection of the row calls up the corresponding Workbook (e.g., to modify the play), including the Play Quality Matrix of figure D.5.

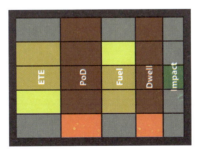

Figure D.5. Play Quality Matrix. The color coding (green, yellow, or red) and deviation of each bar from the center of the matrix indicate whether the corresponding base defense mission parameter is within, above, or below its expected operating range. The parameters depicted in this illustration are: expected time en route, probability of detection, fuel state, how long vehicle(s) can dwell on a site to be surveilled, and impact (degree to which play impacts overall mission success).

The prototype HMIs briefly described above enable a human operator to call plays and for the operator and the AS to work together to specify other play details. HMIs by which the AS communicates its reasoning for play plans and the progress of plays are also illustrated. Other HMIs have been implemented that provide mechanisms by which the human-AS team can track plays that are not active (e.g., waiting for a necessary resource to become available), chain plays together (e.g., a play begins automatically when another play ends), and establish conditions that a play will begin without operator consent (Calhoun et al. 2017b). Improvements have also been initiated to provide a Task Manager HMI that facilitates human-AS coordination and shared awareness.

Appendix E

Toolsets and Datasets

Section 5.3 provided a snippet of available toolsets and datasets that could be used in the implementation of the sample framework presented in section 5.2. In this appendix, we provide a more comprehensive table of potentially useful toolsets and datasets. As stated in section 5.3, the list of toolsets and datasets is not meant to be comprehensive but rather to provide a starting point for future efforts.

Table E.1. List of toolsets and datasets for component functions presented in section 5.3

Section	Primary Function	Secondary Function	Toolset or Dataset Name	Tool or Data	Type	Environment	License	URL
5.3.1	Data-bases		MongoDB	Tool	No SQL Database	Windows/ Linux/ Unix	Open Source	https://www. mongodb.com/
5.3.1	Data-bases		PostgreSQL	Tool	SQL Da-tabase	Windows/ Linux/ Unix	Open Source	https://www. postgresql.org/
5.3.1	Data-bases		CUBRID Manager	Tool	Database Admin		Open Source	http://www. cubrid.org/
5.3.1	Data-bases		Firebird	Tool	SQL Da-tabase	Windows/ Linux/ Unix	Open Source	http://www. firebirdsql.org/ en/start/
5.3.1	Data-bases		Redis	Tool	No SQL Data-base / In-Mem-ory Data Structure Store	Linux, BSD, OS X, Win-64	BSD	https://redis.io/
5.3.1	Data-bases		MariaDB	Tool	SQL Da-tabase		Open Source	https://mariadb. org/
5.3.1	Data-bases		MySQL (Com-munity Edition)	Tool	SQL Da-tabase	Linux, Windows, MacOSX, Free BSD	GNU General Public License	https://mysql. com
5.3.1	Data-bases		SQLite	Tool	SQL Da-tabase	Linux, Windows, MacOSX, Android, Windows Phone	Public Domain	https://www. sqlite.org/

Section	Primary Function	Secondary Function	Toolset or Dataset Name	Tool or Data	Type	Environ-ment	License	URL
5.3.1	Data-bases		SciDB	Tool	High-dimen-sional data	Windows, Linux	Public Domain	http://www.para-digm4.com
5.3.1	Data-bases		McObject eX-tremeDB	Tool	In-memory, embed-ded DB - SQL + NoSQL architec-ture	Windows, Linux, Clusters, iOS	License costs	http://www.mcobject.com/
5.3.1	Data-bases		Hadoop	Tool	Parallel database process-ing	Windows, Linux	Open Source	http://hadoop.apache.com
5.3.1	Data-bases		PERS	Tool	In-memory, embed-ded DB - SQL + NoSQL architec-ture	Java, C#, Android OS, Windows Phone 7, Silverlight	Dual License (Open Source, Propri-etary)	http://www.mcobject.com/perst
5.3.2	Sensors		SUPPRES-SOR	Tool	Elec-tronic Warfare Mission Model-ing	Ubuntu	GOTS	
5.3.2	Sensors		OPNET	Tool	Network Model-ing	Ubuntu, Windows	Com-mercial	www.opnet.com
5.3.2	Sensors		ESAMS	Tool	RadioFre-quency Model-ing	Ubuntu	GOTS	
5.3.2	Sensors		DIADS	Tool	Sensor Model-ing	Ubuntu	GOTS	
5.3.2	Sensors		Vigilant Hammer	Data	Elec-tronic Warfare Data		GOTS	
5.3.2	Sensors		Northern Edge (re-stricted)	Data	Elec-tronic Warfare Data		GOTS	
5.3.2	Sensors		Xpatch	Tool	Sensor Model-ing			

Section	Primary Function	Secondary Function	Toolset or Dataset Name	Tool or Data	Type	Environment	License	URL
5.3.2	Sensors		Raider Tracer	Tool	RF Modeling			
5.3.2	Sensors		Laider Tracer	Tool	Optical Modeling			
5.3.2	Sensors		MZA Toolbox	Tool	Optical Modeling			
5.3.2	Sensors		ScaleME	Tool	Electro-dynamic/acoustic simulator			
5.3.2	Sensors		Mental Ray	Tool	Optical Modeling			
5.3.3.1	Sensor Processing & Fusion		NXP	Tool	Sensor Fusion package using MEMS and magnetic sensors, ARM™ Cortex M0+, M4 and M4F portfolio development boards and Native Android Sensors	Windows, Android OS	Commercial	http://www.nxp.com/products/scensors/nxp-sensor-fusion.XTRSICSNSTLBOX
5.3.3.1	Sensor Processing & Fusion		Sunhillo	Tool	Multi-Sensor Track Fusion Plugin	Sunhillo SGP Product	Commercial (Proprietary)	http://www.sunhillo.com/multi-sensor-track-fuser.html
5.3.3.1	Sensor Processing & Fusion		Vector Informatik: Advanced Driver Assistance Systems (ADAS) development	Tool	vADAS Developer – Multi-Sensor Fusion Development Environment	Windows (Dev in Visual Studio)	Commercial	https://vector.com/vi_vadasdeveloper_en.html

Section	Primary Function	Secondary Function	Toolset or Dataset Name	Tool or Data	Type	Environment	License	URL
5.3.3.1	Sensor Processing & Fusion		OPPORTUNITY Activity Recognition Data Set	Data	Multi-sensor data for human activity recognition	Universal	Open Source	https://archive.ics.uci.edu/ml/datasets/opportunity+activity+recognition
5.3.3.2	Data Mining & Fusion	Sensor Mgmt. & Data Mining	International Business Machine SPSS (CRISP-DM)	Tool	Data Mining Process		COTS	http://www.ibm.com/analytics/us/en/technology/spss/
5.3.3.2	Data Mining & Fusion	Sensor Mgmt. & Data Mining	Advanced-Miner	Tool				
5.3.3.2	Data Mining & Fusion	Sensor Mgmt. & Data Mining	CMRS Data Miner	Tool				
5.3.3.2	Data Mining & Fusion	Sensor Mgmt. & Data Mining	RapidMiner	Tool				
5.3.3.2	Data Mining & Fusion	Sensor Mgmt. & Data Mining	Data Miner Software Kit	Tool				
5.3.3.2	Data Mining & Fusion	Sensor Mgmt. & Data Mining	DBMiner 2.0	Tool				
5.3.3.2	Data Mining & Fusion	Sensor Mgmt. & Data Mining	Delta Miner	Tool				
5.3.3.2	Data Mining & Fusion	Sensor Mgmt. & Data Mining	Exeura Rialto	Tool				

Section	Primary Function	Secondary Function	Toolset or Dataset Name	Tool or Data	Type	Environment	License	URL
5.3.3.2	Data Mining & Fusion	Sensor Mgmt. & Data Mining	KnowledgeMiner 64	Tool				
5.3.4.1	Perception & Event Detection	Learning & Adaption	OpenCV	Tool	Computer Vision		BSD	http://opencv.org/
5.3.4.2	Situation Assessment		NetSA	Tool	Network Situational Awareness	Linux, Solaris, OpenBSD, Mac OS X, and Cygwin	GOTS	http://www.cert.org/netsa/
5.3.4.2	Situation Assessment		Situational Awareness Systems	Tool	Incident management, command & control, BIO-surveillance	Windows 2003 Server	Commercial	http://www.fd-software.com/
5.3.4.2	Situation Assessment		SITAWARE C4I Suite	Tool	C2 and Battle Management		Commercial	https://www.systematicinc.com
5.3.4.2	Situation Assessment		Safe Situational Awareness for Everyone	Tool	Operating Room Situational Awareness		Commercial	https://www.steris.com/healthcare/
5.3.4.2	Situation Assessment		Electricity Infrastructure Operations Center (EIOC)	Tool	Electrical Power Grid Situational Awareness		GOTS	http://eioc.pnnl.gov/research/sitawareness.stm
5.3.4.2	Situation Assessment		Public Health and Medical Emergency Tools	Tool	GIS, Social Medial, and Resource Collections, and Threat-Specific Response Tools	Various		https://www.phe.gov/Preparedness/news/events/NPM2015/Pages/situational-awareness.aspx

Section	Primary Function	Secondary Function	Toolset or Dataset Name	Tool or Data	Type	Environment	License	URL
5.3.4.2	Situation Assessment		California Situational Awareness and Collaboration Tool (SCOUT)	Tool	Incident Command System	Web-based		http://www.caloes.ca.gov/cal-oes-divisions/regional-operations/situation-awareness-and-collaboration-tool
5.3.4.3	Reasoning and Decision-Making		1000Minds	Tool	PAPRIKA method	Online Web Application	COTS	https://www.1000minds.com/
5.3.4.3	Reasoning and Decision-Making		D-Sight	Tool	PROMETHEE, UTILITY	Online Web Application, Mobile	COTS	http://www.d-sight.com/
5.3.4.3	Reasoning and Decision-Making		Decision Lens	Tool	AHP, ANP	Online Web Application	COTS	http://decision-lens.com/
5.3.4.3	Reasoning and Decision-Making		PyKE	Tool	Inference Engine	Python	Open Source	http://pyke.sourceforge.net/using_pyke/index.html
5.3.4.4	Planning & Re-planning		ICAPS Competitions	Tool Data	Planning and Scheduling	N/A	Open Source	http://www.icaps-conference.org/index.php/Main/Competitions
5.3.4.5	Execution Mgmt.		DAM(Decentralized Asset Manager)	Tool			COTS	
5.3.4.5	Execution Mgmt.		M2CS(Multi-vehicle Mission Control System)	Tool			COTS	
5.3.6	Human Computer Interface & Collaboration Environments		OpenVibe	Tool	Brain-Computer Interface	Windows	GNU Affero General Purpose License	http://openvibe.inria.fr/

Section	Primary Function	Secondary Function	Toolset or Dataset Name	Tool or Data	Type	Environment	License	URL
5.3.6	Human Computer Interface & Collaboration Environments	Databases	Defense Information Systems Agency Big Data Platform	Tool	Cloud Infrastructure w/ toolsets	Redhat Linux	GOTS	http://disa.mil/NewsandEvents/2016/Big-Data-Platform
5.3.7	Learning & Adaption		H2O	Tool	General	Java, R, Scala, Python	Open Source	http://www.h2o.ai/#
5.3.7	Learning & Adaption		Caffe	Tool	Deep Learning		BSD-2	http://caffe.berkeleyvision.org/
5.3.7	Learning & Adaption		Azure ML Studio	Tool	General			https://studio.azureml.net/
5.3.7	Learning & Adaption		Apache Singa	Tool	Deep Learning	Apache Spark	Apache License Version 2	https://singa.incubator.apache.org/en/index.html
5.3.7	Learning & Adaption		Amazon Machine Learning	Tool	General	Amazon Web Services		https://aws.amazon.com/machine-learning/
5.3.7	Learning & Adaption		Apache MLlib	Tool	General	Apache Spark	Apache License Version 2	http://spark.apache.org/mllib/
5.3.7	Learning & Adaption		mlpack	Tool	General	C++	BSD	http://mlpack.org/
5.3.7	Learning & Adaption		Pattern	Tool	General			
5.3.7	Learning & Adaption		Scikit-Learn	Tool	General	Python	BSD	http://scikit-learn.org/stable/
5.3.7	Learning & Adaption		Shogun	Tool	General	C++	GNU	http://shogun-toolbox.org/
5.3.7	Learning & Adaption		TensorFlow	Tool	Data flow graphs	Python	Open Source	https://www.tensorflow.org
5.3.7	Learning & Adaption		Veles	Tool	Deep Learning	Python, CUDA, and OpenCL	Open Source	https://velesnet.ml/

Section	Primary Function	Secondary Function	Toolset or Dataset Name	Tool or Data	Type	Environment	License	URL
5.3.7	Learning & Adaption		Torch	Tool	General	GPU-CUDA	BSD	http://torch.ch/
5.3.7	Learning & Adaption		Theano	Tool	General	Python	BSD	http://deeplearning.net/software/theano/index.html
5.3.7	Learning & Adaption	Human Computer Interface	Stanford University Natural Language Processing	Tool	Natural Language Processing		Open Source	
5.3.7	Learning & Adaption		University of Waikato Project (WEKA)	Tool	General	Java	Open Source	http://www.cs.waikato.ac.nz/ml/weka/
5.3.7	Learning & Adaptation		BURLAP	Tool	Reinforcement Learning	Java	Open Source	http://burlap.cs.brown.edu/
5.3.7	Learning & Adaptation		RL-GLUE	Tool	Reinforcement Learning	N/A	Open Source	http://glue.rl-community.org/wiki/Main_Page
5.3.7	Learning & Adaption	Data Mining & Fusion	University of California, Irvine, Machine Learning Repository	Data	ML data	Universal	Open Source	https://archive.ics.uci.edu/ml/datasets.html
5.3.7	Learning & Adaption		United Stated Geological Survey Spectral Library	Data	Hyperspectral signatures	Universal	Open Source	https://speclab.cr.usgs.gov/spectral-lib.html
5.3.7	Learning & Adaption		NASA Jet Propulsion Laboratory - Airborne Visible/Infra-red Imaging Spectrometer (AVIRIS)	Data	Hyperspectral Imagery	Universal	Open Source	https://aviris.jpl.nasa.gov/
5.3.7	Learning & Adaption		ImageNet	Data	Various images	Universal	Open Source	www.image-net.org

Section	Primary Function	Secondary Function	Toolset or Dataset Name	Tool or Data	Type	Environment	License	URL
5.3.8	Domain-Specific Knowledge Base	Learning & Adaption	The Graphical Models Toolkit (GMTk)	Tool	Dynamic Graphical Models & Dynamic Bayesian Networks	Home-brew, Mack-Ports, Linux Remote Package Manager, Source Code	Open Source	http://melodi.ee.washington.edu/gmtk/
5.3.8	Domain-Specific Knowledge Base	Learning & Adaption	Open Markov	Tool	Directed/ Undirected Graphs	Java	Open Source	http://www.openmarkov.org/
5.3.8	Domain-Specific Knowledge Base	Learning & Adaption	UnBBayes	Tool	Bayesian Networks, Influence Diagrams	Java	Open Source	http://unbbayes.sourceforge.net/
5.3.8	Domain-Specific Knowledge Base	Learning & Adaption	WinMine	Tool	Bayesian Networks, Dependency Network Structure	Windows	Open Source (non-commercial)	https://www.microsoft.com/en-us/research/project/winmine-toolkit/

Appendix F

Example Challenge Problems

This appendix provides more details on the domain-independent and domain-dependent challenge problems highlighted in section 6.3.

F.1 Domain-Independent Challenge Problems

Three specific domain-independent challenge problems are described. They include internal representation (how the autonomous system [AS] represents its world model), dynamic agent-to-agent communications, and a theory of consciousness with an accompanying computational framework. They capture three topics relevant to problems that need to be addressed in ASs, but are also relevant for non-AS-based solutions.

F.1.1 Internal Representation

Internal representation, discussed earlier in sections 2.2 and 3.1, concerns how the autonomous system structures what it knows and what it can use to generate meaning about the battlespace. It includes the data structures and algorithms used for learning and for system execution. For example, an artificial neural network (ANN) has nodes that are connected with weights, as well as algorithms for updating the weights and algorithms for predicting an output (e.g., category) based on some observable. Although the application of ANNs and their deep network counterparts can provide significant improvements to machine-based classification of objects and events captured in full motion video, they may not suffice for other classes of problems such as planning and re-planning based on dynamically-changing mission priorities and resources. There is an inherent tension between the internal representation, the stimuli that can be appropriately responded to, and the selection and timing of actions in the response. Accordingly, we believe that internal representation is a fertile area for basic research, applicable across a wide variety of domains and missions.

F.1.2 Dynamic Agent Communications

A central theme in this document is agency of autonomous systems: how they are situated in the world (section 2.1), how bringing multiple ASs together can create emergent behaviors (section 2.3), and how they can learn with the experience of the individual and the group (section 2.4). Communication is key in all these activities. Dynamic agent communications mean that the AS will learn how to communicate its meaning with other ASs using a dynamic, not static or predefined, symbol set. This is necessary in an environment in which agents come and go based on changing tasks being worked on, peer relationships, or cognitive approaches used, so that a fixed set of communications symbols will not suffice.

Historically, a significant amount of effort is spent defining communications protocols and messaging for multiagent systems; this can inadvertently limit the impact the AS has delivering combat effects. Assael suggests that such learned communications protocols can be of the following four forms (2016):

- Centralized learning/centralized execution
- Centralized learning/decentralized execution
- Decentralized learning/centralized execution
- Decentralized learning/decentralized execution

Centralized learning with centralized execution is of little interest for ASs since it can be accomplished by design or through relational learning approaches. Centralized learning with decentralized execution as accomplished in Assael (2016) could be useful in those cases where we know the participants and they remain fixed. Decentralized learning with centralized execution is useful for dynamically learning a protocol, but AS use in operational scenarios is too limited if centralized execution is required. The approach to learned protocols for ASs needs to follow the decentralized learning/decentralized execution paradigm, because not all agent participants will be available at the onset of learning and centralized execution involves too much overhead and inefficiency. As a result, the type of dynamic agent communications we envision for ASs follows the decentralized learning and execution paradigm. There is a need for research on this topic to advance the state of the art with a particular concentration on heterogeneous agents and scalability. There is also a significant need for developing an open and scalable agent communications framework that supports, at least, decentralized learning and execution, although a framework that covers the other communications paradigms could be of significant value to the broader community.

F.1.3 Theory of Consciousness and a Conscious Computing Framework

In section 2.2, we described human and machine consciousness and emphasized that some aspect of the representation of an AS must provide meaning for observations that have never been experienced before. To do this requires a representation that is stable, consistent, useful, and structurally coherent (Lynott and Connell 2010; Yeh and Barsalou 2006; Cardell-Oliver and Liu 2010; Shanahan 1996; Newell 1990; Barsalou 2013). It is a confabulated, yet cohesive, narrative that compliments an experientially learned component (as we described in section 2.4) or subconsciously held component (Vaughan et al. 2014; Lakoff and Narayanan 2010; Rogers 2018). Qualia (plural) are the vocabulary of consciousness (Cowell 2001; Chalmers et al. 1992; Chopra and Kafatos 2014; Hubbard 1996; Rogers 2018) and are the units of conscious cognition. A quale (singular) is the "what" that is evoked in working memory and that is being attended to by the agent as part of its conscious deliberation. A quale could be experienced as a whole when attended to in working memory and is experienced based on how it relates to, and can interact with, other qualia. When the source of the stimulus being attended to is the agent itself, the quale of "self" is evoked to provide self-awareness. An agent that can generate the quale of self can act as an evaluating agent to itself or as a performing agent with respect to some task based on some observable. An agent that can generate the quale of self can determine when it should continue functioning; give itself its own proxy, versus stopping the response and seeking assistance. Ramachandran suggested the existence of three qualia laws: qualia are irrevocable, qualia are flexible with respect to the output, and qualia are buffering (Ramachandran and Hirstein 1997). Since the generation of qualia is used as the defining characteristic of consciousness, it is possible to use the work of Ramachandran as a vector in devising a *theory of consciousness*. Such a theory could also have a set of tenets to define the engineering characteristics for an artificial conscious representation for an AS. An important aspect for a *theory of consciousness* is the construct of Edelman's imagined[1] present, imagined past, and imagined future (1989). That is to say, much of what the agent knows is simulated, seeded by experiences. Qualia represents one approach to developing a *theory of consciousness*, but others exist and should be explored.

We propose that the science and technology (S&T) community flesh out a *theory of consciousness* that can be used to support the development of *artificial consciousness* for an "aware" AS, and equally important, an associated *conscious computing framework*.[2] The ultimate goal of this theory would be a

---

1. In more engineering-focused parlance, simulated.

2. We do not suggest that it is possible at this time to create an artificial general intelligence (Office of Net Assessment 2016). But we do believe that we will achieve a behavioral and computational advantage by pursing a theory of consciousness and an associated computational framework.

simple and elegant set of fundamental laws, analogous to the fundamental laws of physics, something akin to:

- Providing *structural coherence* of an AS's internal representation, ensuring that an AS's interaction with the world is stable, consistent, and useful.

- Providing for *situation-based* processing through the unit of conscious cognition, such as Qualia Theory of Relativity, narratives, and so forth.

- Providing for *conscious representation of situations via simulation*, in which these cognitively decoupled processes can deliberate over an imagined past, imagined present, and imagined future, in the form of a cohesive narrative.

It is anticipated that such a focus would yield a behavioral, and possibly a computational, advantage in AS development, and provide a mechanism to address the unexpected query or event in a fashion that supports true AS flexibility across tasks taken on, peer relationships engaged in, and cognitive approaches used. We believe strongly that this is foundational to achieve the needed AS flexibilities we have discussed earlier.

F.2 Domain-Dependent Challenge Problems

Five specific, domain-dependent challenge problems are described. They include multidomain situational awareness (MDSA); an MDSA operational framework; intelligence, surveillance, and reconnaissance (ISR) processing, exploitation, and dissemination (PED) for narrative generation; multidomain situated consciousness (MDSC); and data-to-decisions air-to-air mission effect chain (MEC).

F.2.1 Multidomain Situational Awareness

The Air Force Chief of Staff recently stated the importance of MDSA and the need for improving current and future Air Force decision-making at the tactical, operational, and strategic levels (Goldfein 2017). SA was covered extensively in section 3.2 and included a common single agent and team SA models based on the perception of the elements in one's environment, comprehending their meaning, and projecting their status into the future (Endsley 1995b). Current SA approaches are linear: data capture, data analysis, knowledge product generation, knowledge product dissemination, followed by decision-making, if called for to affect the environment. Unfortunately, this linear relationship between capture to analysis to decision-making for effects does not scale (Rogers et. al 2014) and is not dynamic enough in battlespaces where the adversary is agile. This approach has too much time lag and is often incapable of supporting timely decision-quality knowledge creation. Current efforts are also stovepiped across domains: they independently handle their domains, fusing across domains after the fact (see earlier discussion in section 5.3.3). The nature of unfolding real-world events across multiple domains (e.g., simultaneous missile and cyberattacks), with an agile and capable adversary, will not be adequately addressed by our linear stovepiped architectures. The key to MDSA is timely decision-quality knowledge. Autonomous systems, working in parallel, across domains, and in a non-linear fashion when called for, can provide the flexible and responsive knowledge management needed by today's decision-making warfighter.

The MDSA problem itself is monolithic and has numerous challenges to overcome. Some of those challenges include understanding the complexities of current operations, changing how we think about SA (that is, moving away from linear-staged processing and domain stovepiping), the span of AS technologies that can be brought to bear on the MDSA problem, and the information technology infrastructure to support the flexible employment of ASs. A key to achieving MDSA will be organizational as well, unifying the relevant functions (e.g., A1, A2, A3, A4, A5/8, A9, A10) in an information-centric environment.

F.2.2 MDSA Operational Framework

Understanding MDSA from one or more relevant operational perspectives would be of great value to operators and to the S&T communities. One perspective could be from the Joint Forces Air Component Commander (JFACC) and his or her staff in an Air Operations Center (AOC). How does the JFACC think about how air, space, and cyber help him/her fight the war? How do the Cyber Operations Centers tightly couple with the J2 to ensure all operational circuits are up and running that support current operations and near-term future operations?

Such a framework should consider MDSA at the strategic, operational, and tactical levels of war and consider an approach that uses experts in both operations (user) and technology (producer) to define that framework (so-called user-producer innovation). This is not a foreign proposition for the operational/technology community; the Defense Information Systems Agency (DISA) has accomplished similar, yet significantly less complex, frameworks for limited cyber SA functions. We believe that the formulation of an MDSA-operational framework would not only be a worthy challenge problem for the S&T community but, if successful, would also serve to deliver additional lethality to the joint fight.

F.2.3 ISR PED for Narrative Generation

A core mission of HAF/A2 is the ISR PED problem. The USAF Distributed Common Ground System (DCGS) is a key component supporting that mission. Much of the processing in ISR PED is manual, and although efforts to bring in automation are currently underway and of great value (e.g., the Air Force Research Lab/RHXB Human Detection & Characterization toolsets), they only begin to scratch the surface when they provide object detection and object characterization.

How ISR PED products should be produced and delivered is different depending on the consumer of those products. For example, a special operator (at the tactical level) on the ground cares about possible immediate threats. A cyber operator (at the operational level) might care about possible threats to his/her infrastructure supporting current operations. A Coalition Forces Commander (at the strategic level) might be concerned about the readiness condition of their air, space, and cyber assets in a particular theater. Thus, ISR PED product resolution, over time and space, will depend on the ultimate consumer of the product.

More concretely, this challenge problem is interested in entity identification (as an example, dogs, cats, horses), a characterization of those entities (German Shepherd, calico, mustang), and their interactions (the animals are playing in the field) that manifest into a comprehensive narrative intended for human-level understanding. Coupled with the context of the mission or task, other objects, their characterizations, and their interactions would be dynamically specified, based on that context. Such an approach would enable one to define and describe events and provide mechanisms to do non-causal event detection where evidence might occur out of order, might be retracted, and might occur over long periods of time. Delivering the resultant narrative to the product consumer at the right level of resolution will be a key challenge.

An event is an important concept; it is any situation that requires attention, and the detection of said events is the focus of this challenge problem. When entity interactions can be understood as a whole, they are situations. When the situation is of interest and requires attention, it is an event. With such a capability, the system could provide an analyst an idea of what to look for next when watching a potential event unfold. A challenge problem of this sort could consider any number of scenarios. A few include: car following and determining who interacts with the car of interest, counting the number of men, women, and children (slant count), characterizing their interactions and possible roles to determine possible impacts of ongoing operations, etc. Many of these will be useful in exploring and defining the general ISR PED problem.

F.2.4 Multidomain Situated Consciousness

We can extend our earlier discussion of artificial consciousness in section F.1.3 from a domain-independent challenge problem to a domain-dependent one by focusing on a new way of thinking about MDSA. Awareness is a measure of the mutual information between reality and the internal representation of some performing agent (human or AS) as deemed by some other evaluating agent. Consciousness is the content of working memory that is being attended to. To illustrate the difference, consider a patient with blind sight, who has lost the visual cortex in both hemispheres of the brain and so has no conscious visual representation (Celesia 2010). These patients, when asked what they see, say they see nothing and that the world is black. However, when they are asked to walk where objects have been placed in their path, they often successfully dodge those objects. What is happening here is that the verbal questioning is calling on the patient to use information that is *consciously available* to the patients, and there is none because of the lack of visual cortex. In contrast, successfully navigating through objects placed their path is calling on the patient to use visual *awareness* of obstacles, something that can be maintained unconsciously. Similarly, body identity integrity disorder and alien hand syndrome (AHS; Blom 2012) are examples of issues that illustrate low awareness while the patient is conscious of the appendages (Sarva 2014). Paraphrasing Albert Einstein, "imagination is more important than knowledge," we claim that consciousness may often more important than awareness. There will always be limitations to how much of reality can be captured in the internal representation of the agent, but there are no limits to imagination.

MDSA focuses on the perception of multidomain elements in the multidomain environment within a volume of time and space, the comprehension of their meaning, and the projection of their status in the near future (modified from Endsley [1995b]). The concept of SA is intimately tied to the mutual information between the internal representation, reality, and awareness. In contrast, MDSC is a stable, consistent, and useful all-source situated simulation that is structurally coherent. This last constraint of being structurally coherent requires the MDSC representation only achieve enough mutual information with reality to maintain stability, consistency, and usefulness. The challenge here is first to establish approaches to SC and then consider the difference in utility to the user compared to a more traditional SA-focused approach.

F.2.5 Data-to-Decisions Air-to-Air Mission Effect Chain

A MEC is defined as: Predict, Prescribe, Find, Fix, Track, Target, Engage, and Assess Anything, Anytime, Anywhere in Any Domain (P2F2T2EA4) (Rogers 2008). An air-to-air MEC executes P2F2T2EA4 in an air-centric mission to enable joint force air superiority in the highly contested environment (USAF ECCT 2016).

Counterair operations are designed to gain control of the air and wrest such control away from an adversary. Air superiority is a condition on the spectrum of air control, which ranges from adversary air supremacy, to air parity, to friendly air supremacy. The air superiority condition is achieved when friendly operations can proceed without prohibitive interference from opposing forces. In modern military operations, achieving this level of control of the air is a critical precondition for success. Air superiority provides freedom from attack, freedom to attack, freedom of action, freedom of access, and freedom of awareness. Importantly, it also precludes adversaries from exploiting similar advantages. As such, air superiority underwrites the full spectrum of joint military operations and provides an asymmetric advantage to friendly forces. A lack of air superiority significantly increases the risk of joint force mission failure as well as the cost to achieve victory both in terms of resources and loss of life. In common discourse, air superiority is often envisioned as a theater-wide condition. In highly contested environments, such a conception may be unrealistic and unnecessary. Air superiority is only needed for the time and over the geographic area required to enable joint operations. The specific amount of time and space required varies significantly across scenarios, mission objectives, and phases of conflict. Accordingly, capability development for air superiority must provide options for commanders to array their forces across a range of durations and geographies.

This challenge problem seeks to employ ASs to gather data from all sources in all domains, rapidly analyzing the data to extract operationally important information and reliably distributing information on the timeline needed to enable critical decisions to create an asymmetric advantage. Here, the goal is to bring to bear all the previously mentioned assets to address any portion of the MEC in support of the counterair mission to the pilot in the cockpit.

Standoff ranges imposed by area denial capabilities degrade the effectiveness of long-range sensors in the highly contested environment, making the counter air mission more volatile. This highlights the importance of all-source intelligence and an ability to bring it to bear inside of the cockpit.

Appendix G

Example Technical Integration Experiments

As part of the prototyping and experimentation process described in section 6.4, we envision conducting a number of technical integration experiments (TIE), illustrated schematically in figure G.1. The goal is to iteratively address one or more of the challenge problems identified in section 6.3 by pushing the boundaries of the creation and representation of the knowledge necessary for autonomous system (AS) flexibility across task, peer, and cognitive dimensions. The diagram shows that there is a minimal amount of development prior to a demonstrable product, as indicated by the blue dashed circle and the blue triangle after TIE_3. Each TIE thereafter may or may not include a demonstration, depending on the goals. The key issue then is to define the TIEs in a meaningful way to deliver operationally useful capabilities as different challenge problems are addressed, while extending the underlying knowledge base and functionality of the AS framework.

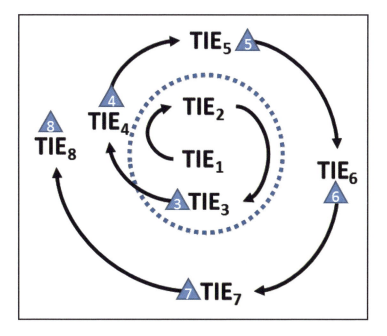

Figure G.1. A series of technical integration experiments with demonstration events (blue triangles) for developing a variety of AS applications

A potential TIE approach for the spiral development of a variety of AS applications—applications that iteratively build on one another—could follow the following development path:

TIE_1—**Fixed Assembly**: Create a set of agents needed to, for example, address the intelligence, surveillance, and reconnaissance (ISR) processing, exploitation, and dissemination (PED) event detection problem (appendix F.2.3) for full-motion video (FMV), and then wire them together in a predefined way. This will necessitate the creation of an initial ontology that can guide future development efforts.

TIE_2—**Cognitive Flexibility**: Establish the basics for cognitive flexibility by enabling some limited flexible assembly of agents for the same video captioning problem. Here, each agent defines its inputs and outputs, which is still an ontology, but they dynamically build their connectivity. The structure could be learned using, for example, relational learning.

TIE_3—**Peer Flexibility**: Add limited peer flexibility by relieving the ontology need by learning communications protocols through centralized learning methods (appendix F.1.2) such as that in Assael (2016). In a fashion similar to that used in TIE_2, the structure could be learned using relational learning.

TIE_4—**Limited Technology Demonstration**: TIE_3 would demonstrate the baseline AS architecture applied to the ISR PED event detection problem (appendix F.2.3) for FMV. It would demonstrate the compartmentalization of different deep-learning networks, add an image-converting agent, and provide alternative network models with different video to text agents used for event detection.

TIE_5—**Task Flexibility**: Begin to explore task flexibility by allowing agents to self-assemble to accomplish a single task.

TIE_6—**Limited Technology Demonstration**: Demonstrate the ability of agent self-assembly by applying it to the ISR PED event detection problem (appendix F.2.3) for FMV where agents assemble to convert different data into a usable form to do the event detection task.

TIE_7—**Data to Decisions, Air-to-Air Mission Effect Chain**: Extract critical processing chains associated with the Air Force Research Lab Sensors Directorate Multi-Source Analytic Development and Evaluation (MAD-E) software (used to reason over multiple data sources to create operationally relevant enemy courses of action [COA]) and the Defense Advanced Research Projects Agency Deep Exploration and Filtering of Text components to agents. This will provide a mechanism to bring all source intelligence based on relevant mission parameters, create enemy and friendly force COAs, and present it to the decision maker for the air-to-air mission-effect chain problem (appendix F.2.5). This TIE requires the addition of a human interface to inject mission intent to serve as a guide for agents to self-assemble for task completion and use multiple cognitive approaches.

TIE_8—**Limited Technology Demonstration**: Process video and textual data, integrate team intent, and, based upon that team intent, demonstrate the ability to provide COAs based on all source intelligence and current mission data.

TIE_9—**Artificial Consciousness**: Add a series of processes that enables the system to simulate over the possible past, possible present, and possible future, so it can evaluate the meaning of stimuli the system has never experienced, where experience comes from the system interacting with the battle space (appendix F.1.3).

TIE_{10}—**Limited Technology Demonstration**: Demonstrate an ability to improve peer, task, and cognitive flexibility for the ISR PED event detection problem (appendix F.2.3) for FMV and the air-to-air mission effect chain problem (appendix F.2.5).

TIE_{11}—**Autonomous Agent**: Demonstrate advances in an ASs internal representation (appendix F.1.1) and artificial consciousness (appendix F.1.3).

TIE_{12}—**Limited Technology Demonstration**: Apply autonomous agents to multidomain situated consciousness (appendix F.2.4).

Appendix H

Autonomous System Vignette

We present here a brief vignette that includes both "at rest" and "in motion" autonomous system (AS) assets and that progresses from an intelligence, surveillance, and reconnaissance (ISR) mission, to an area defense mission, and, finally, to a humanitarian recovery mission. The vignette displays key AS task, peer, and cognitive flexibilities introduced in chapter 1.

Due to political tensions, the UN has established a no-fly zone along the Russian border with the independent country of Crimea. The United States has provided a team of "in motion" autonomous unmanned air vehicles[1] (AUAV) to patrol the border and provide persistent intelligence gathering. The AUAVs can conduct 24-hour surveillance and provide instantaneous reports on border activities, to include troop activities (threat identification and build-up), village/civilian patterns-of-life, and any irregular movement of goods and supplies up to and over the border. The data gathered from these ISR activities feeds into the Global Integrated ISR (GIISR) database envisioned in the Air Force Future Operating Concept (AFFOC; USAF AFFOC 2016), completing the picture of the Russian social, economic, and military status. The border activities in this small contested area, combined with the ISR data repository from many other autonomous AUAV teams along and within the borders, combined with cyber and space monitoring, creates a complete picture that is synthesized and exploited at the Multi-Domain Operations Center (MDOC). A variety of collected data (full-motion video [FMV], Signals Intelligence [SIGINT], etc.) is streamed to the MDOC, where the signals are processed and analyzed for indicators of possible/probable future violations by Russia, calling on a capability for model-based intent inferencing by another AS, this one operating "at rest" at the MDOC. The ISR overwatch is planned and synchronized by the autonomous AUAVs, coordinating overlapping intelligence gathering with each other and other manned assets in the area. The teams of AUAVs can take direction from the MDOC and also query the other AUAVs/manned aircraft in peer-to-peer relationships to verify detected activities and/or contribute sensor information to instances of anomaly detections.

Where a likely but unproved transgression is imminent, another "at rest" AS will notify the MDOC of the probable event and provide several COAs with probability of success (and losses), given the available multidomain ca-

---

1. Unlike today's heavily manned UAVs.

pabilities for human action. On the outset of an unpreventable attack, a part of the AUAV overwatch team would switch from an ISR mission to a defensive posture while other sections of the team would continue to gather, analyze, and stream ISR data, in a fashion that exhibits the kind of AS task flexibility discussed earlier. The AUAV team lead would instantly notify the MDOC and take countermeasures using the right mix of available assets from interconnected domains, calling on the kind of adaptive domain control (ADC) envisioned in the AFFOC. For example, the autonomous AUAV team may call for a cyberattack to disable a nearby radar station, employ electronic warfare countermeasures, and call for an offshore naval missile barrage to counter the incoming transgression. When one countermeasure is defeated or fails, then the AS team can dynamically employ a new mix of tactics across several domains to achieve the desired kinetic results and stop the incursion, exhibiting a form of operationally focused cognitive flexibility in problem-solving. All COAs involving joint assets—and/or the deployment of munitions—would likely require human approval from the MDOC, to keep the human "on-the loop" and duly informed.

In this vignette, the attack includes surface-to-surface missiles, causing a great deal of damage to a nearby city with large civilian casualties. As the attack is being repelled, parts of the AUAV team will conduct a damage assessment, looking for survivors, categorizing triage status, and assessing the environment. As civilian survivor needs are being estimated, the AUAV sensors are also assessing the air quality (detecting airborne contaminants); conducting a chemical, biological, radiological, and nuclear sweep; and testing the water purity. As this information is gathered, the autonomous AUAV team takes inventory of what materials are in the area, which items can be sourced nearby (allied cities, offshore Navy, etc.), and which will need to be brought from the United States. The materials that need to come from the stateside warehouses (tents, medical supplies, water, cots, food, etc.) will be automatically retrieved and gathered from the closest air base (i.e., Dover AFB) using robotic movers, providing the AFFOC's vision of rapid global mobility. As the materials are gathered (think Amazon warehouse) and palletized on the flight line, the next available aircrew is being notified and briefed on the mission (if an aircrew is even required!). The needed materials and support personnel/equipment are sourced from nearby and distant United States installations and those of allied countries, all while keeping the MDOC rapidly informed of the plans and outcomes. As the materials arrive at the recovery scene, teams of AUAVs will continue defensive measure as other teams of AUAVs and autonomous unmanned ground vehicles organize and deliver materials to the areas of need.

Appendix I

Test and Evaluation of Autonomous Systems

The primary objective of test and evaluation (T&E) is to ensure sufficient system safety and performance. Toward this end, the high-level goals of T&E are to capture system safety and performance requirements in the context of relevant operational environments, to use these requirements to guide system development, and to provide compelling evidence that the final system implementation meets these requirements. T&E activities in support of these goals should be performed throughout the entire system lifecycle, starting from the very beginning.

Conceptually, there is no reason these types of T&E activities cannot be performed for autonomous systems (AS), but in practice, ASs have unique features that make these activities challenging. To illustrate this point, consider standard processes for assuring system safety (SAE 1996). These processes involve identifying hazards that could occur in operational environments, identifying combinations of system faults that could lead to them, and designing the system to bring the risk of hazards to within an acceptable probability bound. For physical components, the primary source of evidence that this has been achieved is test. However, test alone would be intractable for very tight probability bounds, and so it is often supplemented by arguments based on design choices. A common example is redundancy; assuming copies of a component fail independently and only one is needed for correct system operation, the probability bound for each individual copy is loosened. Note this argument implicitly relies on first principles of physics, for example a damaging input received by one component has no effect on redundant copies. A major challenge for ASs stems from software, which does not follow the same first principles. In fact, redundant software copies will fail simultaneously under the same conditions, since the fault is due to a logic error rather than a physical failure. Without analogous first principles for software and more generally for ASs, the burden of evidence shifts back to test, which is intractable (Kalra and Paddock 2016). The problem is exacerbated by the complexity of operational environments in which ASs are anticipated to operate and the levels of flexibility they are expected to incorporate.

T&E of autonomous systems will therefore require significant advances in several areas (DOD R&E 2015). Toward first principles for ASs, more expressive mathematical frameworks are needed to formally capture richer requirements and model more diverse system behaviors. To address increased system complexity, these frameworks must enable rapid, transparent, and automated or semi-automated analysis or even design *synthesis* to ensure system behaviors satisfy requirements and that requirements are complete and consistent, e.g. as in *formal methods* approaches in computer science. To support system flexibility and enable analysis of complex systems-of-systems, methods to decompose systems and compositionally reason about high-level behaviors based on component-level interactions are also needed, for example as in *assume-guarantee* contracts and as supported by certain *architecture description languages*. For components that cannot be sufficiently verified through analysis, *runtime assurance* methods provide a possible alternative. For instance, *runtime verification* approaches can be used to detect and help diagnose faults generated by a system during operation, and *runtime enforcement* approaches can additionally override erroneous behaviors, similar to the way in which a Ground Collision Avoidance System takes temporary control of an aircraft when a collision is imminent and returns control to the pilot when it is safe. In addition to these newer approaches, more traditional approaches rooted in modeling and simulation (M&S) and legacy T&E will continue to serve an invaluable role. However, richer and more flexible M&S frameworks are needed to address increasingly complex system behaviors and dynamic operational environments. Furthermore, T&E must be made more efficient by employing a more sequential progressive testing approach throughout the system lifecycle, and improved statistical engineering methods are needed to support developmental and operational testing of nondeterministic systems that operate in dynamic environments (Ahner and Parson 2016). Finally, since T&E of autonomous systems will require a combination of *all* these approaches, new methods are necessary for building and comprehending *assurance cases* that capture networks of interrelated arguments and supporting evidence.

The various facets of autonomous system flexibility, which we have discussed throughout this report, will result in additional challenges for many of these approaches:

- *Peer flexibility* will require new forms of requirements that capture the high-level responsibilities of different roles in an operational environment. For formal methods and compositional reasoning approaches (Bolton et al. 2013; Johnson et al. 2011), it will also require new forms of models that capture system and human capabilities to enable identification of the conditions under which these roles should be performed, to derive more specific low-level responsibilities in response to changes in the environment or the organization, and to verify that certain critical high-level responsibilities are fulfilled under all conditions. Runtime assurance approaches for peer flexibility will also require human and AS models that can be used to determine when one is acting erroneously or is likely to fail and provide feedback on the failure in a timely manner so that the other can address the problem. M&S and T&E approaches will need to be expanded to better address interactions between ASs and humans.

- *Task flexibility* will require new forms of requirements that capture high-level mission goals and constraints. For formal methods, compositional reasoning, and runtime assurance approaches, it will also require models of sensors, system capabilities, and the predicted effects actions have in dynamic environments to enable verification that critical high-level requirements are satisfied. These models will have to address increased levels of uncertainty and reactivity, so methods for formulating and efficiently analyzing models of probabilistic and reactive systems are needed (Kwiatkowska et al. 2002; Manna and Pnueli 1995). Models will also have to address heterogeneous combinations of hardware and software, for example, as in hybrid systems and embedded systems (Alur et al. 1993). M&S and T&E approaches will need to be expanded to support these more complex models. As peer and cognitive flexibility also often involve addressing uncertainty, reactivity, and heterogeneity, these needs likely apply to those facets of autonomy as well.

- *Cognitive flexibility* will require new forms of high-level mission requirements like those needed for task flexibility. Formal methods for cognitive flexibility will additionally require models of the learning process and formal representations of learned knowledge, for example to verify that learned behaviors, decision boundaries, and rules are internally consistent and will not result in the violation of critical requirements based on what is known about the environment. For learning and reasoning approaches that are not amenable to full analysis through formal methods (e.g., artificial neural networks), bounding or proving certain high-level properties such as the range of possible inputs and outputs, or convergence of the learning algorithm, can help provide basic component-level guarantees or characterizations of acceptable behavior that can be used in compositional reasoning and runtime assurance approaches (Schumann et al. 2003; Russell et al. 2015).

Overall, given the complexity of relationships between components in future autonomy systems, assurance cases will play a key role in T&E. From past research on emergent systems and complexity theory, interactions between two systems can result in surprise (Ronald et al. 2007). Since these surprises are not known *a priori* and may be nonreducible—that is, there may be no way to identify a root cause—we will need to be flexible in how we deal with them. Assurance cases will allow us to incrementally build arguments that an AS is sufficiently safe and effective as new surprises are discovered, using whatever approaches are most appropriate given the nature of the surprise.

References

"5 Cloud Computing Trends to Prepare for in 2018." *Network World*, October 15, 2018. https://www.networkworld.com/article/3233134/cloud-computing/5-cloud-computing-trends-to-prepare-for-in-2018.html.

"210-Link-C-3-Flight-Trainer.Pdf." Accessed October 27, 2018. https://www.asme.org/wwwasmeorg/media/ResourceFiles/AboutASME/Who%20We%20Are/Engineering%20History/Landmarks/210-Link-C-3-Flight-Trainer.pdf.

"A Better Thermostat," Carnegie Mellon University (website), October 19, 2018. https://www.cmu.edu/homepage/environment/2012/winter/the-nest-thermostat.shtml.

Aamondt, Agnar, and Enric Plaza. 1994. "Case-Based Reasoning: Foundational Issues, Methodological Variations, and System Approaches." *Artificial Intelligence Communications* 7 (1): 39–59. https://doi.org/10.3233/AIC-1994-7104.

Accenture. 2016. "Intelligent Automation: The Essential New Co-Worker for the Digital Age." *Technology Vision 2016*, n.d. https://www.accenture.com/t20160125T111718__w__/us-en/_acnmedia/Accenture/Omobono/TechnologyVision/pdf/Intelligent-Automation-Technology-Vision-2016.pdf#zoom=50.

Ackley, David H., Geoffrey E. Hinton, and Terrence J. Sejnowski. 1985. "A Learning Algorithm for Boltzmann Machines." *Cognitive* Science 9, no. 1 (January): 147–69. https://doi.org/10.1016/S0364-0213(85)80012-4.

Adolphs, Ralph. 2002. "Trust in the Brain." *Nature Neuroscience* 5, no. 3 (March): 192–93. https://doi.org/10.1038/nn0302-192.

Aggarwal, Charu C. 2016. *Data Mining* - The Textbook. New York: Springer. https://www.springer.com/us/book/9783319141411.

Agrawal, Divyakant, Philip Bernstein, Elisa Bertino, Susan Davidson, Umeshwas Dayal, Michael Franklin, Johannes Gehrke, et al. 2011. "Challenges and Opportunities with Big Data 2011-1." In *Cyber Center Technical Reports,* 2011. https://docs.lib.purdue.edu/cctech/1.

Ahner, Darryl K., and Carl R. Parson. 2016. "Workshop Report: Test and Evaluation of Autonomous Systems." Wright-Patterson AFB, OH: Air Force Institute of Technology, October 22, 2016. https://www.afit.edu/stat/statcoe_files/Workshop%20Report%20Test%20and%20Evaluation%20of%20Autonomous%20Systems%20V2.pdf.

Allen, Greg, and Taniel Chan. 2017. *Artificial Intelligence and National Security*. Cambridge, MA: Belfer Center for Science and International Affairs. https://www.belfercenter.org/sites/default/files/files/publication/AI%20NatSec%20-%20final.pdf.

Allen, James, and James Hendler. 1990. *Readings in Planning*, edited by Austin Tate. San Mateo, CA: Morgan Kaufmann Publishers.

Alonso, Jose-Manuel, and Yao Chen. 2008. "Receptive Field." *Scholarpedia* 4, no. 1 (January): 5393. https://doi.org/10.4249/scholarpedia.5393.

Alur, Rajeev, Costas Courcoubetis, Thomas A. Henzinger, and Pei-Hsin Ho. 1993. "Hybrid Automata: An Algorithmic Approach to the Specification and Verification of *Hybrid Systems*." In Hybrid Systems, edited by Robert L. Grossman, Anil Nerode, Anders P. Ravn, and Hans Rischel, 209–29. Lecture Notes in Computer Science. Berlin: Springer. https://doi.org/10.1007/3-540-57318-6_30.

American Society of Mechanical Engineers (ASME) History and Heritage Committee. 2000. *The Link Flight Trainer: A Historic Mechanical Engineering Landmark*. New York: ASME International Retrieved from ASME: https://www.asme.org/getmedia/d75b81fd-83e8-4458-aba7-166a87d35811/210-Link-C-3-Flight-Trainer.aspx.

Amodei, Dario, Chris Olah, Jacob Steinhardt, Paul Christiano, John Schulman, and Dan Mané. 2016. "Concrete Problems in AI Safety." *ArXiv:1606.06565 [Cs]*, June 21, 2016. http://arxiv.org/abs/1606.06565.

Anaya, Ivan Dario Paez, Viliam Simko, Johann Bourcier, Noël Plouzeau, and Jean-Marc Jézéquel. 2014. "A Prediction-Driven Adaptation Approach for Self-Adaptive Sensor Networks." In *Proceedings of the 9th International Symposium on Software Engineering for Adaptive and Self-Managing Systems - SEAMS 2014*, 145–54. Hyderabad, India: ACM Press, 2014. https://doi.org/10.1145/2593929.2593941.

Anderson, James A. 1972. "A Simple Neural Network Generating an Interactive Memory." *Mathematical Biosciences* 14, no. 3–4 (August): 197–220. https://doi.org/10.1016/0025-5564(72)90075-2.

———. 1988. "Neurocomputing: Foundations of Research." edited by James A. Anderson and Edward Rosenfeld. Cambridge, MA: MIT Press. 181–192. http://dl.acm.org/citation.cfm?id=65669.104398.

Anderson, John R. 1983. *The Architecture of Cognition*. Cambridge, MA: Harvard University Press.

Anderson, John R., Daniel Bothell, Michael D. Byrne, Scott Douglass, Christian Lebiere, and Yulin Qin. 2004. "An Integrated Theory of the Mind." *Psychological Review* 111 (4): 1036–60. https://doi.org/10.1037/0033-295X.111.4.1036.

Anderson, P. W. 1972. "More Is Different." Science 177, no. 4047 (August): 393–96. https://doi.org/10.1126/science.177.4047.393.

Andrews, Robert, Joachim Diederich, and Alan B. Tickle. 1995. "Survey and Critique of Techniques for Extracting Rules from Trained Artificial Neural

Networks." *Knowledge-Based Systems, Knowledge-Based Neural Networks 8,* no. 6 (December): 373–89. https://doi.org/10.1016/0950-7051(96)81920-4.

Anthes, Gary. 2017. "Artificial Intelligence Poised to Ride a New Wave." *Communications of the ACM 60,* no. 7 (July): 19–21.

Aoki, Masanao. 1971. *Introduction to Optimization Techniques: Fundamentals and Applications of Nonlinear Programming.* 1st ed. (Macmillan Series in Applied Computer Sciences. New York: Macmillan.

Arkin, Ronald C. 1989. "Motor Schema-Based Mobile Robot Navigation." *International Journal of Robotics Research 8,* no. 4 (August): 92–112. https://doi.org/10.1177%2F027836498900800406.

———. 1998. *Behavior-Based Robotics.* Cambridge, MA: MIT Press.

Arquilla, John, and Duncan A. Buell. 2015. "The Dangers of Military Robots, the Risks of Online Voting." *Communications of the ACM* 58, no. 7 (July): 12–13.

Ashby, W. Ross. 1956. *An Introduction to Cybernetics.* London: Chapman & Hall, 1957. http://dspace.utalca.cl/handle/1950/6344.

Assael, Y.M., Foerster, J.N., Freitas, N.D., and Whiteson, S. (2016). Learning to Communicate with Deep Multi-Agent Reinforcement Learning. NIPS.

Atahary, Tanvir, Tarek Taha, Fredrick Webber, and Scott Douglass. 2015. "Knowledge Mining for Cognitive Agents through Path Based Forward Checking." In *2015 IEEE/ACIS 16th International Conference on Software Engineering, Artificial Intelligence, Networking and Parallel/Distributed Computing (SNPD),* 1–8. Takamatsu, Japan. https://doi.org/10.1109/SNPD.2015.7176169.

Aumann, R. J. 1989. "*Game Theory.*" In Game Theory, edited by John Eatwell, Murray Milgate, and Peter Newman, 1–53. London: Palgrave Macmillan.

Bäck, Thomas. 1996. *Evolutionary Algorithms in Theory and Practice Evolution Strategies, Evolutionary Programming, Genetic Algorithms.* New York: Oxford University Press.

Bains, Paul. 2006. *The Primacy of Semiosis: An Ontology of Relations.* Toronto Studies in Semiotics and Communication. Toronto: University of Toronto Press.

Bajcsy, Ruzena. 1988. "Active Perception." *Proceedings of the IEEE,* 76(8): 966–1005.

Balch, T., and Ronald C. Arkin. 1998. "Behavior-Based Formation Control for Multirobot Teams." *IEEE Transactions on Robotics and Automation* 14 (6): 926–39. https://doi.org/10.1109/70.736776.

Baldwin, P., and M. Talbert. 2005. "INFOCUS: Information Correlation of UAV Sensors." First International Innovations and Real-time Applications of Distributed Sensor Networks (DSN) Symposium (electronic proceedings). Bethesda, MD.

Balestrini-Robinson, Santiago, Jack M. Zentner, and Tommer R. Ender. 2009. "On Modeling and Simulation Methods for Capturing Emergent Behaviors for Systems of Systems." San Diego, CA: National Defense Industrial Association.

Banks, S. B., and C. S. Lizza. 1991. "Pilot's Associate: A Cooperative, Knowledge-Based System Application." *IEEE Expert 6*, no. 3 (June): 18–29. https://doi.org/10.1109/64.87681.

Baomar, H., and P. J. Bentley. 2016. "An Intelligent Autopilot System That Learns Piloting Skills from Human Pilots by Imitation." In *2016 International Conference on Unmanned Aircraft Systems (ICUAS)*, 1023–31. https://doi.org/10.1109/ICUAS.2016.7502578.

Barlow, Horace B. 1995. The Neuron in Perception." In *The Cognitive Neurosciences*, edited by Michael S. Gazzaniga. Cambridge, MA: MIT Press.

Barlow Jr., R. B., and E. Kaplan. 1977. "Properties of Visual Cells in the Lateral Eye of Limulus in situ: Intracellular recordings." *Journal of General Physiology* 69(2): 203–20.

Barr, Avron, and Edward Albert Feigenbaum. 1981. *The Handbook of Artificial Intelligence*. Vol. 1. Los Altos, CA: Kaufmann.

Barsalou, Lawrence W. 2013. "Mirroring as Pattern Completion Inferences within Situated Conceptualizations." *Cortex* 49, no. 10 (December): 2951–53. https://doi.org/10.1016/j.cortex.2013.06.010.

Bedau, Mark A. 1997. "Weak Emergence." In *Mind, Causation, and World*. Edited by James E. Tomberlin. Philosophical Perspectives 11. 375–99. Boston: Blackwell Publishers.

———. 2008. "Downward Causation and Autonomy in Weak Emergence." In *Emergence: Contemporary Readings in Philosophy and Science*, edited by Mark A. Bedau and Paul Humphreys. Cambridge, MA: MIT Press. http://mitpress.universitypressscholarship.com/view/10.7551/mitpress/9780262026215.001.0001/upso-9780262026215-chapter-8.

Bedau, Mark A., and Paul Humphreys, eds. *Emergence: Contemporary Readings in Philosophy and Science*. Emergence: Contemporary Readings in Philosophy and Science. Cambridge, MA: MIT Press, 2008. https://doi.org/10.7551/mitpress/9780262026215.001.0001.

Bell, Michael Z. 1985. "Why Expert Systems Fail." *The Journal of the Operational Research Society* 36 (7): 613–19. https://doi.org/10.2307/2582480.

Bengio, Yoshua. 2009. "Learning Deep Architectures for AI." *Foundations and Trends in Machine Learning 2*, no. 1 (January): 1–127. https://doi.org/10.1561/2200000006.

———. 2016. "Springtime for AI: The Rise of Deep Learning." *Scientific American* 314, no. 6 (2016): 46–51. https://www.scientificamerican.com/article/springtime-for-ai-the-rise-of-deep-learning/.

Bernoulli, Daniel. 1954. "Exposition of a New Theory on the Measurement of Risk." *Econometrica 22*, no. 1 (January): 23–36. https://doi.org/10.2307/1909829.

Bertoncello, Michele, and Dominik Wee. 2015. "Ten Ways Autonomous Driving Could Redefine the Automotive World." McKinsey (website). June 2015. https://www.mckinsey.com/industries/automotive-and-assembly/our-insights/ten-ways-autonomous-driving-could-redefine-the-automotive-world.

Billman, Lisa, and Marc Steinberg. 2000. "Human System Performance Metrics for Evaluation of Mixed-Initiative Heterogeneous Autonomous Systems." *In Proceedings of the 2007 Workshop on Performance Metrics for Intelligent Systems*, 120–126. PerMIS '07. New York: ACM. https://doi.org/10.1145/1660877.1660893.

Bishop, Christopher M. 2006. *Pattern Recognition and Machine Learning.* New York: Springer.

Blaha, Leslie M., Christopher R. Fisher, Matthew M. Walsh, Bella Z. Veksler, and Glenn Gunzelmann. 2016. "Real-Time Fatigue Monitoring with Computational Cognitive Models." *In Proceedings, Part I, 10th International Conference on Foundations of Augmented Cognition: Neuroergonomics and Operational Neuroscience.* Vol. 9743. New York: Springer-Verlag. https://doi.org/10.1007/978-3-319-39955-3_28.

Blasch, Erik, Guna Seetharaman, Steve Suddarth, Kannappan Palaniappan, Genshe Chen, Haibin Ling, and Arlsan Basharat. 2014. "Summary of Methods in Wide-Area Motion Imagery (WAMI)." edited by Matthew F. Pellechia, Kannappan Palaniappan, Shiloh L. Dockstader, Peter J. Doucette, and Donnie Self, 90890C. Baltimore, MD. https://doi.org/10.1117/12.2052894.

Block, H. D. 1962. "The Perceptron: A Model for Brain Functioning. I." *Reviews of Modern Physics 34*, no. 1 (January): 123–35. https://doi.org/10.1103/RevModPhys.34.123.

Blom, Rianne M., Raoul C. Hennekam, and Damiaan Denys. 2012. "Body Integrity Identity Disorder." *PLoS One*, April 13, 2012. https://journals.plos.org/plosone/article?id=10.1371/journal.pone.0034702.

Bohn, Dieter. 2016. "Google Is Making Its Assistant 'conversational' in Two New Ways." The Verge, May 18, 2016. https://www.theverge.com/2016/5/18/11672938/google-assistant-chatbot-virtual-assistant-io-2016.

Bolton, Matthew L., Ellen J. Bass, and Radu I. Siminiceanu. 2013. "Using Formal Verification to Evaluate Human-Automation Interaction: A Review." *IEEE Transactions on Systems, Man, and Cybernetics: Systems 43*, no. 3 (May): 488–503. https://doi.org/10.1109/TSMCA.2012.2210406.

Bomberger, N. A., C. S. Sahin, S. M. Morales, M. E. Hlasyszyn, and B. J. Rhodes. 2012. "Automated Learning and Visualization of Traffic Patterns from Multi-INT Fused Tracks." Burlington: BAE Systems.

Bonabeau, Eric, Marco Dorigo, and Guy Theraulaz. 1999. *Swarm Intelligence: From Natural to Artificial Isystems*. New York: Oxford University Press.

Bonet, Blai, and Hector Geffner. 2003. "Labeled RTDP: Improving the Convergence of Real-Time Dynamic Programming." *Proceedings of Thirteenth International Conference on Automated Planning and Scheduling*. 12–21.

Bonnet, Philippe. 2000. "Jaguar and Cougar: Next Generation Object-Relational Database Systems." presented at the Second International Workshop on Information Integration and Web-based Applications & Services, Yogyakarta, Indonesia.

Booch, Grady. 2016. "It Is Cold. And Lonely." *IEEE Software*, June 2016.

Booher, Harold. ed. 2003. *Handbook of Human Systems Integration*. Hoboken, NJ: Wiley.

Booth, Serena, James Tompkin, Hanspeter Pfister, Jim Waldo, Krzysztof Gajos, and Radhika Nagpal. 2017. "Piggybacking Robots: Human-Robot Overtrust in University Dormitory Security." In *Proceedings of the 2017 ACM/IEEE International Conference on Human-Robot Interaction*, 426–34. New York. http://www.eecs.harvard.edu/~kgajos/papers/2017/booth17piggybacking.shtml.

Bornstein, Jon. 2015. "DoD Autonomy Roadmap: Autonomy Community of Interest." Aberdeen, MD: Army Research Laboratory Aberdeen Proving Ground. https://apps.dtic.mil/docs/citations/AD1010189.

BostonDynamics. *Hey Buddy, Can You Give Me a Hand?* YouTube, 2018. https://www.youtube.com/watch?v=fUyU3lKzoio.

Bradshaw, Jeffrey M., Alessandro Acquisti, James Allen, Maggie R. Breedy, Larry Bunch, Nathanael Chambers, Lucian Galescu, et al. 2004. "Teamwork-Centered Autonomy for Extended Human-Agent Interaction in Space Applications." Stanford University, Palo Alto, CA. https://pdfs.semanticscholar.org/0e69/030a19e529f857bd0aad94039aff172586b0.pdf?_ga=2.81951126.1253779093.1543262587-2101735069.1541165297.

Bradshaw, Jeffrey M., Paul J. Feltovich, Matthew J. Johnson, Larry Bunch, M. R. Breedy, Tom Eskridge, Hyuckchul Jung, James Lott, and Andrzej Uszok. 2008. "Coordination in Human-Agent-Robot Teamwork." In *2008 International Symposium on Collaborative Technologies and Systems*, 467–76. https://doi.org/10.1109/CTS.2008.4543966.

Braitenberg, Valentino. 1984. *Vehicles: Experiments in Synthetic Psychology*. Cambridge, MA: MIT Press.

Brambilla, Manuele, Eliseo Ferrante, Mauro Birattari, and Marco Dorigo. 2013. "Swarm Robotics: A Review from the Swarm Engineering Perspective." *Swarm Intelligence* 7, no. 1 (March): 1–41. https://doi.org/10.1007/s11721-012-0075-2.

Broadbent, Donald E. 1958. *Perception and Communication*. London: Pergamon Press.

Broca, Paul. 1865. "Sur le siège de la faculté du langage articulé." *Bulletins et Mémoires de la Société d'Anthropologie de Paris* 6 (1): 377–93. https://doi.org/10.3406/bmsap.1865.9495.

Brodie, Scott E., Bruce W. Knight, and Floyd Ratliff. 1978. "The Response of the Limulus Retina to Moving Stimuli: A Prediction by Fourier Synthesis." *Journal of General Physiology* 72, no. 2 (August): 129–66.

Brooks, Rodney A. 1986. "A Robust Layered Control System for a Mobile Robot." *IEEE Journal of Robotics and Automation* 2 (1): 14–23.

———. 1990. "Elephants Don't Play Chess." *Robotics and Autonomous Systems Robotics and Autonomous Systems* 6 (1–2): 3–15.

———. 2017. "The Big Problem with Self-Driving Cars Is People." IEEE Spectrum: Technology, Engineering, and Science News, July 27, 2017. https://spectrum.ieee.org/transportation/self-driving/the-big-problem-with-selfdriving-cars-is-people.

Brown, J. 1995. "Making Robots Conscious of Their Mental States." In *Proceedings of the 1995 AAAI Spring Symposium on Representing Mental States and Mechanisms*. Technical Report SS-95-08. Menlo Park, CA: AAAI Press. 89–98.

Brown, L. 2015. "Deep Learning with GPUs." GEOINT 2016, June 22–25, 2015, Washington, DC.

Bruni, Sylvain, Jessica J. Marquez, Amy Brzezinski, Carl Nehme, and Yves Boussemart. 2007. "Introducing a Human-Automation Collaboration Taxonomy (HACT) in Command and Control Decision-Support Systems." In *12TH ICCRTS "Adapting C2 to the 21st Century,"* 13. Massachusetts Institute of Technology, Cambridge, MA. https://hal.pratt.duke.edu/sites/hal.pratt.duke.edu/files/u13/Introducing%20a%20Human-Automation%20Collaboration%20Taxonomy%20%28HACT%29%20in%20Command%20and%20Control%20DecisionSupport%20Systems%20.pdf.

Bryson, Arthur E., and Yu-Chi Ho. 1969. *Applied Optimal Control Optimization, Estimation, and Control*. Blaisdell Book in the Pure and Applied Sciences. Waltham, MA: Blaisdell.

Buchanan, Bruce G., and Edward H. Shortliffe. 1984. *Rule-Based Expert Systems : The MYCIN Experiments of the Stanford Heuristic Programming*

Project. Addison-Wesley Series in Artificial Intelligence. Reading, MA: Addison-Wesley.

Bui, Ngot, and Vasant Honavar. 2013. "On the Utility of Abstraction in Labeling Actors in Social Networks." In *2013 International Conference on Advances in Social Networks Analysis and Mining (ASONAM)*, 692–98. https://doi.org/10.1109/ASONAM.2013.6785778.

Burlina, Philippe M., Aurora C. Schmidt, and I-Jeng Wang. 2015. "Zero Shot Deep Learning from Semantic Attributes." In *2015 IEEE 14th International Conference on Machine Learning and Applications (ICMLA)*, 871–76. Miami, FL. https://doi.org/10.1109/ICMLA.2015.140.

Butterfield, Rebecca. 2017. "Problem Solving with Tower of Hanoi." *Women in Computing Newsletter* (blog), July 17, 2017. https://medium.com/women -in-computing-newsletter/problem-solving-with-tower-of-hanoi -466210a105e.

Cagnon, Sebastien. "The Roomba Vacuum Robot Explained." About-Robots. com, 2014. http://www.about-robots.com/roomba-vacuum-robot.html.

Calhoun, Gloria L., and Mark H. Draper. 2006. "Multi-Sensory Interfaces for Remotely Operated Vehicles." In *Human Factors of Remotely Operated Vehicles*, edited by Nancy J. Cooke, Heather L. Pringle, Harry K. Pedersen, and Olena Connor, 7:149–63. Advances in Human Performance and Cognitive Engineering Research. Bingley, UK: Emerald Group Publishing Limited. https://doi.org/10.1016/S1479-3601(05)07011-6.

Calhoun, Gloria L., Michael A. Goodrich, John R. Dougherty, and Julie A. Adams. 2016. "Human-Autonomy Collaboration and Coordination Toward Multi-RPA Missions." In *Remotely Piloted Aircraft Systems: A Human Systems Integration Perspective*, edited by Nancy J. Cooke, Leah J. Rowe, Winston Bennett Jr., and DeForest Q. Joralmon, 109–36. Hoboken, NJ: John Wiley & Sons, Ltd, https://doi.org/10.1002/9781118965900.ch5.

Calhoun, Gloria L., Heath A. Ruff, Kyle J. Behymer, and Elizabeth M. Frost. 2018. "Human-Autonomy Teaming Interface Design Considerations for Multi-Unmanned Vehicle Control." *Theoretical Issues in Ergonomics Science* 19, no. 3 (May): 321–52. https://doi.org/10.1080/1463922X.2017.1315751.

Calhoun, Gloria L., Heath Ruff, Kyle Behymer, and Elizabeth M. Mersch. 2017a. "Operator-Autonomy Teaming Interfaces to Support Multi-Unmanned Vehicle Missions." In *Advances in Human Factors in Robots and Unmanned Systems: Proceedings of the AHFE 2016 International Conference on Human Factors in Robots and Unmanned Systems, July 27–31, 2016, Walt Disney World®, Florida, USA*, edited by Pamela Savage-Knepshield and Jessie Chen, 113–36. Cham, Switzerland: Springer International Publishing.

https://www.springerprofessional.de/en/operator-autonomy-teaming-in
terfaces-to-support-multi-unmanned-v/10588676.

Calhoun, Gloria L., Heath A. Ruff, Kyle J. Behymer, and Clayton D. Rothwell. 2017b. "Evaluation of Interface Modality for Control of Multiple Unmanned Vehicles." In HCI. https://doi.org/10.1007/978-3-319-58475-1_2.

Carbonell, Jaime G. 1983. "Derivational Analogy and Its Role in Problem Solving." In *Proceedings of the Third AAAI Conference on Artificial Intelligence*, 64–69. AAAI'83. Washington, DC: AAAI Press. http://dl.acm.org/citation.cfm?id=2886844.2886859.

Card, Stuart K., Thomas P. Moran, and Allen Newell. 1983. *The Psychology of Human-Computer Interaction*. Hillsdale, NJ: Erlbaum.

———. 1986. "The Model Human Processor: An Engineering Model of Human Performance." In *Handbook of Perception and Human Performance*. Vol. 2, Cognitive Processes and Performance edited by K. R. Boff, L. Kaufman, and J. P. Thomas. Oxford, England: John Wiley & Sons.

Cardell-Oliver, Rachel, and Wei Liu. 2010. "Representation and Recognition of Situations in Sensor Networks." *IEEE Communications Magazine 48*, no. 3 (March): 112–117. https://doi.org/10.1109/MCOM.2010.5434382.

Carter, Ash. 2016. Secretary of Defense Ash Carter Submitted Statement to the Senate Appropriations Committee – Defense on the FY 2017 Budget Request for the Department of Defense, § Senate Appropriations Committee. https://www.senate.gov/reference/Years_to_Congress.htm.

Casakin, Hernan, and Gabriela Goldschmidt. 1999. Expertise and the Use of Visual Analogy: Implications for Design Education." *Design Studies* 20, no. 2 (March): 153–75. https://doi.org/10.1016/S0142-694X(98)00032-5.

Casner, Stephen M., Edwin L. Hutchins, and Don Norman. 2016. "The Challenges of Partially Automated Driving." *Commun. ACM 59*, no. 5 (April 2016): 70–77. https://doi.org/10.1145/2830565.

Cassimatis, Nicholas L., J. Gregory Trafton, Magdalena D. Bugajska, and Alan C. Schultz. 2004. "Integrating Cognition, Perception and Action through Mental Simulation in Robots." *Journal of Robotics and Autonomous Systems 49*, no. 1–2 (2004): 12–23.

Catterall, William A. 2012. "Voltage-Gated Sodium Channels at 60: Structure, Function and Pathophysiology." *Journal of Physiology* 590, no. 11 (June): 2577–89. https://doi.org/10.1113/jphysiol.2011.224204.

Celesia, Gastone G. 2010. "Visual Perception and Awareness: A Modular System." *Journal of Psychophysiology* 24, no. 2 (2010): 62–67. https://doi.org/10.1027/0269-8803/a000014.

Cenamor, Isabel, Tomás de la Rosa, and Fernando Fernández. 2014. "IBACOP and IBACOP2 Planner." In *International Planning Competition*, 35–38.

"Center for Space Studies - Military Force Enhancement," October 16, 2018. http://www.au.af.mil/au/awc/space/enhance.htm.

Cerf, Vinton G. 2018. "A Comprehensive Self-Driving Car Test." *Communications of the ACM 61*, no. 2 (February): 7.

Chabod, Luc, and Philippe Galaup. 2012. "Shared Resources for Airborne Multifunction Sensor Systems." In *IET International Conference on Radar Systems (Radar 2012)*, 1–4, 2012. https://doi.org/10.1049/cp.2012.1669.

Chai, Joyce Y., Rui Fang, Changsong Liu, and Lanbo She. 2016. "Collaborative Language Grounding Toward Situated Human-Robot Dialogue." *AI Magazine 37* (4): 32–45. https://doi.org/10.1609/aimag.v37i4.2684.

Chakraborti, Tathagata, Yu Zhang, David E. Smith, and Subbarao Kambhampati. 2016. "Planning with Resource Conflicts in Human-Robot Cohabitation." In *Proceedings of the 2016 International Conference on Autonomous Agents & Multiagent Systems*, 1069–1077. AAMAS '16. Richland, SC: International Foundation for Autonomous Agents and Multiagent Systems. http://dl.acm.org/citation.cfm?id=2936924.2937081.

Chalkiadakis, Georgios, Edith Elkind, and Nicholas R. Jennings. 2009. "Simple Coalitional Games with Beliefs." In *Proceedings of the 21st International Jont Conference on Artifical Intelligence*, 85–90. IJCAI'09. San Francisco, CA: Morgan Kaufmann Publishers Inc. http://dl.acm.org/citation.cfm?id=1661445.1661460.

Chalmers, David J. 1996. *The Conscious Mind: In Search of a Fundamental Theory*. New York: Oxford University Press.

Chalmers, David J., Robert M. French, and Douglas R. Hofstadter. 1992. "High-Level Perception, Representation, and Analogy: A Critique of Artificial Intelligence Methodology." *Journal of Experimental & Theoretical Artificial Intelligence* 4, no. 3 (July): 185–211. https://doi.org/10.1080/09528139208953747.

Chapman, Pete, Julian Clinton, Randy Kerber, Thomas Khabaza, Thomas Reinartz, Colin Shearer, and Rüdiger Wirth. 2000. "CRISP-DM 1.0 Step-by-Step Data Mining Guide," 2000. http://www.crisp-dm.org/CRISPWP-0800.pdf.

Chatty, Abdelhak, Philippe Gaussier, Ilhem Kallel, Philippe Laroque, Florence Pirard, and Adel M. Alimi. 2013. "Evaluation of Emergent Structures in a 'Cognitive' Multi-Agent System Based on On-Line Building and Learning of a Cognitive Map." In *5th International Conference on Agents and Artificial Intelligence (ICAART)*, 269–75. Barcelona, Spain. https://hal.archives-ouvertes.fr/hal-00955928.

Chella, Antonio, and Riccardo Manzotti. 2007. "Artificial Consciousness." In *Perception-Action Cycle: Models, Architectures, and Hardware*, edited by Vassilis Cutsuridis, Amir Hussain, and John G. Taylor, 637–71. Springer

Series in Cognitive and Neural Systems. New York: Springer. https://doi
.org/10.1007/978-1-4419-1452-1_20.

Chen, Jessie Y. C., and Michael J. Barnes. 2014. "Human–Agent Teaming for
Multirobot Control: A Review of Human Factors Issues." *IEEE Transactions on Human-Machine Systems* 44, no. 1 (February): 13–29. https://
doi.org/10.1109/THMS.2013.2293535.

Cheng, Jie, David A. Bell, and Weiru Liu. 1997. "An Algorithm for Bayesian
Belief Network Construction from Data." In *Proceedings of Artificial Intelligence and Statistics*, 83-90.

Chin, Josh. 2018. "Chinese Police Add Facial-Recognition Glasses to Surveillance Arsenal." *Wall Street Journal*, February 7, 2018, sec. Tech. https://
www.wsj.com/articles/chinese-police-go-robocop-with-facial-recognition
-glasses-1518004353.

"China's Xi Pledges Billions in Loans, Aid to Arab Nations." AP News, July 12,
2018. https://apnews.com/70016e1098bb46b887019a34577830e3.

Choong, William. 2018. "Quad Goals: Wooing ASEAN." The Strategist, July
11, 2018. https://www.aspistrategist.org.au/quad-goals-wooing-asean/.

Chopra, Deepak, and Menas C. Kafatos. 2014. "From Quanta to Qualia: How
a Paradigm Shift Turns into Science." *Philosophy Study 4* (4). https://doi
.org/10.17265/2159-5313/2014.04.005.

Chun, Wendell H., Thomas Spura, Fank C. Alvidrez, and Randy J. Stiles. 2006.
"Spatial Dialog and Unmanned Aerial Vehicles." *In Human Factors of
Remotely Operated Vehicles*, edited by Nancy J. Cooke, Heather L. Pringle,
Harry K. Pedersen, and Olena Connor, 7:193–206. Advances in Human
Performance and Cognitive Engineering Research. Bingley, UK: Emerald
Group Publishing Limited. https://www.emeraldinsight.com/doi
/abs/10.1016/S1479-3601(05)07014-1.

Ciavarelli, Anthony. 1997. "Cockpit Control and Display Design Hazard
Analysis." Retrieved from CNET: https://www.cnet.navy.mil/nascweb
/sas/files/cockpit_control_and_display_hazard_checklist.pdf.

Clark, Herbert H. 1996. *Using Language*. 1st ed. New York: Cambridge University Press.

Clark, Matthew, Jim Alley, Paul Deal, Jeffrey DePriest, Eric Hansen, Connie
Heitmeyer, Richard Nameth, et al. 2015. "Autonomy Community of Interest (COI) Test and Evaluation, Verification and Validation (TEVV)
Working Group: Technology Investment Strategy 2015-2018." Office of
the Assistant Secretary of Defense (Research and Engineering), May 1,
2015. https://apps.dtic.mil/docs/citations/AD1010194.

Cochrane, Joe. 2018. "Indonesia, Long on Sidelines, Starts to Confront China's
Territorial Claims." *New York Times*, August 7, 2018, sec. World. https://

www.nytimes.com/2017/09/10/world/asia/indonesia-south-china-sea-military-buildup.html.

Cohen, Philip R., and Sharon L. Oviatt. 1995. "The Role of Voice Input for Human-Machine Communication." *Proceedings of the National Academy of Sciences 92*, no. 22 (October): 9921–27. https://doi.org/10.1073/pnas.92.22.9921.

Colombini, Esther Luna, Alexandre da Silva Simões, and Carlos Henrique Costa Ribeiro. 2017. "An Attentional Model for Autonomous Mobile Robots." *IEEE Systems Journal* 11, no. 3 (September): 1308–19. https://doi.org/10.1109/JSYST.2015.2499304.

Condliffe, Jamie. 2017. "A New Sensor Gives Driverless Cars a Human-like View of the World." MIT Technology Review, December 11, 2017. https://www.technologyreview.com/s/609718/a-new-sensor-gives-driverless-cars-a-human-like-view-of-the-world/.

Cook, Maia B., Cory A. Reith, and Mary K. Ngo. 2015. "Displays for Effective Human-Agent Teaming: The Role of Information Availability and Attention Management." In *Virtual, Augmented and Mixed Reality: 7th International Conference, VAMR 2015*, edited by Randall Shumaker and Stephanie Lackey. Cham, Switzerland: Springer International Publishing.

Cooke, Nancy J., H. Pringle, H. Pedersen, and O. Connor, eds. 2006. *Human Factors of Remotely Operated Vehicles*. New York: Elsevier.

Cooke, Nancy J., Eduardo Salas, Janis A. Cannon-Bowers, and Renée J. Stout. 2000. "Measuring Team Knowledge." *Human Factors* 42, no. 1 (March): 151–73. https://doi.org/10.1518/001872000779656561.

Cooper, Gregory F. 1990. "The Computational Complexity of Probabilistic Inference Using Bayesian Belief Networks (Research Note)." *Artificial Intelligence* 42, no. 2–3 (March): 393–405. https://doi.org/10.1016/0004-3702(90)90060-D.

Corey, M., C. Ford, J. Marx, I. MacLeod, G. van Osterom, P. VanDenBroeke, and W. Whitman. 2016. "Autonomy in DoD Cyberspace Operations, AFRL/RIGA Cross-Functional Discussion Group Meeting Notes." Air Force Research Laboratory.

Couture, Mario. 2007. "Complexity and Chaos - State-of-the-Art; Glossary." Technical note. Quebec: Defence R&D Canada - Valcartier, September. http://www.dtic.mil/dtic/tr/fulltext/u2/a475275.pdf.

Coward, L. Andrew, and Ron Sun. 2004. Criteria for an Effective Theory of Consciousness and Some Preliminary Attempts." *Consciousness and Cognition* 13, no. 2 (June): 268–301. https://doi.org/10.1016/j.concog.2003.09.002.

Cowell, Christopher Williams. 2001. "Minds, Machines and Qualia: A Theory of Consciousness" (dissertation). Berkeley: University of California.

Cox, Michael T. 2005. "Metacognition in Computation: A Selected Research Review." *Artificial Intelligence*, Special Review Issue, 169, no. 2 (December): 104–41. https://doi.org/10.1016/j.artint.2005.10.009.

Crevier, Daniel. 1993. *AI: The Tumultuous History of the Search for Artificial Intelligence*. New York: BasicBooks.

Crick, Francis, and Christof Koch. 1990. "Towards a Neurobiological Theory of Consciousness." *Seminars in the Neurosciences* 2 (1990): 263–75.

Cropper, Andrew, Alireza Tamaddoni-Nezhad, and Stephen H. Muggleton. 2016. "Meta-Interpretive Learning of Data Transformation Programs." In *Inductive Logic Programming*, edited by Katsumi Inoue, Hayato Ohwada, and Akihiro Yamamoto, 9575:46–59. Cham: Springer International Publishing. https://doi.org/10.1007/978-3-319-40566-7_4.

"CSDL." IEEE Computer Society. Accessed November 2, 2018. https://www.computer.org/csdl/magazine/ex/2000/06/x6016/13rRUy3gn1R.

Cummings, Mary L., and Amy S. Brzezinski. 2010. "Global vs. Local Decision Support for Multiple Independent UAV Schedule Management." *International Journal of Applied Decision Sciences* 3, no. 3 (January): 188–205. https://doi.org/10.1504/IJADS.2010.036098.

Curtis, Jon, Gavin Matthews, and David Baxter. 2005. "On the Effective Use of Cyc in a Question Answering System." In *Proceedings of the IJCAI Workshop on Knowledge and Reasoning for Answering Questions*, 61–70.

Cybenko, George. 1989. "Approximation by Superpositions of a Sigmoidal Function." *Mathematics of Control, Signals, and Systems* 2 (4): 303–14. https://doi.org/10.1007/BF02551274.

"Cyber Workforce Retention (Book, 2016) [WorldCat.Org]," January 31, 2018. http://www.worldcat.org/title/cyber-workforce-retention/oclc/957696601&referer=brief_results.

"Cybersecurity: Public Sector Threats and Responses," October 15, 2018. http://web.a.ebscohost.com.aufric.idm.oclc.org/ehost/ebookviewer/ebook/bmxlYmtfXzQxMTk1OF9fQU41?sid=e3b85e5c-32cc-41dd-8165-a52f65bf5476@sessionmgr4008&vid=1&format=EB&rid=1.

Dahm, Werner J. A. 2011. *Technology Horizons: A Vision for Air Force Science and Technology 2010-30*. Maxwell AFB, AL: Air University Press. https://media.defense.gov/2017/Apr/07/2001728430/-1/-1/0/B_0126_TECHNOLOGYHORIZONS.PDF.

Dall'Igna Jr., Alcino, Renato S. Silva, Kleber C. Mundim, and Laurent E. Dardenne. 2004. "Performance and Parameterization of the Algorithm Simplified Generalized Simulated Annealing." *Genetics and Molecular Biology* 27 (4): 616–22. https://doi.org/10.1590/S1415-47572004000400024.

Davidson, Janet E., Rebecca Deuser, and Robert J. Sternberg. 1995. "The Role of Metacognition in Problem Solving." In *Metacognition: Knowing about Knowing*, edited by Janet Metcalfe and Arthur P Shimamura, 208–25. Cambridge, MA: MIT Press.

Davis, Ernest, and Gary Marcus. 2015. "Commonsense Reasoning and Commonsense Knowledge in Artificial Intelligence." *Communications of the ACM* 58, no. 9 (September): 92–103. https://doi.org/10.1145/2701413.

De Weck, Olivier L., Daniel Roos, and Christopher L Magee. 2011. *Engineering Systems: Meeting Human Needs in a Complex Technological World.* Cambridge, MA: MIT Press.

De Weerdt, Mathijs, and Brad Clement. 2009. "Introduction to Planning in Multiagent Systems." *Multiagent and Grid Systems: An International Journal* 5 (4): 345–55. http://doi.org/10.3233/MGS-2009-0133.

"DeepMind." DeepMind, October 21, 2018. https://deepmind.com/.

Demir, Mustafa, Nathan J. McNeese, and Nancy J. Cooke. 2016. "Team Situation Awareness within the Context of Human-Autonomy Teaming." *Cognitive Systems Research* 46 (December): 3–12. https://doi.org/10.1016/j.cogsys.2016.11.003.

Dennett, Daniel C. 1991. *Consciousness Explained.* New York: Little, Brown and Company.

———. 2017. *From Bacteria to Bach and Back.* New York: W. W. Norton and Company.

Department of Defense. 1991. Data Fusion Subpanel of the Joint Directors of Laboratories, Technical Panel for C3. Data Fusion Lexicon.

———. 2012. "Autonomy in Weapons Systems, Department of Defense Directive No. 3000.09. Washington, DC.

Deptula, David A. 2006. "Effects-Based Operations." Air and Space Power Journal 20, no. 1 (Spring): 4-6.

———. 2016. "Beyond JSTARS: Rethinking the Combined Airborne Battle Management and Ground Surveillance System." Mitchell Institute Policy Papers. Vol. 2. September 2016.

DHPI (Directorate of Human Performance Integration). 2008. *Air Force Human Systems Integration Handbook: Planning and Execution of Human Systems Integration.* Brooks City-Base, TX: USAF DHPI Human Performance Optimization Division. http://www.acqnotes.com/Attachments/Air%20Force%20Human%20System%20Integration%20Handbook.pdf.

Diamantides, N. D. 1958. "A Pilot Analog for Airplane Pitch Control." *Journal of the Aerospace Sciences* 25 (6): 361–70. https://doi.org/10.2514/8.7687.

Dietterich, Thomas G. 2000. "Ensemble Methods in Machine Learning." In *Proceedings of the First International Workshop on Multiple Classifier*

Systems, 1–15. London: Springer-Verlag. https://dl.acm.org/citation
.cfm?id=743935.

Dietterich, Thomas G., and Eric J. Horvitz. 2015. "Rise of Concerns About AI: Reflections and Directions." *Communications of the ACM* 58, no. 10 (September): 38–40.

Dix, Jürgen, Thomas Eiter, Michael Fink, Axel Polleres, and Yingqian Zhang. 2003. "Monitoring Agents Using Declarative Planning." *Fundamenta Informaticae* 57, no. 2–4 (April): 345–70.

Dixon, Raymond A., and Scott D. Johnson. 2011. "Experts vs. Novices: Differences in How Mental Representations Are Used in Engineering Design." *Journal of Technology Education* 23 (1): 47–65.

DOD ONA (Department of Defense, Office of Net Assessment). "(Artificial) Intelligence: What Questions Should DoD Be Asking?" Summer Study. Washington, DC: Office of Net Assessment, 2016.

DOD R&E (Department of Defense, Research and Engineering). 2015. Autonomy Community of Interest (COI) Test and Evaluation, Verification and Validation (TEVV) Working Group. 2015. "Technology Investment Strategy 2015–2018." Washington, DC: Office of the Assistant Secretary of Defense. https://apps.dtic.mil/dtic/tr/fulltext/u2/1010194.pdf.

Domingos, Pedro. 2015. *The Master Algorithm: How the Quest for the Ultimate Learning Machine Will Remake Our World*. London: Penguin Books.

Dong, Guangheng, Xiao Lin, and Marc N. Potenza. 2015. "Decreased Functional Connectivity in an Executive Control Network Is Related to Impaired Executive Function in Internet Gaming Disorder." *Progress in Neuro-Psychopharmacology & Biological Psychiatry* 57 (March): 76–85. https://doi.org/10.1016/j.pnpbp.2014.10.012.

Dorneich, Michael C., Patricia May Ververs, Santosh Mathan, Stephen Whitlow, and Caroline C. Hayes. 2012. "Considering Etiquette in the Design of an Adaptive System." *Journal of Cognitive Engineering and Decision Making* 6, no. 2 (June): 243–65. https://doi.org/10.1177/15553434124 41001.

Dougherty, Llewellyn S., and Thomas F. Saunders. 2003. "Technology for Machine-to-Machine Intelligence, Surveillance, and Reconnaissance Integration, Vol. 1, Executive Summary and Brief." Washington, DC: United States Air Force Scientific Advisory Board.

Draper, Mark, Gloria Calhoun, Michael Hansen, Scott Douglass, Sarah Spriggs, Michael Patzek, Allen Rowe, et al. 2017. "Intelligent Multi-Unmanned Vehicle Planner with Adaptive Collaborative/Control Technologies (Impact)." In *19th International Symposium on Aviation Psychology*, 226–31. Dayton, OH. https://corescholar.libraries.wright.edu/isap_2017/85.

Dray, Susan. 1995. "The Importance of Designing Usable Systems." *Interactions* 2, no. 1 (January): 17–20. https://doi.org/10.1145/208143.208152.

DSB (Defense Science Board). 2012. "DSB Autonomy Study: The Role of Autonomy in DoD Systems." Task Force Report. Washington, DC: Office of the Undersecretary of Defense for Acquisition, Technology, and Logistics. https://www.acq.osd.mil/dsb/reports/2010s/AutonomyReport.pdf.

———. 2016. "Defense Science Board Summer Study on Autonomy." Report of the Defense Science Board Summer Study on Autonomy." Washington, DC: Office of the Undersecretary of Defense for Acquisition, Technology, and Logistics, June 2016. http://www.dtic.mil/dtic/tr/fulltext/u2/1017790.pdf.

DTIC (Defense Technical Information Center). 2005. "The Application of Category Theory and Analysis of Receiver Operating Characteristics to Information Fusion" (DTIC ADA450338).

Duda, Richard O., Peter E. Hart, and David G. Stork. 2012. *Pattern Classification*. New York: Wiley.

Duda, Richard O., and Edward H. Shortliffe. 1983. "Expert Systems Research." *Science* 220, no. 4594 (April): 261–68. https://doi.org/10.1126/science.6340198.

Duffy, Brian R. 2003. "Anthropomorphism and the Social Robot." *Robotics and Autonomous Systems* 42, no. 3–4 (March): 177–90. https://doi.org/10.1016/S0921-8890(02)00374-3.

Durso, Francis T., Katherine A. Rawson, and Sara Girotto. 2007. "Comprehension and Situation Awareness." In *Handbook of Applied Cognition*, edited by Francis T. Durso, Raymond S. Nickerson, Susan T. Dumais, Stephan Lewandowsky, and Timothy J. Perfect, 163–93. Hoboken, NJ: John Wiley & Sons, Ltd. https://doi.org/10.1002/9780470713181.ch7.

Dyer, Dave. (n.d.). "A Brief History of Lisp Machines" (website). Accessed January 5, 2019. http://real-me.net/ddyer/lisp/.

Dyer, Jean L. 1984. Team Research and Team Training: A State of the Art Review." In *Human Factors Review*, edited by Frederick A. Muckler, 285–323. Santa Monica, CA: Human Factors Society.

Dzindolet, Mary T., Scott A. Peterson, Regina A. Pomranky, Linda G. Pierce, and Hall P. Beck. 2003. "The Role of Trust in Automation Reliance." *International Journal of Human-Computer Studies* 58, no. 6 (June): 697–718. https://doi.org/10.1016/S1071-5819(03)00038-7.

Dzindolet, Mary T., Linda G. Pierce, Hall P. Beck, Lloyd A. Dawe, and B. Wayne Anderson. 2001. "Predicting Misuse and Disuse of Combat Identification Systems." *Military Psychology* 13, no. 3 (January): 147–64. https://doi.org/10.1207/S15327876MP1303_2.

Ecemis, M. İhsan, James Wikel, Christopher Bingham, and Eric Bonabeau. 2008. "A Drug Candidate Design Environment Using Evolutionary Computation." *IEEE Transactions on Evolutionary Computation* 12, no. 5 (October): 591–603. https://doi.org/10.1109/TEVC.2007.913131.

Edelman, Gerald M. 1989. The Remembered Present: *A Biological Theory of Consciousness*. New York: Basic Books.

Edwards, Chris. 2015. "Growing Pains for Deep Learning." *Communications of the ACM*, 58(7).

Eldar, Yonina C., and Gitta Kutyniok. 2012. *Compressed Sensing: Theory and Applications*. New York: Cambridge University Press.

Eliasmith, Chris, Terrence C. Stewart, Xuan Choo, Trevor Bekolay, Travis DeWolf, Yichuan Tang, and Daniel Rasmussen. 2012. "A Large-Scale Model of the Functioning Brain." *Science*, 338(6111): 1202–1205. doi:10.1126/science.1225266.

Ellis, Kevin, Armando Solar-Lezama, and Joshua B. Tenenbaum. 2016. Sampling for Bayesian Program Learning." *In Proceedings of the 30th International Conference on Neural Information Processing Systems*, 1297–1305. NIPS'16. Curran Associates Inc. http://dl.acm.org/citation.cfm?id=3157096.3157241.

Endsley, Mica R. 1987. "SAGAT: A Methodology for the Measurement of Situation Awareness (NOR DOC 87-83)." Northrop Corporation, 1987.

———. 1995a. "Toward a Theory of Situation Awareness in Dynamic Systems." *Human Factors* 37 (1): 32–64.

———. 1995b. "Measurement of Situation Awareness in Dynamic Systems." *Human Factors* 37, no. 1 (March): 65–84. https://doi.org/10.1518/001872095779049499.

———. 2015a. "Final Reflections: Situation Awareness Models and Measures." *Journal of Cognitive Engineering and Decision Making* 9, no. 1 (March): 101–11. https://doi.org/10.1177/1555343415573911.

———. 2015b. "Situation Awareness Misconceptions and Misunderstandings." *Journal of Cognitive Engineering and Decision Making* 9, no. 1 (March): 4–32. https://doi.org/10.1177/1555343415572631.

———. 2015c. *Autonomous Horizons: Autonomy in the Air Force—A Path to the Future*. Vol. 1, *Human Autonomy Teaming* (AF/ST TR 15-01). Washington, DC.

———. 2016. *Designing for Situation Awareness: An Approach to User-Centered Design*. Boca Raton, FL: CRC Press. https://www.crcpress.com/Designing-for-Situation-Awareness-An-Approach-to-User-Centered-Design/Endsley/p/book/9781420063554.

Etzioni, Amitai, and Oren Etzioni. 2016. "Designing AI Systems That Obey Our Laws and Values." *Communications of the ACM 59*, no. 9 (September): 29–31.

Evans, Jonathan St. B. T., and Keith E. Stanovich. 2013. "Dual-Process Theories of Higher Cognition: Advancing the Debate." *Perspectives on Psychological Science* 8, no. 3 (May): 223–41. https://doi.org/10.1177/1745691612460685.

Everstine, Brian W. 2018. "The F-22 and the F-35 Are Struggling to Talk to Each Other. . . and to the Rest of the USAF." *Air Force Magazine*, March 2018. http://www.airforcemag.com/MagazineArchive/Pages/2018/March%20 2018/The-F-22-and-the-F-35-Are-Struggling-to-Talk-to-Each-Other-And-to-the-Rest-of-USAF.aspx.

Farooq, Umer, and Jonathan Grudin. 2016. "Human-Computer Integration." *Interactions* 23, no. 6 (October): 26–32. https://doi.org/10.1145/3001896.

Federal Aviation Administration. (n.d.). "FAA Human Factors Awareness Web Course." Retrieved from http://www.hf.faa.gov/Webtraining/index.htm.

Fehrenbacher, Katie. 2015. How Tesla's Autopilot Learns." *Fortune*, October 16, 2015. http://fortune.com/2015/10/16/how-tesla-autopilot-learns/.

Feigenbaum, Edward A., Pamela McCorduck, and Penny Nii. 1988. *The Rise of the Expert Company: How Visionary Companies Are Using Artificial Intelligence to Achieve Higher Productivity and Profits*. New York: Times Books.

Ferrucci, David, Eric Brown, Jennifer Chu-Carroll, James Fan, David Gondek, Aditya A. Kalyanpur, Adam Lally, et al. 2010. "Building Watson: An Overview of the DeepQA Project." *AI Magazine* 31, no. 3 (July): 59–79. https://doi.org/10.1609/aimag.v31i3.2303.

Festinger, Leon. 1957. *A Theory of Cognitive Dissonance*. Stanford, CA: Stanford University Press.

Fikes, Richard E., and Nils J. Nilsson. 1971. "Strips: A New Approach to the Application of Theorem Proving to Problem Solving." *Artificial Intelligence* 2, no. 3 (December): 189–208. https://doi.org/10.1016/0004 -3702(71)90010-5.

Fisher, C. R., M. M. Walsh, L. M. Blaha, G. Gunzelmann, and B. Z. Veksler. 2016. "Efficient Parameter Estimation of Cognitive Models for Real-Time Performance Monitoring and Adaptive Interfaces." In *Proceedings of the 14th International Conference on Cognitive Modeling (ICCM 2016)*, edited by D. Reitter and F. E. Ritter. University Park, PA: Penn State.

Fitts, Paul M., and R. E. Jones. 1947. *Psychological Aspects of Instruments Display*. Washington: Department of Commerce.

Flach, John M., and Cynthia O. Dominguez. 1995. "Use-Centered Design: Integrating the User, Instrument, and Goal." *Ergonomics in Design* 3, no. 3 (July): 19–24. https://doi.org/10.1177/106480469500300306.

Flavell, John H. 1979. "Metacognition and Cognitive Monitoring: A New Area of Cognitive–Developmental Inquiry." *American Psychologist* 34 (10): 906–11. https://doi.org/10.1037/0003-066X.34.10.906.

Fletcher, Justin R. 2016. "Synaptic Annealing: Anisotropic Simulated Annealing and Its Application to Neural Network Synaptic Weight Selection." Thesis, Air Force Institute of Technology. https://scholar.afit.edu/etd/459.

"Flight of the Drones." *The Economist*, October 8, 2011. https://www.economist.com/briefing/2011/10/08/flight-of-the-drones.

Foerster, Jakob N., Yannis M. Assael, Nando de Freitas, and Shimon Whiteson. 2016. "Learning to Communicate with Deep Multi-Agent Reinforcement Learning." *Advances in Neural Information Processing Systems*, 2145–53.

Fogel, David. 2002. *Blondie24*. San Francisco: Morgan Kaufmann Pub.

Forrest, Stephanie, and Melanie Mitchell. 2016. "Adaptive Computation: The Multidisciplinary Legacy of John H. Holland." *Communications of the ACM* 59, no. 8 (July): 58–63. https://doi.org/10.1145/2964342.

Fowler, Martin. 2002. *Patterns of Enterprise Application Architecture*. 1 edition. Boston: Addison-Wesley Professional.

Franklin, Stan, and Art Graesser. 1996. "Is It an Agent, or Just a Program?: A Taxonomy for Autonomous Agents." In *Proceedings of the Workshop on Intelligent Agents III, Agent Theories, Architectures, and Languages*, 21–35. ECAI '96. London, UK, UK: Springer-Verlag. http://dl.acm.org/citation.cfm?id=648203.749270.

Fredslund, Jakob, and Maja J. Matarić. 2002. "A General Algorithm for Robot Formations Using Local Sensing and Minimal Communication." *IEEE Transactions on Robotics and Automation* 18, no. 5 (October): 837–46. https://doi.org/10.1109/TRA.2002.803458.

Freedman, Reva, Syed S. Ali, and Susan McRoy. 2000. "Links: What Is an Intelligent Tutoring System?" *Intelligence* 11, no. 3 (September): 15–16. https://doi.org/10.1145/350752.350756.

Friedman, Nir, Lise Getoor, Daphne Koller, and Avi Pfeffer. 1999. "Learning Probabilistic Relational Models." In *Proceedings of the 16th International Joint Conference on Artificial Intelligence*. Vol. 2, 1300–1307. IJCAI'99. San Francisco: Morgan Kaufmann Publishers Inc. http://dl.acm.org/citation.cfm?id=1624312.1624404.

Frommer, Dan. 2016. "Tesla Explains How Its Entire Fleet Is Learning to Be Better Self-Driving Cars Together." Recode, September 12, 2016. https://www.recode.net/2016/9/12/12889358/tesla-autopilot-data-fleet-learning.

Fuster, Joaquín M., Mark Bodner, and James K. Kroger. 2000. "Cross-Modal and Cross-Temporal Association in Neurons of Frontal Cortex." *Nature* 405 (6784): 347–51. https://doi.org/10.1038/35012613.

Gaifman, Haim. 1964. "Concerning Measures on Boolean Algebras." *Pacific Journal of Mathematics* 14, no. 1 (March): 61–73. https://doi.org/10.2140/pjm.1964.14.61.

Galster, Scott M., and Erica M. Johnson. 2013. "Sense-Assess-Augment: A Taxonomy for Human Effectiveness." Wright-Patterson AFB, OH: Air Force Research Laboratory, May 2013. http://www.dtic.mil/docs/citations/ADA585921.

Galvani, Luigi. (1791) 1953. *Commentary on the Effect of Electricity on Muscular Motion.* Translated by R. M. Green. Cambridge, MA: Licht.

GAO (Government Accountability Office). 2015. "Defense Advanced Research Projects Agency: Key Factors Drive Transition of Technologies, but Better Training and Data Dissemination Can Increase Success" (GAO-16-5). Washington, DC. http://www.gao.gov/assets/680/673746.pdf.

Garcez, Artur S. d'Avila, Dov M. Gabbay, and Krysia B. Broda. 2012. Neural-Symbolic Learning System: Foundations and Applications. Berlin: Springer-Verlag.

Gat, Erann. 1998. "On Three-Layer Architectures." In *Artificial Intelligence and Mobile Robots*, edited by David Kortenkamp, R. Peter Bonasso, and Robin Murphy, 195–210. Cambridge, MA: MIT Press. http://dl.acm.org/citation.cfm?id=292092.292130.

Gelb, Arthur. 1974. *Applied Optimal Estimation.* Cambridge, MA: MIT Press.

Gelepithis, Petros A. M. 2014. "A Novel Theory of Consciousness." *International Journal of Machine Consciousness* 6 (2): 125–39.

Genter, Katie, and Peter Stone. 2016. "Adding Influencing Agents to a Flock." In *Proceedings of the 2016 International Conference on Autonomous Agents & Multiagent Systems*, 615–623. AAMAS '16. Richland, SC: International Foundation for Autonomous Agents and Multiagent Systems. http://dl.acm.org/citation.cfm?id=2936924.2937015.

Gentner, Dedre, and Albert L. Stevens. 1983. *Mental Models.* Hillsdale, NJ: L. Erlbaum Associates.

Georgeff, Michael P., Barney Pell, Martha E. Pollack, Milind Tambe, and Michael Wooldridge. 1998. "The Belief-Desire-Intention Model of Agency." In *Proceedings of the 5th International Workshop on Intelligent Agents V, Agent Theories, Architectures, and Languages*, 1–10. ATAL '98. London: Springer-Verlag. http://dl.acm.org/citation.cfm?id=648205.749450.

Gerkey, Brian P., and Maja J. Mataric. 2004. "A Formal Analysis and Taxonomy of Task Allocation in Multi-Robot Systems." *International Journal of Robotics Research* 23, no. 9 (September): 939–54. https://doi.org/10.1177/0278364904045564.

Getoor, Lise, and Ben Taskar. 2007. *Introduction to Statistical Relational Learning*. Cambridge, MA: MIT Press.

Ghallab, Malik, Dana S. Nau, and Paolo Traverso. 2004. Automated Planning Theory and Practice. Amsterdam: Elsevier.

———. 2014. "The Actor's View of Automated Planning and Acting: A Position Paper." *Artificial Intelligence* 208 (March): 1–17. https://doi.org/10.1016/j.artint.2013.11.002.

———. 2016. *Automated Planning and Acting*. New York: Cambridge University Press.

Gibson, James J. 1947." Motion Picture Testing and Research." Army Air Forces Aviation Psychology Program Research Report. Washington, DC: US Government Printing Office. http://www.dtic.mil/dtic/tr/fulltext/u2/651783.pdf.

———. 1966. *The Senses Considered as Perceptual Systems*. Boston: Houghton Mifflin.

Goldberg, David E. 1989. *Genetic Algorithms in Search, Optimization & Machine Learning*. Reading, MA: Addison-Wesley.

Goldfein, David. 2017. "Enhancing Multi-Domain Command and Control. . . Tying It All Together." Department of Defense, March 2017. https://www.af.mil/Portals/1/documents/csaf/letter3/Enhancing_Multi-domain_CommandControl.pdf.

Gomes, Lee. 2017. "Can We Copy the Brain? - The Neuromorphic Chip's Make-or-Break Moment." *IEEE Spectrum* 54, no. 6 (June): 52–57. https://doi.org/10.1109/MSPEC.2017.7934233.

"Google Is Opening a China-Based Research Lab Focused on Artificial Intelligence." TechCrunch, October 12, 2018. https://techcrunch.com/2017/12/12/google-opening-an-office-focused-on-artificial-intelligence-in-china/.

"Google Translate Is Getting Really, Really Accurate." Boston Globe, October 12, 2018. https://www.bostonglobe.com/business/2016/10/03/google-translate-getting-really-really-accurate/L1FuGTeV3JocnsWVTVTdaP/story.html.

Goose, Stephen. 2015. "The Case for Banning Killer Robots: Point." *Communications of the ACM* 58, no. 12 (December): 43–45. https://doi.org/10.1145/2835963.

Goth, Gregory. 2016. "Deep or Shallow, NLP Is Breaking Out." *Communications of the ACM* 59, no. 3 (February): 13–16. https://doi.org/10.1145/2874915.

Gouin, D., R. Evdokiou, and R. Vernik. 2004. "A Showcase of Visualization Approaches for Military Decision Makers" (ADA427996). Defence

RandD Canada. Edinburgh, Australia: Defence Science and Technology Organization.

Graves, Alex, Greg Wayne, and Ivo Danihelka. 2014. "Neural Turing Machines." London: Google DeepMind. Retrieved from *arXiv:1410.5401v2 [cs.NE]*.

Green, Spence, Jeffrey Heer, and Christopher D. Manning. 2015. "Natural Language Translation at the Intersection of AI and HCI." *Communications of the ACM* 58, no. 9 (September): 46–53. https://doi.org/10.1145/2767151.

Greenblatt, Nathan A. 2016. "Self-Driving Cars and the Law." *IEEE Spectrum* 53, no. 2 (February 2016): 46–51. https://doi.org/10.1109/MSPEC.2016.7419800.

Greengard, Samuel. 2017. "Gaming Machine Learning." *Communications of the ACM* 60, no. 12 (November): 14–16. https://doi.org/10.1145/3148817.

Grice, H. Paul. 1975. "Logic and Conversation." In *Syntax and Semantics*, edited by Peter Cole and Jerry L. Morgan, 3:41–58. Speech Acts. New York: Academic Press.

Gross, Charles G. 2002. "Genealogy of the 'Grandmother Cell.' " *Neuroscientist* 8, no. 5 (October): 512–18. https://doi.org/10.1177%2F107385802237175.

Guerlain, Stephanie A., Philip J. Smith, Jodi Heinz Obradovich, Sally Rudmann, Patricia Strohm, Jack W. Smith, John Svirbely, and Larry Sachs. 1999. "Interactive Critiquing as a Form of Decision Support: An Empirical Evaluation." *Human Factors* 41, no. 1 (March): 72–89. https://doi.org/10.1518/001872099779577363.

Guizzo, Erico, and Evan Ackerman. 2015. "The Hard Lessons of DARPA's Robotics Challenge [News]." *IEEE Spectrum* 52, no. 8 (August): 11–13. https://doi.org/10.1109/MSPEC.2015.7164385.

Gunning, David. 2016. "Explainable Artificial Intelligence (XAI)," 36. Dayton, OH. https://www.darpa.mil/attachments/XAIProgramUpdate.pdf.

Guttal, Vishwesha, and Iain D. Couzin. 2010. "Social Interactions, Information Use, and the Evolution of Collective Migration." *Proceedings of the National Academy of Sciences* 107, no. 37 (September): 16172–77. https://doi.org/10.1073/pnas.1006874107.

Hackos, JoAnn T, and Janice Redish. 1998. *User and Task Analysis for Interface Design*. New York: Wiley. http://www.books24x7.com/marc.asp?bookid=11249.

Haikonen, Pentti O. A. 2013. "Consciousness and Sentient Robots." *International Journal of Machine Consciousness* 5 (1): 11–26.

Hall, David L., and James Llinas. 1997. "An Introduction to Multisensor Data Fusion." *Proceedings of the IEEE* 85, no. 1 (January): 6–23. https://doi.org/10.1109/5.554205.

Hammond, Kenneth R., Robert M. Hamm, Janet Grassia, and Tamra Pearson. 1987. "Direct comparison of the Efficacy of Intuitive and Analytical Cognition in Expert Judgment." *IEEE Transactions on Systems, Man, and Cybernetics* 17 (5): 753–70. https://doi.org/10.1109/TSMC.1987.6499282.

Hansen, Michael, Gloria Calhoun, Scott Douglass, and Dakota Evans. 2016. "Courses of Action Display for Multi-Unmanned Vehicle Control: A Multi-Disciplinary Approach." In *2016 AAAI Fall Symposium Series.* Arlington, VA. https://aaai.org/ocs/index.php/FSS/FSS16/paper/view/14104.

Hardman, Nicholas, John Colombi, David Jacques, and Ray Hill. 2008. "What Systems Engineers Need to Know About Human - Computer Interaction." *INSIGHT* 11, no. 2 (April): 19–23. https://doi.org/10.1002/inst.200811219.

Harnad, Stevan. 1990. "The Symbol Grounding Problem." *Physica D: Nonlinear Phenomena* 42, no. 1–3 (June): 335–46.

Hastie, Trevor, Robert Tibshiriani, and Jerome Friedman. 2009. *The Elements of Statistical Learning: Data Mining, Inference, and Prediction.* New York: Springer.

Haxby, J. V., C. L. Grady, B. Horwitz, L. G. Ungerleider, M. Mishkin, R. E. Carson. P. Herscovitch, M. B. Schapiro, and S. I. Rapoport. 1991. "Dissociation of Object and Spatial Visual Processing Pathways in Human Extrastriate Cortex." *Proceedings of the National Academy of Science* 88, no. 5 (March): 1621–1625. https://www.ncbi.nlm.nih.gov/pmc/articles/PMC51076/.

Hayes-Roth, Barbara. 1995. "An Architecture for Adaptive Intelligent Systems." *Artificial Intelligence* 72, no. 1–2 (January): 329–65. https://doi.org/10.1016/0004-3702(94)00004-K.

Hayes-Roth, Frederick, Donald A. Waterman, and Douglas B. Lenat. 1983. *Building Expert Systems.* Boston: Addison-Wesley Longman Publishing Co., Inc.

Haykin, Simon. 2009. *Neural Networks and Learning Machines.* 3rd ed. Upper Saddle River, NJ: Pearson Education.

Hebb, Donald. 1949. *The Organization of Behavior.* New York: Wiley.

Helie, Sebastien, and Ron Sun. 2010. "Incubation, Insight, And Creative Problem Solving: A Unified Theory and a Connectionist Model." *Psychological Review* 117 (3): 994–1024. https://doi.org/10.1037/a0019532.

Helmert, Malte, Gabriele Röger, and Erez Karpas. 2011. "Fast Downward Stone Soup: A Baseline for Building Planner Portfolios." 21st International Conference on Automated Planning and Scheduling, 28–35. Freibrug, Germany.

Hendler, James, and Jennifer Golbeck. 1997. "Metcalfe's Law, Web 2.0, and the Semantic Web." *Web Semantics: Science, Services and Agents on the World Wide Web* 6, no. 1 (February): 14–20. https://doi.org/10.1016/j .websem.2007.11.008.

Hergeth, Sebastian, Lutz Lorenz, Roman Vilimek, and Josef F. Krems. 2016. "Keep Your Scanners Peeled: Gaze Behavior as a Measure of Automation Trust During Highly Automated Driving." *Human Factors* 58, no. 3 (May): 509–19. https://doi.org/10.1177/0018720815625744.

Herman, David. 2013. *Storytelling and the Sciences of Mind*. Cambridge, MA: MIT Press.

Hernandez, Daniela. 2014. "Meet the Man Google Hired to Make AI a Reality." Wired, January 16, 2014. https://www.wired.com/2014/01/geoffrey-hinton-deep-learning/

———. 2017. "Can Robots Learn to Improvise?" *Wall Street Journal*, December 15, 2017. https://www.wsj.com/articles/can-robots-learn-to -improvise-1513368041.

Hesslow, Germund, and Dan-Anders Jirenhed. 2007. "Must Machines Be Zombies? Internal Simulation as a Mechanism for Machine Consciousness." *In Proceedings of the AAAI Symposium*, 6. Washington, DC. http:// www.aaai.org/Papers/Symposia/Fall/2007/FS-07-01/FS07-01-014.pdf.

Hill, Andrew and Gregg Thompson. 2016. "Five Giant Leaps for Robotkind: Expanding The Possible in Autonomous Weapons." War on the Rocks (blog). December 28, 2016. Retrieved from https://warontherocks .com/2016/12/five-giant-leaps-for-robotkind-expanding-the-possible -in-autonomous-weapons/.

Hilyard, Steven A. 1993. "Electrical and Magnetic Brain Recordings: Contributions to Cognitive Neuroscience." *Current Opinion in Neurobiology* 3, no. 2 (April): 217–24. https://doi.org/10.1016/0959-4388(93)90213-I.

Hilyard, Steven A., Wolfgang A. Teder-Salejarvi, and Thomas F. Munte. 1998. "Temporal Dynamics of Early Perceptual Processing." *Current Opinion in Neurobiology* 8, no. 2 (April): 202–210. https://doi.org/10.1016 /S0959-4388(98)80141-4.

Hinman, Michael L. 2002. "Some Computational Approaches for Situation Assessment and Impact Assessment." *Proceedings of the Fifth International Conference on Information Fusion 2002*. Vol. 1, 687–93. http://dx.doi .org/10.1109/ICIF.2002.1021221.

Hinton, Geoffrey E., Li Deng, Dong Yu, George Dahl, Abdel-rahman Mohamed, Navdeep Jaitly, Andrew Senior, et al. 2012. "Deep Neural Networks for Acoustic Modeling in Speech Recognition: The Shared Views of Four

Research Groups." *IEEE Signal Processing Magazine* 29, no. 6 (November): 82–97. https://doi.org/10.1109/MSP.2012.2205597.

Hinton, Geoffrey E., Simon Osindero, and Yee-Whye Teh. 2006. "A Fast Learning Algorithm for Deep Belief Nets." *Neural Computation* 18 (7): 1527–54.

Hodgkin, A. L., and A. F. Huxley. 1952. "A Quantitative Description of Membrane Current and its Application to Conduction and Excitation in Nerves." *Journal of Physiology* 117 (4): 500–544. https://www.ncbi.nlm.nih.gov/pmc/articles/PMC1392413/pdf/jphysiol01442-0106.pdf.

Hoey, Jesse, Robert St-Aubin, Alan Hu, and Craig Boutilier. 1999. "SPUDD: Stochastic Planning Using Decision Diagrams." Conference on Uncertainty in Artificial Intelligence. Stockholm. http://www.cs.toronto.edu/~cebly/Papers/spudd.pdf.

Hoff, Kevin Anthony, and Masooda Bashir. 2015. "Trust in Automation: Integrating Empirical Evidence on Factors That Influence Trust." *Human Factors* 57, no. 3 (May): 407–34. https://doi.org/10.1177/0018720814547570.

Hoffman, Robert R., and Laura G. Militello. 2009. *Perspectives on Cognitive Task Analysis.* New York: Taylor and Francis Group.

Holland, John H. 1975. *Adaptation in Natural and Artificial Systems.* Cambridge, MA: MIT Press.

———. 1996. *Hidden Order: How Adaptation Builds Complexity*, Reading, MA: Addison-Wesley.

———. 2006. "Studying Complex Adaptive Systems," *Journal of Systems Science and Complexity* 19, no. 1 (March): 1–8. doi:10.1007/s11424-006-0001-z.

Hooper, Daylond James, Jeffrey P. Duffy, Thomas C. Hughes, and Gloria L. Calhoun. 2015. "A Taxonomy for Improving Dialog between Autonomous Agent Developers and Human-Machine Interface Designers." In AAAI Fall Symp. Tech. Rep. *AAAI Fall Symposium - Technical Report, FS-15-01:81–88.* Arlington, VA.

Hoopes, James, ed. 1991. *Peirce on Signs.* Chapel Hill: University of North Carolina Press.

Hopfield, J. J. 1982. "Neural Networks and Physical Systems with Emergent Collective Computational Abilities." *Proceedings of the National Academy of Sciences of the United States of America* 79 (8): 2554–58.

Horne, Bill. 2016. "Trust Me. Trust Me Not." *IEEE Security Privacy* 14, no. 3 (May): 3–5. https://doi.org/10.1109/MSP.2016.56.

Horswill, Ian. 2000. "A Laboratory Course in Behavior-Based Robotics." *IEEE Intelligent Systems* 15, no. 6 (November): 16–21. https://doi.org/10.1109/5254.895852.

"How Google's AlphaGo Beat Lee Sedol, a Go World Champion." *The Atlantic*, October 12, 2018. https://www.theatlantic.com/technology/archive/2016 /03/the-invisible-opponent/475611/.

Hubbard, Timothy L. 1996. "The Importance of a Consideration of Qualia to Imagery and Cognition." *Consciousness and Cognition* 5 (3): 327–58.

Hubel, D. H., and T. N. Wiesel. 1962. "Receptive Fields, Binocular Interaction and Functional Architecture in the Cat's Visual Cortex." *The Journal of Physiology* 160 (1): 106–54. https://www.ncbi.nlm.nih.gov/pmc/articles /PMC1359523/.

Hudlicka, Eva. 2003. "To Feel or Not to Feel: The Role of Affect in Human-Computer Interaction." *International Journal of Human-Computer Studies* 59, no. 1–2 (July): 1–32. https://doi.org/10.1016/S1071-5819(03)00047-8.

Hudlicka, Eva, and Greg Zacharias. 2005. "Requirements and Approaches for Modeling Individuals within Organizational Simulations." In *Organizational Simulation*, edited by William B. Rouse, and Kenneth R. Boff, 79–137. Hoboken, NJ: John Wiley and Sons. http://dx.doi.org/10.1002/0471739448 .ch5.

Humphreys, Paul. 1997. "How Properties Emerge." *Philosophy of Science* 64, no. 1 (March): 1–17. https://doi.org/10.1086/392533.

Hutchins, Andrew R., M. L. Cummings, Mark Draper, and Thomas Hughes. 2015. "Representing Autonomous Systems' Self-Confidence through Competency Boundaries." *Proceedings of the Human Factors and Ergonomics Society Annual Meeting* 59, no. 1 (September): 279–83. https:// doi.org/10.1177/1541931215591057.

Hytla, Patrick, Kevin S. Jackovitz, Eric J. Balster, Juan R. Vasquez, and Michael L. Talbert. 2012. "Detection and Tracking Performance with Compressed Wide Area Motion Imagery." *2012 IEEE National Aerospace and Electronics Conference*. Dayton, OH. https://doi.org/10.1109/NAE CON.2012.6531049.

iamcarolfierce13. "Google Improves Voice Recognition, Hits 95% Accuracy | Android News." *AndroidHeadlines.Com* (blog), June 2, 2017. https:// www.androidheadlines.com/2017/06/google-improves-voice-recognition -hits-95-accuracy.html.

"IBM Archives: Deep Blue." TS200, January 23, 2003. www-03.ibm.com/ibm /history/exhibits/vintage/vintage_4506VV1001.html.

"IBM Research – Home." www.research.ibm.com/.

Ingrand, Felix, and Malik Ghallab. 2017. "Deliberation for Autonomous Robots: A Survey." *Artificial Intelligence* 247 (June): 10–44. https://doi.org/10.1016 /j.artint.2014.11.003.

International Conference on Simulation of Adaptive Behavior. 1993. Herbert L Roitblat, Jean-Arcady Meyer, and Stewart W Wilson, eds. *From Animals to Animats 2: Proceedings of the Second International Conference on Simulation of Adaptive Behavior*. Cambridge, Mass.; London: MIT Press.

Ishtiaq, Khizer. 2012. "Thesis – Emergent Design Systems in Architecture." *Emergent Design Systems* (blog), June 19, 2012. https://khizerishtiaq .wordpress.com/2012/06/19/undergrad-thesis-emergent-design -systems-in-architecture/.

Islam, Monirul M., and K. Murase. 2005. "Chaotic Dynamics of a Behavior-Based Miniature Mobile Robot: Effects of Environment and Control Structure." *Neural Networks* 18, no. 2 (March): 123–44. https://doi .org/10.1016/j.neunet.2004.09.002.

Jacobs, Robert A., Michael I. Jordan, Steven J. Nowlan, and Geoffrey E. Hinton. 1991. "Adaptive Mixtures of Local Experts." *Neural Computation* 3, no. 1 (Spring): 79–87.

Janakiram, M. S. 2017. "In the Era of Artificial Intelligence, GPUs Are the New CPUs." *Forbes*, August 7, 2017. https://www.forbes.com/sites/jana kirammsv/2017/08/07/in-the-era-of-artificial-intelligence-gpus-are -the-new-cpus/#2265d7fb5d16.

Jennings, Nick R. 1993. "Specification and Implementation of a Belief-Desire-Joint-Intention Architecture for Collaborative Problem Solving." *International Journal of Intelligent and Cooperative Information Systems* 2 (3): 289–318. https://doi.org/10.1142/S0218215793000137.

Jensen, K. (Wikimedia Commons contributors). 2012. File: CRISP-DM Process Diagram. Retrieved from https://commons.wikimedia.org/w/index .php?title=File:CRISP-DM_Process_Diagram.pngandoldid= 231928936.

Joe, Jeffrey C., John O'Hara, Heather D. Medema, and Johanna H. Oxstrand. 2014. "Identifying Requirements for Effective Human-Automation Teamwork." In PSAM 2014, *Probabilistic Safety Assessment and Management*. Honolulu, HI. https://www.researchgate.net/publication/282722194_Iden tifying _requirements_for_effective_human-automation_teamwork.

Johansson, Petter, Lars Hall, Sverker Sikström, and Andreas Olsson. 2005. "Failure to Detect Mismatches between Intention and Outcome in a Simple Decision Task." *Science* 310, no. 5745 (October): 116–19. https:// doi.org/10.1126/science.1111709.

John, Bonnie E., and David E. Kieras. 1996. "The GOMS Family of User Interface Analysis Techniques: Comparison and Contrast." *ACM Transactions on Computer-Human Interaction* 3, no. 4 (December): 320–51. http://www.di.ubi.pt/~agomes/ihc/artigos/john2.pdf.

Johnson, Chris W. 2006. "What are Emergent Properties and How Do They Affect the Engineering of Complex Systems?" *Reliability Engineering and System Safety* 91, no. 12 (December): 1475–81. http://dx.doi.org/ 10.1016 /j.ress.2006.01.008.

Johnson, Matthew, Jeffrey M. Bradshaw, Paul J. Feltovich, Catholijn M. Jonker, Birna van Riemsdijk, and Maarten Sierhuis. 2011. "The Fundamental Principle of Coactive Design: Interdependence Must Shape Autonomy." In *Coordination, Organizations, Institutions, and Norms in Agent Systems VI*, edited by M. De Vos, N. Fornara, J. V. Pitt, and G. Vouros, 172–91. Berlin: Springer.

Johnson, Matt, Jeffrey M. Bradshaw, Robert R. Hoffman, Paul J. Feltovich, and David D. Woods. 2014. "Seven Cardinal Virtues of Human-Machine Teamwork: Examples from the DARPA Robotic Challenge." *IEEE Intelligent Systems* 29, no. 6 (December): 74–80. https://doi.org/10.1109 /MIS.2014.100.

Johnson-Laird, Philip Nicholas. 1983. *Mental Models: Towards a Cognitive Science of Language, Inference, and Consciousness*. Cambridge, MA: Harvard University Press.

Kaber, David B., and Mica R. Endsley. 2003. "The Effects of Level of Automation and Adaptive Automation on Human Performance, Situation Awareness and Workload in a Dynamic Control Task." *Theoretical Issues in Ergonomics Science* 5, no. 2 (March): 113–53. https://doi.org/10.1080 /1463922021000054335.

Kaebling, Leslie, Michael L. Littman, and Anthony R. Cassandra. 1998. "Planning and Acting in Partially Observable Stochastic Domains." *Artificial Intelligence* 101, no. 1–2 (May): 99–134. https://doi.org/10.1016 /S0004-3702(98)00023-X.

Kahneman, Daniel. 2011a. "Don't Blink! The Hazards of Confidence." *New York Times*, October 19, 2011. https://www.nytimes.com/2011/10/23 /magazine/dont-blink-the-hazards-of-confidence.html.

———. 2011b. *Thinking, Fast and Slow*. New York: Farrar, Straus, and Giroux.

Kahneman, Daniel, Paul Slovic, and Amos Tversky. 1982. *Judgement Under Uncertainty: Heuristics and Biases*. Cambridge, UK: Cambridge University Press.

Kahneman, Daniel, and Amos Tversky. 1979. "Prospect Theory: Analysis of Decision Making Under Risk." *Econometrica* 47, no. 2 (March): 263–92. https://www.jstor.org/stable/1914185.

Kalman, R. E. 1960. "A New Approach to Linear Filtering and Prediction Problems." *Journal of Basic Engineering* 82, no. 1 (March): 35–45. https:// doi:10.1115/1.3662552.

Kalra, Nidhi, and Susan M. Paddock. 2016. "Driving to Safety: How Many Miles of Driving Would It Take to Demonstrate Autonomous Vehicle Reliability?" Technical Report RR-1478-RC. Santa Monica, CA: RAND. https://www.rand.org/pubs/research_reports/RR1478.html.

Kandel, Abraham. 1992. *Fuzzy Expert Systems*. Boca Raton, FL: CRC Press.

Kandel, Eric R., James H. Schwartz, and Thomas M. Jessell. 2013. *Principles of Neural Science*. 5th ed. New York: McGraw Hill Professional.

Kaplan, Jerry. 2017. "Artificial Intelligence: Think Again." *Communications of the ACM* 60, no. 1 (January): 36–38.

Kaplan, Ehud, and Robert B. Barlow Jr. 1975. "Properties of Visual Cells in the Lateral Eye of Limulus in Situ." *Journal of General Physiology* 66 (3): 303–26.

Kearns, Kristin, C. Schumacher, and J. Imhoff. 2016. Update to AFRL Directors: Autonomy for Loyal Wingman. Air Force Research Laboratory, Airman Systems Directorate.

Keeney, Ralph L., and Howard Raiffa. 1993. *Decisions with Multiple Objectives: Preferences and Value Trade-Offs*. Cambridge, UK: Cambridge University Press.

Keim, Brandon. 2015. "Dr. Watson Will See You . . . Someday." *IEEE Spectrum* 52, no. 6 (June): 76–77. https://doi.org/10.1109/MSPEC.2015.7115575.

Keller, James M., Derong Liu, and David B Fogel. 2016. *Fundamentals of Computational Intelligence: Neural Networks, Fuzzy Systems, and Evolutionary Computation*. Hoboken, NJ: Wiley.

Keller, Thomas, and Malte Helmert. 2013. "Trial-based Heuristic Tree Search for Finite Horizon MDPs." *Proceedings of the 23rd International Conference on Automated Planning and Scheduling* 23 (June): 135–43.

Kelly, Christopher, M. Boardman, P. J. Goillau, and Emmanuella Jeannot. 2003. "Guidelines for Trust in Future ATM Systems: A Literature Review." Brussels, Belgium: EUROCONTROL, May 5, 2003. https://www.research gate.net/publication/311065869_Guidelines_for_Trust_in_Future_ATM _Systems_A_Literature_Review.

Kennedy, William G., and Robert E. Patterson. 2012. "Modeling of Intuitive Decision Making in ACT-R." *Proceedings of the 11th International Conference on Cognitive Modeling*. Berlin, Germany.

Kieras, David E., and David E. Meyer. 1997. "An Overview of the EPIC Architecture for Cognition and Performance with Application to Human-Computer Interaction." *Journal of Human-Computer Interaction* 12, no. 4 (December): 391–438. https://doi.org/10.1207/s15327051hci1204_4.

Kieras, David E., Scott D. Wood, and David E. Meyer. 1997. "Predictive Engineering Models Based on the EPIC Architecture for a Multimodal High-

Performance Human-Computer Interaction Task." *ACM Transactions on Computer-Human Interaction* 4, no. 3 (September): 230–75. https://doi.org/10.1145/264645.264658.

Kilgore, Ryan, and Martin Voshell. 2014. "Increasing the Transparency of Unmanned Systems: Applications of Ecological Interface Design." *In Virtual, Augmented and Mixed Reality.* Applications of Virtual and Augmented Reality, edited by Randall Shumaker and Stephanie Lackey, 378–89. Lecture Notes in Computer Science. Springer International Publishing.

Kim, Eugene. 2016. "Amazon's $775 Million Deal for Robotics Company Kiva Is Starting to Look Really Smart." *Business Insider*, June 15, 2016. https://www.businessinsider.com/kiva-robots-save-money-for-amazon-2016-6.

Kim, Jaegwon. 2006. "Being Realistic About Emergence." In *The Re-Emergence of Emergence*, edited by Philip Clayton and Paul Sheldon Davies, 189. New York: Oxford University Press.

Kim, Yeo Keun, Kitae Park, and Jesuk Ko. 2003. "A Symbiotic Evolutionary Algorithm for the Integration of Process Planning and Job Shop Scheduling." *Computers and Operations Research* 30, no. 8 (July): 1151–71. https://doi.org/10.1016/S0305-0548(02)00063-1.

Kirk, Benjamin H., Jonathan W. Owen, Ram M. Narayanan, Shannon D. Blunt, Anthony F. Martone, and Kelly D. Sherbondy. 2017. "Cognitive Software Defined Radar: Waveform Design for Clutter and Interference Suppression." Proceedings SPIE 10188, Radar Sensor Technology XXI, 1018818. https://doi.org/10.1117/12.2262305.

Kirkpatrick, Keith. 2015. "The Moral Challenges of Driverless Cars." *Communications of the ACM* 58, no. 8 (July): 19–20. https://doi.org/10.1145/2788477.

———. 2016. "Can We Trust Autonomous Weapons?" *Communications of the ACM* 59, no. 12 (December): 27–29. https://doi.org/10.1145/3005678.

Kirlik, Alex. 1993. "Modeling Strategic Behavior in Human-Automation Interaction: Why an 'Aid' Can (and Should) Go Unused." *Human Factors* 35, no. 2 (June): 221–42. https://doi.org/10.1177/001872089303500203.

Klarreich, Erica. 2016. "Learning Securely." *Communications of the ACM* 59, no. 11 (October): 12–14. https://doi.org/10.1145/2994577.

Klein, Gary. 1997. "The Recognition-Primed Decision (RPD) Model: Looking Back, Looking Forward." In *Naturalistic Decision Making*, edited by Caroline E. Zsambok and Gary Klein, 285–92. Expertise: Research and Applications. Hillsdale, NJ: Lawrence Erlbaum Associates, Inc.

———. 1998. *Sources of Power*. Cambridge, MA: MIT Press, 1998.

———. 2008. "Naturalistic Decision Making." *Human Factors* 50, no. 3 (June): 456–60. https://doi.org/10.1518/001872008X288385.

Klein, Gary, Roberta Calderwood, and Donald G. MacGregor. 1989. "Critical Decision Method for Eliciting Knowledge." *IEEE Transactions on Systems, Man, and Cybernetics* 19, no. 3 (May): 462–72. https://doi.org/10.1109/21.31053.

Klein, Gary, Brian Moon, and Robert R. Hoffman. 2006. "Making Sense of Sensemaking 2: A Macrocognitive Model." *IEEE Intelligent Systems* 21, no. 5 (September): 88–92. https://doi.org/10.1109/MIS.2006.100.

Klein, Gary, David D. Woods, Jeffrey M. Bradshaw, and Paul J. Feltovich. 2004. "Ten Challenges for Making Automation a 'Team Player' in Joint Human-Agent Activity." *IEEE Journals & Magazine* 19, no. 6 (December): 91–95.

Kleinman, D., Baron, S., and Levison, W. 1970). "An Optimal Control Model of Human Response: Theory and Validation." *Automatica* 6, no. 3 (May): 357–69. https://doi.org/10.1016/0005-1098(70)90051-8.

Knight, Will. 2016. "AI's Language Problem." *Technology Review* 119, no. 5 (September): 28–37. https://www.technologyreview.com/s/602094/ais-language-problem/.

———. 2017a. "Paying with Your Face." *Technology Review* 120, no. 2 (March/April). https://www.technologyreview.com/s/603494/10-breakthrough-technologies-2017-paying-with-your-face/.

———. 2017b. "The Dark Secret at the Heart of AI." Intelligent Machines (blog), *Technology Review* (website), April 11, 2017. https://www.technologyreview.com/s/604087/the-dark-secret-at-the-heart-of-ai/.

Koch, Christof, and Giulio Tononi. 2008. "Can Machines Be Conscious?" IEEE Spectrum, June 1, 2008. https://spectrum.ieee.org/biomedical/imaging/can-machines-be-conscious.

———. 2017. "Can We Quantify Machine Consciousness?" *IEEE Spectrum*, May 25, 2017. https://spectrum.ieee.org/computing/hardware/can-we-quantify-machine-consciousness.

Kocsis, Levente, and Csaba Szepesvari. 2006. "Bandit Based Monte-Carlo Planning." In *Proceedings of the 17th European Conference on Machine Learning*, Lecture Notes in Computer Science 4212, edited by J. Furnkranz, T. Scheffer, and M. Spiliopoulou, 282–93. https://doi.org/10.1007/11871842_29.

Kohonen, Teuvo. 1972. "Correlation Matrix Memories." *IEEE Transactions on Computers* 21, no. 4 (April): 353–59. https://doi.org/10.1109/TC.1972.5008975.

———. 1981. "Automatic Formation of Topological Maps of Patterns in a Self-Organizing System." Proceedings of the 2nd Scandinavia Conference on Image Analysis, 214–20. Espoo, Finland.

———. 1995. "Learning Vector Quantization." *Self-Organizing Maps. Springer Series in Information Sciences.* Vol. 30. Berlin: Springer. https://doi.org/10.1007/978-3-642-97610-0_6.

———. 2001. *Self-Organizing Maps.* 3rd ed. Berlin: Springer Verlag.

Kohonen, Teuvo, and Erkki Oja, eds. 1997. Workshop on Self-Organizing Maps WSOM'97. Helsinki University of Technology, Finland.

Kolobov, Andrey, M. Mausam, and Daniel S. Weld. 2012. "LRTDP versus UCT for Online Probabilistic Planning." *Proceedings of the Twenty-Sixth AAAI Conference on Artificial Intelligence.* 1786–92.

Kolodner, Janet. 1993. *Case-Based Reasoning.* San Mateo, CA: Morgan Kaufmann Publishers

———. 1992. "An Introduction to Case-Based Reasoning." *Artificial Intelligence Review* 6 no. 1 (March): 3–34. https://link.springer.com/article/10.1007/BF00155578.

Komenda, A., P. Novák, V. Lisý, B. Bošanský, M. Čáp, and M. Pěchouček. 2012. "Tactical Operations of Multi-robot Teams in Urban Warfare (Demonstration)." *Proceedings of the 11th International Conference on Autonomous Agents and Multiagent Systems* (AAMAS 2012), 3, 1473–74.

Korf, Jakob. 2014. "Emergence of Consciousness and Qualia from a Complex Brain." *Folia Medica* 56, no. 4 (December): 289–96. https://doi.org/10.1515/folmed-2015-0010.

Kott, Alexander, David S. Alberts, and Cliff Wang. 2015. "Will Cybersecurity Dictate the Outcome of Future Wars?" *Computer* 48, no. 12 (December): 98–101. https://doi.org/10.1109/MC.2015.359.

Kozlowski, Steve W. J., and Katherine J. Klein. 2000. "A Multi-Level Approach to Theory and Research in Organizations: Contextual, Temporal, and Emergent Processes." In *Multilevel Theory, Research, and Methods in Organizations: Foundations, Extensions, and New Directions*, edited by Katherine J. Klein and Steve W. J. Kozlowski, 3–90. San Francisco: Jossey-Bass.

Krakovsky, Marina. 2016. "Reinforcement Renaissance." *Communications of the ACM* 59 no. 8 (August): 12–14. https://doi.org/10.1145/2949662.

Krauthamer, George. 1968. "Form Perception across Sensory Modalities." *Neuropsychologia* 6, no. 2 (June): 105–13. https://doi.org/10.1016/0028-3932(68)90052-3.

Kroo, Ilan. 2004. "Collectives and Complex System Design." Lecture presented at the VKI Lecture Series on Optimization Methods & Tools for Multicriteria/Multidisciplinary Design, von Karman Institute, Brussels, Belgium, November 15, 2004.

Kuipers, Benjamin. 2018. "How Can We Trust a Robot?" *Communications of the ACM* 61, no. 3 (March): 86–95.

Kwiatkowska, Marta, Gethin Norman, and David Parker. 2002. "PRISM: Probabilistic Symbolic Model Checker." In *Computer Performance Evaluation: Modelling Techniques and Tools. TOOLS 2002. Lecture Notes in Computer Science.* Vol. 2324, 200–204. Edited by T. Field, P. G. Harrison, J. Bradley, and U. Harder. Berlin: Springer Verlag.

Lachman, Roy, Janet L. Lachman, and Earl C. Butterfield. 1979. *Cognitive Psychology and Information Processing: An Introduction.* Hillsdale, NJ: Lawrence Erlbaum.

Laird, John. 2012. *The Soar Cognitive Architecture.* Cambridge, MA: MIT Press.

Laird, John, and Shiwali Mohan. 2018. "Learning, Fast and Slow: Levels of Learning in General Autonomous Intelligent Agents." Association for the Advancement of Artificial Intelligence. https://soar.eecs.umich.edu /pubs/Laird_Mohan_FastAndSlow_AAAI18.pdf.

Laird, J. E., A. Newell, and P. S. Rosenbloom. 1993. *The Soar Papers: Research on Integrated Intelligence.* Cambridge, MA: MIT Press.

Lakoff, George, and Srini Narayanan. 2010. "Toward a Computational Model of Narrative." *AAAI Fall Symposium: Computational Models of Narrative.*

Langley, Derrick, Ronald A. Coutu Jr., LaVern A. Starman, and Peter J. Collins. 2011. "MEMS Integrated Metamaterial Structure Having Variable Resonance for RF Applications." *MEMS and Nanotechnology* 2:115–20. https://doi.org/10.1007/978-1-4419-8825-6_17.

Larson, Christina. 2018. "China's Massive Investment in Artificial Intelligence Has an Insidious Downside." *Science*, February 7, 2018. https://www.sciencemag.org/news/2018/02/china-s-massive-investment-artificial -intelligence-has-insidious-downside.

Launchbury, John. 2016. "JASON Study on Future of AI."

LeCun, Yann. 1988. "A Theoretical Framework for Back-Propogation." In *Proceedings of the 1988 Connectionist Models Summer School.* Edited by D. Touretzky, G. Hinton, and T. Segjowski, 21–28. Pittsburgh, PA: Carnegie Mellon University, Morgan Kaufmann.

LeCun, Yann, Yoshua Bengio, and Geoffrey Hinton. 2015. "Deep Learning." *Nature* 521:436–44. https://www.nature.com/articles/nature14539.

LeCun, Yann, L. Bottou, Y. Bengio, and P. Haffner. 1998. "Gradient-Based Learning Applied to Document Recognition." *Proceedings of the IEEE* 86, no. 11 (November): 2278–2324.

Lee, John D., and Katrina A. See. 2004. "Trust in Automation: Designing for Appropriate Reliance." *Human Factors* 46, no. 1 (March): 50–80. https:// doi.org/10.1518/hfes.46.1.50_30392.

Lenat, Doug B. 1998. "From 2001 to 2001: Common Sense and the Mind of HAL," in *Hal's Legacy: 2001's Computer as Dream and Reality*, edited by David G. Stork. Cambridge, MA: MIT Press.

Lettvin, J., H. Maturana, W. McCulloch, and W. Pitts. 1959. "What the Frog's Eye Tells the Frog's Brain." *Proceedings of the Institute of Radio Engineers* 47, no. 11 (November): 1940–51. https://doi.org/10.1109/JRPROC.1959.287207.

Liao, Wenzhi, Aleksandra Pizurica, Rik Bellens, Sidharta Gautama, and Wilfried Philips. 2015. "Generalized Graph-Based Fusion of Hyperspectral and LiDAR Data Using Morphological Features." *IEEE Geoscience and Remote Sensing Letters* 12 (3): 552–56.

Lindsay, Robert K., Bruce G. Buchanan, Edward A. Feigenbaum, and Joshua Lederberg. 1980. *Applications of Artificial Intelligence for Organic Chemistry: The Dendral Project*. New York: McGraw-Hill Book Company.

Litman, Diane, Satinder Singh, Michael Kearns, and Marilyn Walker. 2000. "NJFun: A Reinforcement Learning Spoken Dialogue System." *Proceedings of the 2000 ANLP/NAACL Workshop on Conversational Systems*.

Luckham, David. 2002. *The Power of Events: An Introduction to Complex Event Processing in Distributed Enterprise Systems*. Boston: Addison-Wesley.

Lynott, Dermot, and Louise Connell. 2010. "Embodied Conceptual Combination." *Frontiers in Psychology* 1:212. https://doi.org/10.3389/fpsyg.2010.00212.

Lyons, Joseph B., Kolina S. Koltai, Nhut T. Ho, Walter B. Johnson, David E. Smith, and R. Jay Shively. 2016a. "Engineering Trust in Complex Automated Systems." *Ergonomics in Design* 24, no. 1 (January): 13–17. https://doi.org/10.1177/1064804615611272.

Lyons, Joseph B., Nhut T. Ho, William E. Fergueson, Garrett G. Sadler, Samantha D. Cals, Casey E. Richardson, and Mark A. Wilkins. 2016b. "Trust of an Automatic Ground Collision Avoidance Technology: A Fighter Pilot Perspective." *Military Psychology* 28, no. 4 (May): 271–77. https://doi.org/10.1037/mil0000124.

Madhaven, P., and D. A. Wiegmann. 2007. "Similarities and Differences between Human–Human and Human–Automation Trust: An Integrative Review." *Theoretical Issues in Ergonomics Science* 8 (4): 277–301.

Maes, Pattie. 1995. "Artificial Life Meets Entertainment: Lifelike Autonomous Agents." *Communications of the ACM* 38, no. 11 (November): 108–114. https://doi.org/10.1145/219717.219808.

Maetz, Cade. 2017. "A Mystery AI Just Crushed the Best Human Players at Poker." Wired (website), January 31, 2017. https://www.wired.com/2017/01/mystery-ai-just-crushed-best-human-players-poker/.

Manna, Zohar, and Amir Pnueli. 1995. *Temporal Verification of Reactive Systems: Safety*. New York: Springer Science and Business Media.

Manolopoulos, Yannis, Yannis Theodoridis, and Vassilis Tsotras. 2000. *Advanced Database Indexing*. 17. New York: Springer Science and Business Media.

MarketsandMarkets. 2017. "Internet of Things Market by Software Solution & Application—Global Forecast 2022," June 2017. https://www.marketsandmarkets.com/Market-Reports/internet-of-things-market-573.html.

Markoff, John. 2011. "Computer Wins on 'Jeopardy!': Trivial, It's Not." *New York Times*, February 16, 2011. https://www.nytimes.com/2011/02/17/science/17jeopardy-watson.html.

Markus, M. Lynne, Ann Majchrzak, and Les Gasser. 2002. "A Design Theory for Systems That Support Emergent Knowledge Processes." *MIS Quarterly* 26, no. 3 (September): 179–212.

Masiello, Thomas J. 2013. "Air Force Research Laboratory Autonomy Science and Technology Strategy." Wright-Patterson Air Force Base, OH: Air Force Research Laboratory, December 2, 2013. https://web.archive.org/web/20170125102447/http://www.defenseinnovationmarketplace.mil/resources/AFRL_Autonomy_Strategy_DistroA.PDF.

Mataric, Maja J. 1993. "Designing Emergent Behaviors: From Local Interactions to Collective Intelligence." In *Proceedings of the Second International Conference on from Animals to Animats 2: Simulation of Adaptive Behavior*. Honolulu. Edited by J. A. Meyer, H. Roitblat, and S. Wilson, 432–41. Cambridge, MA: MIT Press.

———. 1995. "Issues and Approaches in the Design of Collective Autonomous Agents." *Robotics and Autonomous Systems* 16, no. 2 (December): 321–31. https://doi.org/10.1016/0921-8890(95)00053-4.

———. 1997. "Behaviour-Based Control: Examples from Navigation, Learning, and Group Behaviour." *Journal of Experimental & Theoretical Artificial Intelligence* 9, no. 2–3 (April): 323–36. https://doi.org/10.1080/095281397147149.

Mayer, Roger C., James H. Davis, and F. David Schoorman. 1995. "An Integrative Model of Organizational Trust." *The Academy of Management Review* 20, no. 3 (July): 709–34. https://doi.org/10.2307/258792.

Mayoh, Brian, Enn Tyugu, and Jaan Penjam, editors. 2013. *Constraint Programming*. Berlin: Springer Science and Business Media. http://dx.doi.org/10.1007/978-3-642-85983-0.

McCarthy, John. 1960. "Recursive Functions of Symbolic Expressions and Their Computation by Machine, Part I." *Communications of the ACM* 3, no. 4 (April): 184–95. https://doi.org/10.1145/367177.367199.

———. 1968. "Programs with Common Sense." In *Semantic Information Processing*, edited by M. L. Minsky, 403–418. Cambridge, MA: MIT Press.

McCarthy, John, and Vinay Chaudhri. 2004. DARPA Workshop on Self-Aware Computer Systems, SRI Headquarters, Arlington, VA, April 2004.

McCarthy, John, Marvin Minsky, N. Rochester, and C. E. Shannon. 1955. "A Proposal for the Dartmouth Summer Research Project on Artificial Intelligence." August 31, 1955. http://www-formal.stanford.edu/jmc/history/dartmouth/dartmouth.html.

McCarthy, John, Marvin Minsky, Aaron Sloman, Leiguang Gong, T. Lau, Leora Morgenstern, Erik T. Mueller, Doug Riecken, Moninder Singh, and Push Singh. 2002. "An Architecture of Diversity for Commonsense Reasoning [Technical Forum]." *IBM Systems Journal* 41 (3): 530–39. https://doi.org/10.1147/SJ.2002.5386871.

McCorduck, Pamela. 2004. *Machines Who Think*. 2nd ed. Natick, MA: A. K. Peters, Ltd.

McCulloch, Warren S., and Walter Pitts. 1943. "A Logical Calculus of Ideas Immanent in Nervous Activity." *Bulletin of Mathematical Biophysics* 5, no. 4 (December): 115–33. https://doi.org/10.1007/BF02478259.

McDaniel, Patrick, Nicolas Papernot, and Z. Berkay Celik. 2016. "Machine Learning in Adversarial Settings." *IEEE Security Privacy* 14, no. 3 (May): 68–72. https://doi.org/10.1109/MSP.2016.51.

McDonnell, John R., Robert G. Reynolds, and David B. Fogel. 1995. *Evolutionary Programming IV: Proceedings of the Fourth Annual Conference on Evolutionary Programming*, San Diego. Cambridge, MA: A Bradford Book.

McEwan, Desmond, Geralyn R. Ruissen, Mark A. Eys, Bruno D. Zumbo, and Mark R. Beauchamp. 2017. "The Effectiveness of Teamwork Training on Teamwork Behaviors and Team Performance: A Systematic Review and Meta-Analysis of Controlled Interventions." *PLoS One* 12, no. 1 (January). https://journals.plos.org/plosone/article?id=10.1371/journal.pone.0169604.

McGettigan, Carolyn, and Sophie K. Scott. 2012. Cortical Asymmetries in Speech Perception: What's Wrong, What's Right, and What's Left?" *Trends in Cognitive Science* 16, no. 5 (May): 269–76. http://doi.org/10.1016/j.tics.2012.04.006.

McGinn, Conor, Michael Francis Cullinan, George Walsh, Cian Donavan, and Kevin Kelly. 2015. "Towards an Embodied System-Level Architecture for Mobile Robots." In *2015 International Conference on Advanced Robotics (ICAR)*, 536–42. https://doi.org/10.1109/ICAR.2015.7251508.

McGuirl, John M., and Nadine B. Sarter. 2006. "Supporting Trust Calibration and the Effective Use of Decision Aids by Presenting Dynamic System Confidence Information." *Human Factors* 48, no. 4 (September): 656–65. https://doi.org/10.1518/001872006779166334.

McPhee, Michele, K. C. Baker, and Corky Siemaszko. 2015. "Deep Blue, IBM's Supercomputer, Defeats Chess Champion Garry Kasparov in 1997." *New York Daily News*. May 10, 2015. http://www.nydailynews.com/news/world/kasparov-deep-blues-losingchess-champ-rooke-article-1.762264.

McRuer, Duane, Dunstan Graham, Ezra Krendel, and William Reisener Jr. 1965. "Human Pilot Dynamics in Compensatory Systems" (AFFDL-TR-65-15). Wright-Patterson AFB, OH: Air Force Systems Command, Flight Dynamics Laboratory. Retrieved from http://www.dtic.mil/get-tr-doc/pdf?AD=AD0470337.

Merényi, Erzsébet, William H. Farrand, James V. Taranik, and Timothy B. Minor. 2014. "Classification of Hyperspectral Imagery with Neural Networks: Comparison to Conventional Tools." *EURASIP Journal on Advances in Signal Processing* 2014:71. https://doi.org/10.1186/1687-6180-2014-71.

Merenyi, Erzsébet, Michael J. Mendenhall, and Patrick O'Driscoll, eds. 2016. "Advances in Self-Organizing Maps and Learning Vector Quantization." *Proceedings of the 11th International Workshop WSOM 2016*, Houston, TX, January 6–8, 2016. New York: Springer Nature. http://doi.org/czfd.

Metz, Cade. 2018. "Bets on AI Open a New Chip Frontier." *New York Times*, January 14, 2018. https://www.nytimes.com/2018/01/14/technology/artificial-intelligence-chip-start-ups.html.

Metzinger, Thomas, and David J. Chalmers. 1995. "Selected Bibliography: Consciousness in Philosophy, Cognitive Science and Neuroscience: 1970-1995." In *Conscious Experience*, edited by Thomas Metzinger. Paderborn, Germany: Schöningh.

Meyer, David E., and David E. Kieras. 1997. "A Computational Theory of Executive Cognitive Processes and Multiple-Task Performance: Parts 1 and 2." *Psychological Review* 104 (1): 3–65. http://doi.org/fwrwd8.

Michie, Donald. 1972. "Programmer's Gambit." *New Scientist* 55, no. 809 (August): 329–32.

Militello, Laura G., Robert J. B. Hutton, Rebecca M. Pliske, Betsy J. Knight, Gary Klein, and Josephine Randel. 1997. "Applied Cognitive Task Analysis (ACTA) Methodology." NPRDS TN-98-4. San Diego: Navy Personnel Research and Development Center.

Mill, John Stuart. 1843. *System of Logic*. London: Longmans, Green, Reader, and Dyer.

Miller, Christopher A., and Raja Parasuraman 2003. "Beyond Levels of Automation: An Architecture for More Flexible Human-Automation Collaboration." *Proceedings of the Human Factors and Ergonomics Society Annual Meeting* 47, no. 1 (October): 182–86. https://doi.org/10.1177/154193120304700138.

———. 2007. "Designing for Flexible Interaction between Humans and Automation: Delegation Interfaces for Supervisory Control." *Human Factors* 49, no. 1 (February): 57–75. https://doi.org/10.1518/001872007779598037.

Miller, Duncan C., and Jack A. Thorpe. 1995. "SIMNET: The Advent of Simulator Networking." *Proceedings of the IEEE* 83, no. 8 (August): 1114–23. https://doi.org/10.1109/5.400452.

Miner, Gary, John Elder IV, Andrew Fast, Thomas Hill, and Dursun Delen. *Practical Text Mining and Statistical Analysis for Non-Structured Text Data Applications.* Waltham, MA: Academic Press, 2012.

Minsky, Marvin. 1968. "Matter, Mind, and Models." In *Semantic Information Processing*, edited by Marvin L. Minsky, 425–32. Cambridge, MA: MIT Press.

———. 1975. "A Framework for Representing Knowledge." In *The Psychology of Computer Vision*, edited by P. Winston. New York: McGraw-Hill.

———. 1986. *The Society of Mind.* New York: Simon and Schuster.

———. 1991. "Logical vs. Analogical or Symbolic vs. Connectionist or Neat vs. Scruffy." *AI Magazine* 12, no. 2 (Summer): 34–51.

———. 2006. *The Emotion Machine.* New York: Simon and Schuster.

Minsky, Marvin, and Seymour A. Papert. 1969. *Perceptrons: An Introduction to Computational Geometry.* Cambridge, MA: MIT Press.

Mislevy, Robert, Geneva Haertel, Michelle Riconscente, Daisy Wise Rutstein, and Cindy Ziker. 2017. "Model-Based Inquiry. In Assessing Model-Based Reasoning using Evidence-Centered Design." Springer Briefs in Statistics. New York: Springer.

Mitchell, Melanie. 2006. "Complex Systems: Network Thinking." *Artificial Intelligence*, Special Review Issue, 170, no. 18 (December): 1194–1212. https://doi.org/10.1016/j.artint.2006.10.002.

Mitchell, Russ. 2018. "Tesla Crash Highlights a Problem: When Cars Are Partly Self-Driving, Humans Don't Feel Responsible." *Los Angeles Times*, January 25, 2018. http://www.latimes.com/business/autos/la-fi-hy-tesla-autopilot-20180125-story.html.

Mitchell, Tom. 1997. *Machine Learning.* Boston: McGraw-Hill.

Mittal, Saurabh, Margery J. Doyle, and Eric Watz. 2013. "Detecting Intelligent Agent Behavior with Environment Abstraction in Complex Air Combat Systems." In *2013 IEEE International Systems Conference (SysCon)*, 662–70. Orlando, FL. https://doi.org/10.1109/SysCon.2013.6549953.

Miyake, Akira, and Priti Shah, eds. 1999. *Models of Working Memory: Mechanisms of Active Maintenance and Executive Control.* Cambridge, UK: Cambridge University Press.

Miyake, A., Friedman, N., Emerson, M., Witzki, A., Howerter, A., and Wager, T. 2000. "The Unity and Diversity of Executive Functions and Their Contributions to Complex 'Frontal Lobe' Tasks: A Latent Variable Analysis." *Cognitive Psychology* 41, no. 1 (August): 49–100. https://doi .org/10.1006/cogp.1999.0734.

Mnih, Volodymyr, Koray Kavukcuoglu, David Silver, Alex Graves, Ioannis Antonoglou, Daan Wierstra, and Martin Riedmiller. 2013. "Playing Atari with Deep Reinforcement Learning." *ArXiv:1312.5602 [Cs]*, NIPS Deep Learning Workshop, December 19, 2013. http://arxiv.org /abs/1312.5602.

Mnih, Volodymyr, Koray Kavukcuoglu, David Silver, Andrei A. Rusu, Joel Veness, Marc G. Bellemare, Alex Graves, et al. 2015. "Human-Level Control through Deep Reinforcement Learning." *Nature* 518:529–42. https://doi.org/10.1038/nature14236.

Modayil, Joseph, and Benjamin Kuipers. 2008. "The Initial Development of Object Knowledge by a Learning Robot." *Robotics and Autonomous Systems* 56, no. 11 (November): 879–90. https://doi.org/10.1016/j.robot .2008.08.004.

Monirul Islam, Md., and Kazuyuki Murase. 2005. "Chaotic Dynamics of a Behavior-Based Miniature Mobile Robot: Effects of Environment and Control Structure." Neural Networks: *The Official Journal of the International Neural Network Society* 18, no. 2 (March): 123–44. https://doi.org /10.1016/j.neunet.2004.09.002.

Montalbano, A., G. Baj, D. Papadia, E. Tongiorgi, and M. Sciancalepore. 2013. "Blockade of BDNF Signaling Turns Chemically-Induced Long-Term Potentiation into Long-Term Depression." *Hippocampus* 23, no. 10 (October): 879–89. https://doi.org/10.1002/hipo.22144.

Moore, Gordon E. 1965. "Cramming More Components onto Integrated Circuits, Reprinted from Electronics, Volume 38, Number 8, April 19, 1965, pp.114 Ff." *IEEE Solid-State Circuits Society Newsletter* 11, no. 3 (September 2006): 33–35. https://doi.org/10.1109/N-SSC.2006.4785860.

Moorman, Christine, Rohit Deshpandé, and Gerald Zaltman. 1993. "Factors Affecting Trust in Market Research Relationships." *Journal of Marketing* 57, no. 1 (1993): 81–101. https://doi.org/10.2307/1252059.

Moravec, Hans. 1988. *Mind Children*. Cambridge, MA: Harvard University Press.

Morgan, Jacob. 2014. *The Future of Work: Attract New Talent, Build Better Leaders, and Create a Competitive Organization*. Hoboken, NJ: Wiley.

————. 2015. "The 5 Types of Organizational Structures: Part 1, The Hierarchy." Forbes.com, July 6, 2015. https://www.forbes.com/sites/jacobmorgan

/2015/07/06/the-5-types-of-organizational-structures-part-1-the-hierarchy /#2516b3225252.

Morvan, Laurence, Francis Hintermann, and Madhu Vazirani. 2017. "Five Ways to Win with Digital Platforms: Executive Summary." Accenture. https://www.accenture.com/t20180705T112755Z__w__/us-en/_acnmedia /PDF-29/Accenture-Five-Ways-To-Win-With-Digital-Platforms-Executive -Summary.pdf.

Mosier, Kathleen L. 2009. "Searching for Coherence in a Correspondence World." *Judgment and Decision Making* 4, no. 2 (March): 154–63.

Mosier, Kathleen L., Linda J. Skitka, Susan Heers, and Mark Burdick. 1998. "Automation Bias: Decision Making and Performance in High-Tech Cockpits." *International Journal of Aviation Psychology* 8 (1): 47–63.

Moyer, Christopher. 2016. "How Google's AlphaGo Beat a Go World Champion." *The Atlantic*, March 28, 2016. https://www.theatlantic.com/technology /archive/2016/03/the-invisible-opponent/475611/.

Mozur, Paul. 2017. "Bejing Wants AI to Be Made in China by 2030." *New York Times*, July 20, 2017. https://www.nytimes.com/2017/07/20/business /china-artificial-intelligence.html?_r=0 –.

Muir, Bonnie M. 1994. "Trust in Automation: Part I. Theoretical Issues in the Study of Trust and Human Intervention in Automated Systems." *Ergonomics* 37, no. 11 (November): 1905–22. https://doi.org/10.1080/00140139408964957.

Mura, M. D., A. Villa, J. A. Benediktsson, J. Chanussot, and L. Bruzzone. 2011. "Classification of Hyperspectral Images by Using Extended Morpho-logical Attribute Profiles And Independent Component Analysis." *IEEE Geoscience and Remote Sensing Letters* 8 (3):542–46. https://doi.org/ 10.1109/LGRS.2010.2091253.

Murphy, Robin R. 2002. *Introduction to AI Robotics*. Cambridge, MA: MIT Press.

Murphy, Robin R., Christine L. Lisetti, Russell Tardif, Liam Irish, and Aaron Gage. 2002. "Emotion-Based Control of Cooperating Heterogeneous Mobile Robots." *IEEE Transactions on Robotics and Automation* 18, no. 5 (October): 744–57. https://doi.org/10.1109/TRA.2002.804503.

Myers, Karen L. 1998. "Towards a Framework for Continuous Planning and Execution." *Proceedings of the AAAI Fall Symposium on Distributed Continual Planning*.

Najafabadi, Maryam M., Flavio Villanustre, Taghi M. Khoshgoftaar, Naeem Seliya, Randall Wald, and Edin Muharemagic. 2015. "Deep Learning Applications and Challenges in Big Data Analytics." *Journal of Big Data* 2, no. 1 (February): 1. https://doi.org/10.1186/s40537-014-0007-7.

NASA. 2012. "Technology Readiness Level." October 28, 2012, https://www
.nasa.gov/directorates/heo/scan/engineering/technology/txt_accordion1
.html.

National Research Council. 1990. *Quantitative Modeling of Human Performance in Complex, Dynamic Systems*. Washington, DC: The National
Academies Press. https://doi.org/10.17226/1490.

———. 1998. *Modeling Human and Organizational Behavior: Application to
Military Simulations*. Washington, DC: The National Academies Press.
https://doi.org/10.17226/6173.

———. 2006. *Defense Modeling, Simulation, and Analysis: Meeting the Challenge*. Washington, DC: The National Academies Press. https://doi
.org/10.17226/11726.

National Transportation Safety Board (NTSB). 2017. "Collision Between a
Car Operating with Automated Vehicle Control Systems and a Tractor-
Semitrailer Truck Near Williston, Florida May 7, 2016." Washington,
DC: NTSB, September 12, 2017. https://www.ntsb.gov/investigations
/AccidentReports/Pages/HAR1702.aspx.

Nehme, Carl E., Stacey D. Scott, Mary Cummings, and Carina Furusho.
2006. "Generating Requirements for Futuristic Hetrogenous Unmanned Systems." *Proceedings of the Human Factors and Ergonomics
Society Annual Meeting* 50, no. 3 (October): 235–39. https://doi
.org/10.1177 /154193120605000306.

———. 1976. *Cognition and reality: Principles and implications of cognitive
psychology*. San Francisco: W. H. Freeman and Company.

Nelson-Miller. 2016. "User Interface vs Human Machine Interface: What's the
Difference?" Nelson-Miller, Inc. (blog), November 12, 2015. http://
www.nelson-miller.com/user-interface-vs-human-machine-interface
-whats-the-difference/.

Newell, Allen. 1990. *Unified Theories of Cognition*. Cambridge, MA: Harvard
University Press.

Newell, Allen, and Herbert A. Simon. 1963. "GPS: A Program that Simulates
Human Thought." In *Computers and Thought*, edited by Edward A. Feigenbaum, and Jerome A. Feldman. New York: McGraw-Hill.

Newell, Allen, Herbert A. Simon, and J. C. Shaw. 1959. "Report on a General
Problem-Solving Program." Santa Monica, CA: RAND. http://www
.worldcat.org/title/report-on-a-general-problem-solving-program
/oclc/2271713&referer=brief_results.

New York Times. 2018. "On 'Jeopardy!' Watson Win Is All but Trivial." October
12, 2018. https://www.nytimes.com/2011/02/17/science/17jeopardy-watson
.html.

Nielsen, Jakob, and Robert L Mack, eds. 1994. *Usability Inspection Methods*. New York: Wiley, 1994.

Nikolopoulos, Chris. 1997. *Expert Systems: Introduction to First and Second Generation and Hybrid Knowledge Based Systems*. New York: Marcel Decker.

Nilsson, Nils J. 1986. "Probabilistic Logic." *Artificial Intelligence* 28, no. 1 (February): 71–87. https://doi.org/10.1016/0004-3702(86)90031-7.

Nisbett, Richard E., and Timothy D. Wilson. 1977. "Telling More Than We Can Know: Verbal Reports on Mental Processes." *Psychological Review* 84:231–59. http://hdl.handle.net/2027.42/92167.

Nkambou, Roger, Riichiro Mizoguchi, and Jacqueline Bourdeau. 2010. *Advances in Intelligent Tutoring Systems*. Heidelberg: Springer.

Noda, Kuniaki, Yuki Yamaguchi, Kazuhiro Nakadai, Hiroshi G. Okuno, and Tetsuya Ogata. 2015. "Audio-Visual Speech Recognition Using Deep Learning." *Applied Intelligence* 42, no. 4 (June): 722–37. http://doi.org/10.1007/s10489-014-0629-7.

Norman, Donald A., and Stephen W. Draper, eds. 1986. *User Centered System Design: New Perspectives on Human-Computer Interaction*. Hillsdale, NJ: Lawrence Erlbaum.

Norris, Guy. 2017. "Inside View: Automatic Integrated Collision Avoidance System." Aviation Week & Space Technology, November 10, 2017. http://aviationweek.com/icas.

Norvig, Peter. 2016. "On Chomsky and the Two Cultures of Statistical Learning." Personal website, 2016. http://norvig.com/chomsky.html.

Nunez, Sergio, Daniel Borrajo, and Carlos Linares Lopez. 2012. "Performance Analysis of Planning Portfolios." *Proceedings of the Fifth Annual Symposium Combinatorial Search*, 65–71. http://www.aaai.org/ocs/index.php/SOCS/SOCS12/paper/view/5377.

Nvidia. 2018. "GPU vs CPU? What Is GPU Computing?" Nvidia, April 3, 2018. https://web.archive.org/web/20180403233933/http://www.nvidia.com/object/what-is-gpu-computing.html.

Nwana, Hyacinth S. 1990. "Intelligent Tutoring Systems: An Overview." *Artificial Intelligence Review* 4:251–77. http://doi.org/10.1007/BF00168958.

O'Connor, Timothy. 1994. "Emergent Properties." *American Philosophical Quarterly* 31 (2): 91–104.

O'Hara, John M., and James Higgins. 2010. "Human-System Interfaces to Automatic Systems: Review Guidance and Technical Basis." BNL Technical Report 91017-2010. Upton, NY: Brookhaven National Laboratory.

Oizumi, Masafumi, Larissa Albantakis, and Giulio Tononi. 2014. "From the Phenomenology to the Mechanisms of Consciousness: Integrated Information Theory 3.0." *PLoS Computation Biology* 10, no. 5 (May).

https://journals.plos.org/ploscompbiol/article?id=10.1371/journal
.pcbi.1003588.

Olavsrud, Thor. 2014. "10 IBM Watson-Powered Apps That Are Changing
Our World | CIO." CIO, November 6, 2014. https://www.cio.com/article
/2843710/big-data/10-ibm-watson-powered-apps-that-are-changing
-our-world.html.

Onwubolu, Godfrey, and B. V. Babu. 2004. *New Optimization Techniques in
Engineering*. Berlin: Springer-Verlag.

Opitz, David, and Richard Maclin. 1999. "Popular Ensemble Methods: An
Empirical Study." *Journal of Artificial Intelligence Research* 11 (August):
169–98. https://doi.org/10.1613/jair.614.

Ortiz, Carlos Enrique. 2008. "On Self-Modifying Code and the Space Shuttle
OS." Personal website, August 18, 2007. http://weblog.cenriqueortiz
.com/computing/2007/08/18/on-self-modifying-code-and-the-space
-shuttle-os/.

Osmundson, John S., Thomas V. Huynh, and Gary O. Langford. 2008. "Emergent
Behavior in Systems of Systems." INCOSE *International Symposium*,
KR14, 18, no. 1 (June): 1557–68. https://doi.org/10.1002/j.2334-5837.2008
.tb00900.x.

Oviatt, Sharon. 1999. "Ten Myths of Multimodal Interaction." *Communications of the ACM* 42, no. 11 (November): 74–81. https://doi.org/10.1145
/319382.319398.

Pais, Darren. 2012. "Emergent Collective Behavior in Multi-Agent Systems:
An Evolutionary Perspective." Dissertation, Princeton University, 2012.
http://www.princeton.edu/~naomi/theses/Pais_thesis_main.pdf.

Palacios, Hector, Alexandre Albore, and Hector Geffner. 2014. "Compiling
Contingent Planning into Classical Planning: New Translations and
Results." Sixth *European Conference on Planning*, 1–8.

Pan, Sinno J., and Qiang Yang. 2010. "A Survey on Transfer Learning." *IEEE
Transactions on Knowledge and Data Engineering* 22, no. 10 (October):
1345–59. https://doi.org/10.1109/TKDE.2009.191.

Panait, Liviu, and Sean Luke. 2005. "Cooperative Multi-Agent Learning: The
State of the Art." *Autonomous Agents and Multi-Agent Systems* 11, no. 3
(November): 387–434. https://doi.org/10.1007/s10458-005-2631-2.

Papernot, Nicolas, Patrick McDaniel, Somesh Jha, Matt Fredrikson, Z. Berkay
Celik, and Ananthram Swami. 2016. The Limitations of Deep Learning
in Adversarial Settings." In *ArXiv:1511.07528 [Cs, Stat]*. Saarbrucken,
Germany. http://arxiv.org/abs/1511.07528.

Parasuraman, Raja, and Scott Galster. 2013. "Sensing, Assessing, and Augmenting Threat Detection: Behavioral, Neuroimaging, and Brain Stimulation

Evidence for the Critical Role of Attention." *Frontiers in Human Neuro-science* 7 (June): 1–10. https://doi.org/10.3389/fnhum.2013.00273.

Parasuraman, Raja, and Dietrich H. Manzey. 2010. "Complacency and Bias in Human Use of Automation: An Attentional Integration." *Human Factors* 52, no. 3 (June): 381–410. https://doi.org/10.1177/0018720810376055.

Parasuraman, Raja, and Christopher A. Miller. 2004. "Trust and Etiquette in High-Criticality Automated Systems." *Communications of the ACM* 47, no. 4 (April): 51–55. https://doi.org/10.1145/975817.975844.

Parasuraman, Raja, and Victor Riley. 1997. "Humans and Automation: Use, Misuse, Disuse, Abuse." *Human Factors* 39, no. 2 (June): 230–53. https://doi.org/10.1518/001872097778543886.

Parasuraman, Raja, Thomas B. Sheridan, and Christopher D. Wickens. 2000. "A Model for Types and Levels of Human Interaction with Automation." *IEEE Transactions on Systems, Man, and Cybernetics - Part A: Systems and Humans* 30, no. 3 (May): 286–97. https://doi.org/10.1109/3468.844354.

———. 2008. "Situation Awareness, Mental Workload and Trust in Automation: Viable Empirically Supported Cognitive Engineering Constructs." *Journal of Cognitive Engineering and Decision Making* 2, no. 2 (2008): 140–60. https://doi.org/10.1518/155534308X284417.

Parker, D. B. 1985. "Learning Logic: Casting the Human Brain in Silicon," Tech. Rep. TR-47. Center for Computational Research in Economics and Management Science. Cambridge, MA: Massachusetts Institute of Technology.

Parker, Geoffrey, and Marshall W. van Alstyne. 2016. " 'The Platform Revolution'—An Interview with Geoffrey Parker and Marshall Van Alstyne." Interview by Sangeet Paul Choudhary. *Marketing Journal*, March 30, 2016. http://www.marketingjournal.org/the-platform-revolution-an-interview-with-geoffrey-parker-and-marshall-van-alstyne/.

Patrascu, Monica, Monica Dragoicea, and Andreea Ion. 2014. "Emergent Intelligence in Agents: A Scalable Architecture for Smart Cities." In 2014 *18th International Conference on System Theory, Control and Computing (ICSTCC)*. Sinaia, Romania: IEEE. https://doi.org/10.1109/ICSTCC.2014.6982412.

Patterson, Robert Earl. 2012. "Cognitive Engineering, Cognitive Augmentation, and Information Display." *Journal of the Society for Information Display* 20, no. 4 (April): 208–13. https://doi.org/10.1889/JSID20.4.208.

———. 2017. "Intuitive Cognition and Models of Human-Automation Interaction." *Human Factors* 59 (1): 101–15. https://doi.org/10.1177/0018720816659796.

Patterson, Robert Earl, and Robert G. Eggleston. 2017. "Intuitive Cognition." *Journal of Cognitive Engineering and Decision Making* 11, no. 1 (March): 5–22. https://doi.org/10.1177/1555343416686476.

Paugam-Mosiy, Helene. 2006. "Spiking Neuron Networks: A Survey" (IDIAP-RR 06-11). IDIAP Research Institute.

Pearl, Judea. 1988. *Probabilistic Reasoning in Intelligent Systems*. San Francisco: Morgan Kaufmann.

Peirce, Charles S. 1960. *Collected Papers of Charles Sanders Pierce*. Vol. 2. Cambridge, MA: Belknap Press of Harvard University Press.

———. 1991. *Peirce on Signs: Writings on Semiotic*. Edited by James Hoopes. Chapel Hill: University of North Carolina Press.

Pellerin, Cheryl. 2016. "Deputy Secretary: Third Offset Strategy Bolsters America's Military Deterrence." DoD News, October 31, 2016. https://dod.defense.gov/News/Article/Article/991434/deputy-secretary-third-offset-strategy-bolsters-americas-military-deterrence/.

Penfield, Wilder, and Theodore Rasmussen. 1950. "The Cerebral Cortex of Man: A Clinical Study of Localization of Function." *Journal of the American Medical Association* 144, no. 16 (December): 1412. https://doi.org/10.1001/jama.1950.02920160086033.

Pettersson, Ola. 2005. "Execution Monitoring in Robotics: A Survey." *Robotics and Autonomous Systems* 53, no. 2 (November): 73–88. https://doi.org/10.1016/j.robot.2005.09.004.

Pew, Richard W., and Anne S. Mavor, eds. 1998. *Modeling Human and Organizational Behavior: Application to Military Simulations*. Washington, DC: National Research Council, The National Academy Press.

Pfeffer, Avi. 2016. *Practical Probabilistic Programming*. Shelter Island, NY: Manning Publications.

Pfeifer, Rolf, and Josh Bongard. 2007. *How the Body Shapes the Way We Think*. Cambridge, MA: MIT Press.

Piatetsky, Gregory. 2014. "CRISP-DM, Still the Top Methodology for Analytics, Data Mining, or Data Science Projects." *KDnuggets* (website). October 28, 2014. http://www.kdnuggets.com/2014/10/crisp-dm-top-methodology-analytics-data-mining-data-science-projects.html.

Picard, Rosalind W. 1997. *Affective Computing*. Vol. 252. Cambridge, MA: MIT Press.

———. 2003. "Affective Computing: Challenges." *International Journal of Human-Computer Studies* 59, no. 1–2 (July): 55–64. https://www.sciencedirect.com/science/article/pii/S1071581903000521.

Piccinini, Gualtiero. 2006. "Computational Explanation in Neuroscience." *Synthese* 153, no. 3 (December): 343–53. https://doi.org/10.1007/s11229-006-9096-y.

Picoh, Nicholas J., Daniel Hunter, James V. White, Amy Kao, Daniel Bostwick, and Eric K. Jones. 2004. "Multi-Hypothesis Abductive Reasoning for Link Discovery." *Proceedings of KDD-2004.*

Pierce, David. 2018. "Inside the Lab Where Amazon's Alexa Takes Over the World." *WIRED*, January 8, 2018. https://www.wired.com/story/amazon-alexa-development-kit/.

Pitts, Walter, and Warren S. McCulloch. 1947. "On How We Know Universals: The Perception of Auditory and Visual Forms." *Bulletin of Mathematical Biophysics* 9, no. 3 (September):127–47. https://doi.org/10.1007/BF02478291.

Polk, Thad A., and Colleen Seifert, eds. 2002. *Cognitive Modeling.* Cambridge, MA: MIT Press.

Pomerleau, Mark. 2015. "Drone Innovators Showcase Single-Operator Capability." Defense Systems, March 4, 2015. https://defensesystems.com/articles/2015/03/04/drone-showcase-single-operator-control.aspx.

Pontefract, Dan. 2015. "What Is Happening at Zappos?" Forbes.com, May 11, 2015. https://www.forbes.com/sites/danpontefract/2015/05/11/what-is-happening-at-zappos/.

Posner, M. I. 2012. "Imaging Attention Networks." *Neuroimage* 61, no. 2 (June): 450–56. https://doi.org/10.1016/j.neuroimage.2011.12.040.

Potember, R. 2017. "Perspectives on Research in Artificial Intelligence and Artificial General Intelligence Relevant to DoD" (JSR-16-Task-003). McLean, VA: The MITRE Corporation. https://fas.org/irp/agency/dod/jason/ai-dod.pdf.

Praeger Security International. 2018. "Should Cyber Attacks on the United States by Other Nations Be Considered an Act of War?" October 15, 2018. https://psi-praeger-com.aufric.idm.oclc.org/Search/Display/2174719?fromhome =true.

"Predix Platform | GE Digital," October 12, 2018. https://www.ge.com/digital/iiot-platform.

Prokhorov, Danil, ed. 2008. *Computational Intelligence in Automotive Applications.* Berlin: Springer-Verlag.

Puranik, Marty. 2017. "5 Cloud Computing Trends to Prepare for in 2018." *Network World* (website), October 18, 2017. https://www.networkworld.com/article/3233134/cloud-computing/5-cloud-computing-trends-to-prepare-for-in-2018.html.

"Putin: Leader in Artificial Intelligence Will Rule World." CNBC, October 12, 2018. https://www.cnbc.com/2017/09/04/putin-leader-in-artificial-intelligence-will-rule-world.html.

Quiroga, R. Quian, L. Reddy, G. Kreiman, C. Koch, and I. Fried. 2005. "Invariant Visual Representation by Single Neurons in the Human Brain." *Nature* 435:1102–1107. https://doi.org/10.1038/nature03687.

Quora Session. 2016. Quora Session with Yann LeCun. (Quora, Interviewer) Retrieved from https://www.quora.com/session/Yann-LeCun/1.

Ramachandran, V. S. and W. Hirstein. 1997. "Three Laws of Qualia: What Neurology Tells Us about the Biological Functions of Consciousness." *Journal of Consciousness Studies* 4 (5–6): 429-457.

Rao, Anand, and Michael Georgeff. 1995. "BDI Agents: From Theory to Practice." *Proceedings of the First International Conference on Multiagent Systems.* Menlo Park, CA: AAAI.

Read, Dwight W. 2002. "A Multitrajectory, Competition Model of Emergent Complexity in Human Social Organization." *Proceedings of the National Academy of Sciences of the United States of America* 99, no. 10 (May): 7251–56. https://www.jstor.org/stable/i354634.

Reed, Daniel A., and Jack Dongarra. 2015. "Exascale Computing and Big Data." *Communications of the ACM* 58, no. 7 (July): 56–68. http://doi.org/f7h2dw.

Reese, Hope. 2016. "AI Experts Weigh in on Microsoft CEO's 10 New Rules for Artificial Intelligence." *TechRepublic*, June 30, 2016. https://www.techrepublic.com/article/ai-experts-weigh-in-on-microsoft-ceos-10-new-rules-for-artificial-intelligence/.

Reggia, James A. 2013. "The Rise of Machine Consciousness: Studying Consciousness with Computational Models." *Neural Networks: The Official Journal of the International Neural Network Society* 44 (August): 112–31. https://doi.org/10.1016/j.neunet.2013.03.011.

Reim, Garrett. 2018. "ANALYSIS: US Air Force Eyes Adoption of 'Loyal Wingman' UAVs." FlightGlobal, September 13, 2018. https://www.flightglobal.com/news/articles/analysis-us-air-force-eyes-adoption-of-loyal-wingm-451383/.

Reyna, Valerie F., Christina F. Chick, Jonathan C. Corbin, and Andrew N. Hsia. 2014. "Developmental Reversals in Risky Decision Making: Intelligence Agents Show Larger Decision Biases Than College Students." *Psychological Science* 25, no. 1 (January): 76–84. https://doi.org/10.1177/0956797613497022.

Reyna, Valerie F., and Farrell J. Lloyd. 2006. "Physician Decision Making and Cardiac Risk: Effects of Knowledge, Risk Perception, Risk Tolerance, and Fuzzy Processing." *Journal of Experimental Psychology: Applied* 12, no. 3 (September): 179–95. https://doi.org/10.1037/1076-898X.12.3.179.

Riek, Laurel D. 2012. "Wizard of Oz Studies in HRI: A Systematic Review and New Reporting Guidelines." *Journal of Human-Robot Interaction* 1 (1): 119–36.

Riley, Victor. 1989. "A General Model of Mixed-Initiative Human-Machine Systems." *Proceedings of the Human Factors Society Annual Meeting* 33, no. 2 (October): 124–28. https://doi.org/10.1177/154193128903300227.

Ritter, Frank E., Nigel R. Shadbolt, David G. Elliman, Richard Young, Fernand Gobet, and Gordon D. Baxter. 2003. *Techniques for Modeling Human Performance in Synthetic Environments: A Supplementary Review.* Wright-Patterson Air Force Base, OH: Human Systems Information Analysis Center.

Rochester, N., J. H. Holland, L. H. Habit, and W. L. Duda. 1956. "Tests on a Cell Assembly Theory of the Action of the Brain, Using a Large Digital Computer." *IRE Transactions on Information Theory* 2, no. 3 (September), 80–93. https://doi.org/10.1109/TIT.1956.1056810.

Roger, Gabriele, Florian Pommerening, and Jendrik Seipp. 2014. "Fast Downward Stone Soup 2014." *Eighth International Planning Competition*, 28–31. https://ai.dmi.unibas.ch/papers/roeger-et-al-ipc2014.pdf.

Rogers, Steven, David Ryer, and Jeffrey Eggers. 2014. "Sensing as a Service: An ISR Horizons Future Vision." Washington, DC: Headquarters US Air Force. https://defenseinnovationmarketplace.dtic.mil/wp-content /uploads/2018/02/20140612_SensingAsAService-signed_PAR.pdf.

Rogers, Steven K., Charles Sadowski, Kenneth W. Bauer, Mark E. Oxley, Matthew Kabrisky, Adam Rogers, and Stephen D. Mott. 2008. "The Life and Death of ATR/Sensor Fusion and the Hope for Resurrection." *Proceedings Volume 6967, Automatic Target Recognition XVIII*; SPIE Defense and Security Symposium, 2008, Orlando, Florida. https://doi.org /10.1117/12.783012.

Ronald, Edmund M. A., Moshe Sipper, and Mathieu S. Capcarrère. 2007. "Design, Observation, Surprise! A Test of Emergence." *Artificial Life* 5, no. 3 (July): 225–39. https://doi.org/10.1162/106454699568755.

Rosenblatt, F. 1958. "The Perceptron: A Probabilistic Model for Information Starage and Organization in The Brain." *Psychological Review* 65 (6): 386–408.

Rothwell, Clayton D., and Valerie L. Shalin. 2017. "Human-Machine Articulation Work: Functional Dependency Dialogue for Human-Machine Teaming." In 19th *International Symposium on Aviation Psychology*,

270–75. Dayton, OH, 2017. https://corescholar.libraries.wright.edu/isap_2017/20.

Rouse, Margaret. 2008. "Data Analytics." In "How to Solve Your TMI Problem: Data Science Analytics to the Rescue." TechTarget Network (website). Last updated December 2016. http://searchdatamanagement.techtarget.com/definition/data-analytics.

Rubenstein, Joshua S., David E. Meyer, and Jeffrey E. Evans. 2001. "Executive Control of Cognitive Processes in Task Switching." *Journal of Experimental Psychology: Human Perception and Performance* 27 (4): 763–97. http://dx.doi.org/10.1037/0096-1523.27.4.763.

Rumelhart, David E., Geoffrey E. Hinton, and Ronald J. Williams. 1986. "Learning Internal Representations by Error Propagation." In *Parallel Distributed Processing: Explorations in the Microstructure of Cognition*, edited by David E. Rumelhart and James L. McClelland, 1, 318–62. Cambridge, MA: MIT Press.

Russell, Stuart. 2015. "Recent Developments in Unifying Logic and Probability." *Communications of the ACM* 58, no. 7 (July): 88–97. http://dl.acm.org/citation.cfm?id=2797100.2699411.

———. 2016. "Should We Fear Supersmart Robots?" *Scientific American* 314, no. 6 (June): 58–59. https://doi.org/10.1038/scientificamerican0616-58.

Russell, Stuart, Daniel Dewey, and Max Tegmark. 2015. "Research Priorities for Robust and Beneficial Artificial Intelligence." *AI Magazine* 36, no. 4 (Winter): 105–14.

Russell, Stuart J., and Peter Norvig. 1995. *Artificial Intelligence: A Modern Approach*. Englewood Cliffs, NJ: Prentice-Hall.

———. 2010. *Artificial Intelligence: A Modern Approach*. 3rd ed. Upper Saddle River, NJ: Pearson Education Limited.

Russo, J. Edward, Eric J. Johnson, and Debra L. Stephens. 1989. "The Validity of Verbal Protocols." *Memory and Cognition* 17, no. 6 (November): 759–69. https://doi.org/10.3758/BF03202637.

SAE International. 1996. *Guidelines and Methods for Conducting the Safety Assessment Process on Civil Airborne Systems and Equipment*. ARP4761. S18 Aircraft and System Development and Safety Assessment Committee. https://doi.org/10.4271/ARP4761.

Salas, Eduardo, Nancy J. Cooke, and Michael A. Rosen. 2008. "On Teams, Teamwork, and Team Performance: Discoveries and Developments." *Human Factors* 50, no. 3 (June): 540–47. https://doi.org/10.1518/001872008X288457.

Salas, Eduardo, Carolyn Prince, David P. Baker, and Lisa Shrestha. 1995. "Situation Awareness in Team Performance: Implications for Measurement and Training." *Human Factors* 37, no. 1 (March): 123–36. https://doi.org/10.1518/001872095779049525.

Salas, Eduardo, Dana E. Sims, and C. Shawn Burke. 2005. "Is There a 'Big Five' in Teamwork?" *Small Group Research* 36, no. 5 (October): 555–99. https://doi.org/10.1177/1046496405277134.

Salmon, Merrilee H. 2013. *Introduction to Logic and Critical Thinking.* 6th ed. Boston: Wadsworth.

Samsonovich, Alexei, Kamilla R. Johannsdottir, Antonio Chella, and Ben Goertzel, eds. 2010. *Biologically Inspired Cognitive Architectures: Proceedings of the First Annual Meeting of the BICA Society.* Amsterdam: IOS Press.

Sanz, Ricardo, Ignacio López, and Julita Bermejo-Alonso. 2007. "A Rationale and Vision for Machine Consciousness in Complex Controllers." In *Artificial Consciousness*, edited by Antonio Chella and Riccardo Manzotti, 141–55. New York: Springer.

Sarter, Nadine B., and David D. Woods. 1995. "How in the World Did We Ever Get into That Mode? Mode Error and Awareness in Supervisory Control." *Human Factors* 37, no. 1 (March): 5–19. https://doi.org/10.1518/001872095779049516.

Sarva, Harini, Andres Deik, and William Lawrence Severt. 2014. "Pathophysiology and Treatment of Alien Hand Syndrome." *Tremor and Other Hyperkinetic Movements* 4:241. https://dx.doi.org/10.7916%2FD8VX0F48.

Savage, Neil. 2016. "Seeing More Clearly." *Communications of the ACM* 59, no. 1 (January): 20–22. https://doi.org/10.1145/2843532.

Sawtelle, Jonathan D. 2016. *Resilient Effective Adaptable Leadership.* Maxwell AFB, AL: Air University Press.

Schank, Roger C., and Robert P. Abelson. 1977. Scripts, Plans, Goals and *Understanding: An Inquiry into Human Knowledge Structures.* Hillsdale, NJ: Lawrence Erlbaum Associates.

Schank, Roger C. and Larry Tesler. 1969. "A Conceptual Dependency Parser for Natural Language." *Proceedings of the 1969 Conference on Computational Linguistics*, September 1–4, 1969, Sång-Säby, Sweden. Stockholm: Research Group for Quantitative Linguistics. https://doi.org/10.3115/990403.990405.

Scharre, Paul. 2015. "Between a Roomba and a Terminator: What Is Autonomy?" War on the Rocks (blog), February 18, 2015. https://warontherocks.com/2015/02/between-a-roomba-and-a-terminator-what-is-autonomy/.

———. 2016. "Autonomous Weapons and Operational Risk." Center for a New American Security, February 29, 2016. https://www.cnas.org/publications/reports/autonomous-weapons-and-operational-risk.

Schechner, Sam, Douglas MacMillan, and Liza Lin. 2017. "U.S. and Chinese Companies Race to Dominate AI." *Wall Street Journal*. Updated January 18, 2018. https://www.wsj.com/articles/why-u-s-companies-may-lose -the-ai-race-1516280677.

Schmidhuber, Juergen. 2007. "Godel Machines: Fully Self-referential Optimal Universal Self-Improvers." In *Artificial General Intelligence*, edited by Ben Goertzel and Cassio Pennachin, 199–226. New York: Springer.

———. 2014. "Deep Learning in Neural Networks: An Overview." *Neural Networks* 61 (January): 85–117. https://doi.org/10.1016/j.neunet.2014.09.003.

Schneider, David. 2017. "Deeper and Cheaper Machine Learning." *IEEE Spectrum* 54, no. 1 (January): 42–43. https://doi.org/10.1109/MSPEC.2017.7802746.

Schölkopf, Bernhard, and Alexander J. Smola. 2002. *Learning with Kernels: Support Vector Machines, Regularization, Optimization*. Cambridge, MA: MIT Press.

Schooler, Jonathan W., Stellan Ohlsson, and Kevin Brooks. 1993. "Thoughts Beyond Words: When Language Overshadows Insight." *Journal of Experimental Psychology: General* 122 (2): 166–83. http://psycnet.apa.org /doi/10.1037/0096-3445.122.2.166.

Schumann, Johann, Pramod Gupta, and Stacy Nelson. 2003. "On Verification and Validation of Neural Network Based Controllers." International Conference on Engineering Applications of Neural Networks, 3.

Schumann, Johann, Kristin Y. Rozier, Thomas Reinbacher, Ole J. Mengshoel, Timmy Mbaya, and Corey Ippolito. 2013. "Towards Real-time, On-board, Hardware-supported Sensor and Software Health Management for Unmanned Aerial Systems." Annual Conference of the Prognostics and Health Management Society, New Orleans, 2013. https://www .phmsociety.org/sites/phmsociety.org/files/phm_submission/2013 /phmc_13_053.pdf.

Schupp, Harald T., Tobias Flaisch, Jessica Stockburger, and Markus Junghofer. 2006. "Emotion and Attention: Event-Related Brain Potential Studies." In *Progress in Brain Research*. Vol. 156, edited by S. Anders, G. Ende, M. Junghofer, J. Kissler, and D. Wildgruber. Amsterdam: Elsevier.

Schwartz, Casey. 2011. "Data-to-Decisions: S&T Priority Initiative." Slide presentation. NDIA Disruptive Technologies Conference, Washington, DC, November 8–9, 2011. https://ndiastorage.blob.core.usgovcloudapi .net/ndia/2011/disruptive/Schwartz.pdf.

Sears, Andrew, and Ben Shneiderman. 1991. "High Precision Touchscreens: Design Strategies and Comparisons with a Mouse." *International Journal of Man-Machine Studies* 34, no. 4 (April): 593–613. https://doi .org/10.1016/0020-7373(91)90037-8.

Sentz, Kari, and Scott Ferson. 2002. "Combination of Evidence in Dempster-Shafer Theory." Sandia Report SAND2002-0835. Albuquerque, NM: Sandia National Labs. https://doi.org/10.2172/800792.

Shanahan, Murray. 1996. "Robotics and the Common Sense Informatic Situation." In *Proceedings of the 12th European Conference on Artificial Intelligence*, edited by Wolfgang Wahlster, 684–88. New York: Wiley.

Shang, Lifeng, Zhengdong Lu, and Hang Li. 2015. "Neural Responding Machine for Short-Text Conversation." In *Proceedings of the 53rd Annual Meeting of the Association for Computational Linguistics and the 7th International Joint Conference on Natural Language Processing*. Vol. 1, *Long Papers*, 1577–86. Beijing, China. http://aclweb.org/anthology/P15-1000.

Shapiro, Ehud Y. 1983. "Logic Programs with Uncertainties: A Tool for Implementing Rule-Based Systems." In *Proceedings of the 8th International Joint Conference on Artificial Intelligence*. Vol. 1, 529–32. Karlsruhe, West Germany, August 8–12, 1983.

Shearer, Colin. 2000. "The CRISP-DM Model: The New Blueprint for Data Mining." *Journal of Data Warehousing* 22, no. 4 (Fall): 13–22. https://mineracaodedados.files.wordpress.com/2012/04/the-crisp-dm-model-the-new-blueprint-for-data-mining-shearer-colin.pdf.

Sheridan, Thomas B. 1976. "Toward a General Model of Supervisory Control." In *Monitoring Behavior and Supervisory Control*, edited by Thomas B. Sheridan and Gunnar Johannsen, 271–81. NATO Conference Series. Vol. 1. Boston: Springer.

Sheridan, Thomas B., and William Lawrence Verplank. 1978. *Human and Computer Control of Undersea Teleoperators*. Cambridge, MA: MIT Press.

Shladover, Steven. 2016. "What 'Self-Driving' Cars will Really Look Like." *Scientific American* (June). https://www.scientificamerican.com/article/what-self-driving-cars-will-really-look-like/.

Shneiderman, Ben. 1983. "Direct Manipulation: A Step Beyond Programming Languages." *Computer* 16, no. 8 (August): 57–69. https://doi.org/10.1109/MC.1983.1654471.

———. 1992. *Designing the User Interface: Strategies for Effective Human-computer Interactions*. 2nd ed. Reading, MA: Addison-Wesley.

Shneiderman, Ben, and C. Plaisant. 2005. *Designing the User Interface*. 4th ed. Boston: Addison-Wesley.

Shu, Yufei, and Kazuo Furuta. 2005. "An Inference Method of Team Situation Awareness Based on Mutual Awareness." *Cogn. Technol. Work* 7, no. 4 (November): 272–87. https://doi.org/10.1007/s10111-005-0012-x.

Silberstein, Michael, and John McGeever. 1999. "The Search for Ontological Emergence." *Philosophical Quarterly* 49, no. 195 (April): 201–14. https://doi.org/10.1111/1467-9213.00136.

Silver, David, Aja Huang, Chris J. Maddison, Arthur Guez, Laurent Sifre, George van den Driessche, Julian Schrittwieser, et al. 2016. "Mastering the Game of Go with Deep Neural Networks and Tree Search." *Nature* 529:484-489. https://doi.org/10.1038/nature16961.

Silver, David, Julian Schrittwieser, Karen Simonyan, Ioannis Antonoglou, Aja Huang, Arthur Guez, Thomas Hubert, et al. 2017. "Mastering the Game of Go without Human Knowledge." Nature 550:354. https://www.nature.com/articles/nature24270.

Simmons, Chris B., Sajjan Shiva, HarKeerat Bedi, and Dipankar Dasgupta. 2014. "AVOIDIT: A Cyber Attack Taxonomy." In *Proceedings of the 9th Annual Symposium on Information Assurance* (ASIA '14), 2–12. Albany, NY.

Simoes, Alexandre da Silva, Esther Luna Colombini, and Carlos Henrique Ribeiro. 2017. "CONAIM: A Conscious Attention-Based Integrated Model for Human-Like Robots." *IEEE Systems Journal* 11, no. 3 (September): 1296–1307. https://doi.org/10.1109/JSYST.2015.2498542.

Simons, John. 2016. "Tomorrow's Business Leaders Learn How to Work with AI." *Wall Street Journal*, November 30, 2016. https://www.wsj.com/articles/tomorrows-business-leaders-learn-how-to-work-with-a-i-1480517287.

Singh, Push. 2003. "Examining the Society of Mind." *Computing and Informatics* 22, no. 6 (February): 521–43.

Singh, Pushpinder. 2005. "EM-ONE: An Architecture for Reflective Commonsense Thinking." Thesis, Massachusetts Institute of Technology. http://dspace.mit.edu/handle/1721.1/33926.

Sloman, Aaron. 2001. "Beyond Shallow Models of Emotion." In *Cognitive Processing: International Quarterly of Cognitive Science* 2:177–198.

Sloman, Aaron, and Ron Chrisley. 2003. "Virtual Machines and Consciousness." *Journal of Consciousness Studies* 10 (4–5): 133–72. https://pdfs.semanticscholar.org/a039/0e02e8e0cd23a3f908cb9248c111faabe1a1.pdf.

Smith, David Canfield, Allen Cypher, and Jim Spohrer. 1994. "KidSim: Programming Agents without a Programming Language." *Communications of the ACM* 37, no. 7 (July): 54–67. https://doi.org/10.1145/176789.176795.

Smith, Philip J. 2017. "Making Brittle Technologies Useful." In *Cognitive Systems Engineering: The Future for a Changing World*, edited by Philip J. Smith and Robert R. Hoffman. Boca Raton, FL: CRC Press.

Socher, Richard, Milind Ganjoo, Christopher D. Manning, and Andrew Y. Ng. 2013. "Zero-Shot Learning through Cross-Modal Transfer." In *Proceedings of the 26th International Conference on Neural Information Processing Systems.* Vol. 1, 935–43. Lake Tahoe, NV.

Soto-Faraco, S., J. Lyons, M. Gazzaniga, C. Spence, and A. Kingstone. 2002. "The Ventriloquist in Motion: Illusory Capture of Dynamic Information across Sensory Modalities." *Cognitive Brain Research* 14, no. 1 (June): 139–46.

Spears, William M., Diana F. Spears, Jerry C. Hamann, and Rodney Heil. 2004. "Distributed, Physics-Based Control of Swarms of Vehicles." *Autonomous Robots* 17, no. 2 (September): 137–62. https://doi.org/10.1023/ B:AURO .0000033970.96785.f2.

Squire, Larry R. 2004. "Memory Systems of the Brain: A Brief History and Current Perspective." *Neurobiology of Learning and Memory* 82, no. 3 (November): 171–77. https://doi.org/10.1016/j.nlm.2004.06.005.

———. 2009. "Memory and Brain Systems: 1969-2009." *Journal of Neuroscience: The Official Journal of the Society for Neuroscience* 29, no. 41 (October): 12711–16. https://doi.org/10.1523/JNEUROSCI.3575-09.2009.

Stanley, Kenneth O., and Risto Mikkulainen. 2002. "Evolving Neural Networks through Augmenting Topologies." *Evolutionary Computation* 10 (2): 99–127.

Stanovich, Keith E. 2011. *Rationality and the Reflective Mind.* New York: Oxford University Press, 2010.

St-Aubin, Robert, Jesse Hoey, and Craig Boutilier. 2001. "APRICODD: Approximate Policy Construction Using Decision Diagrams." *Advances in Neural Information Processing Systems* (NIPS).

Steele, Julia, and Noah Iliinsky. 2010. *Beautiful Visualization: Looking at Data through the Eyes of Experts.* Sebastopol, CA: O'Reilly Media.

Stevens, Michael J., and Michael A. Campion. 1994. "The Knowledge, Skill, and Ability Requirements for Teamwork: Implications for Human Resource Management." *Journal of Management* 20, no. 2 (April): 503–30. https://doi.org/10.1177/014920639402000210.

Stuckenbruck, Linn C. 1979. "The Matrix Organization." *Project Management Quarterly* 10, no. 3 (September): 21–33.

Sun, Ron. 2002. *Duality of the Mind: A bottom-up Approach Toward Cognition.* Mahwah, NJ: Lawrence Erlbaum Associates.

———. 2003. "A Tutorial on CLARION 5.0." July 22, 2003. http://www.cogsci .rpi.edu/~rsun/sun.tutorial.pdf.

———. 2004. "Desiderata for Cognitive Architectures." *Philosophical Psychology* 17 (3): 341–73. https://doi.org/10.1080/0951508042000286721.

———. 2007. "The Importance of Cognitive Architectures: An Analysis Based on CLARION." *Journal of Experimental and Theoretical Artificial Intelligence* 19 (2): 159–93. https://doi.org/10.1080/09528130701191560.

Sun, Ron, ed. 2006. *Cognition and Multi-Agent Interaction: From Cognitive Modeling to Social Simulation*. Cambridge, UK: Cambridge University Press, 2006.

Sun, Ron, and Xi Zhang. 2006. "Accounting for a Variety of Reasoning Data within a Cognitive Architecture." *Journal of Experimental and Theoretical Artificial Intelligence* 18 (2): 169–91. https://doi.org/10.1080/09528130600557713.

Sun, Ron, Xi Zhang, and Robert Matthews. 2006. "Modeling Meta-Cognition in a Cognitive Architecture." *Cognitive Systems Research* 7, no. 4 (December): 327–38. https://doi.org/10.1016/j.cogsys.2005.09.001.

Swanson, H. Lee. 1990. "Influence of Metacognitive Knowledge and Aptitude on Problem Solving." *Journal of Educational Psychology* 82 (2): 306–14. https://doi.org/10.1037/0022-0663.82.2.306.

Szu, Harold, and Ralph Hartley. 1987. "Fast Simulated Annealing." *Physics Letters A* 122 (3/4): 157–62. https://doi.org/10.1016/0375-9601(87)90796-1.

Talbert, Michael L., Patrick Baldwin, and Guna Seetharaman. 2006. "Information Expectation from Unmanned Aircraft Swarms." In *Next-Generation Communication and Sensor Networks 2006*, 6387:63870A. International Society for Optics and Photonics. https://doi.org/10.1117/12.686315.

Tate, Austin, John Levine, Peter Jarvis, and Jeff Dalton. 2000. "Using AI Planning Technology for Army Small Unit Operations." *Proceedings of the Fifth International Conference on Artificial Intelligence Planning Systems* (AIPS_2000), edited by Steve Chien, Subbarao Kambhampati, and Craig A. Knoblock, 379–86. https://www.aaai.org/Papers/AIPS/2000/AIPS00-042.pdf.

Tauer, Gregory, Rakesh Nagi, and Moises Sudit. 2013. "The Graph Association Problem: Mathematical Models and a Lagrangian Heuristic." *Naval Research Logistics (NRL)* 60, no. 3 (April): 251–68. https://doi.org/10.1002/nav.21532.

Taylor, Glenn, Ben Purman, Paul Schermerhorn, Guillermo Garcia-Sampedro, Robert Hubal, Kathleen Crabtree, Allen Rowe, and Sarah Spriggs. 2015. "Multi-Modal Interaction for UAS Control." In *SPIE Preceedings*, edited by Robert E. Karlsen, Douglas W. Gage, Charles M. Shoemaker, and Grant R. Gerhart, Vol. 946802. Baltimore: SPIE, 2015. https://doi.org/10.1117/12.2180020.

The Economist. 2011. "Flight of the Drones." October 8, 2011. https://www.economist.com/node/21531433.

"The History of the Modern Graphics Processor." TechSpot, October 15, 2018. https://www.techspot.com/article/650-history-of-the-gpu/.

"The Role of Autonomy in DoD Systems." (n.d.). 125.

Thelen, Esther, and Linda B. Smith. 1996. *A Dynamic Systems Approach to Development of Cognition and Action*. Cambridge, MA: MIT Press, A Bradford Book.

Thompson, William R. 1999. *Great Power Rivalries*. Columbia, S.C: University of South Carolina Press, 1999.

Thorsen, Steven N. 2018. "The Application of Category Theory and Analysis of Receiver Operating Characteristics to Information Fusion." Dissertation, Air Force Institute of Technology, 2005.

Tononi, Giulio, Melanie Boly, Marcello Massimini, and Christof Koch. 2016. "Integrated Information Theory: From Consciousness to Its Physical Substrate." *Nature Reviews Neuroscience* 17:450–61.

Torralba, A., V. Alcazar, and D. Borrajo. (n.d.). "SymBA*: A Symbolic Bidirectional A* Planner." The 2014 International Planning Competition. 105–9.

Torres y Quevedo, Leonardo. 1903. Patente de Invención por un Sistema Denominado Telekino para Gobernar a Distancia un Movimiento Mecánico (Patent Application for a System, Called Telekine, to Steer a Mechanical Movement at Distance). Madrid, Spain: Spanish Patents and Trademarks Office, June 10, 1903.

Toulmin, Stephen E. 1959. *The Uses of Argument*. Cambridge, UK: Cambridge University Press.

Tran, Trung. 2016. "Evolution of Machine Learning: From Expert to Exploratory Systems." Seminar presented at the DARPA SAGA Workshop, Dayton, OH.

Treisman, Anne M. 1964. "Selective Attention in Man." *British Medical Bulletin* 20, no. 1 (January): 12–16. https://doi.org/10.1093/oxfordjournals.bmb.a070274.

Tsallis, Constantino, and Daniel Adrian Stariolo. 1996. "Generalized Simulated Annealing." *Physica A: Statistical and Theoretical Physics* 233 (1-2): 395–406. http://dx.doi.org/10.1016/S0378-4371(96)00271-3.

Turing, Alan M. 1950. "Computing Machinery and Intelligence." *Mind* 49:433–60. https://doi.org/10.1093/mind/LIX.236.433.

Turner, Karen. 2016. "Google Translate Is Getting Really, Really Accurate." *Washington Post*, October 3, 2016. https://www.washingtonpost.com/news/innovations/wp/2016/10/03/google-translate-is-getting-really-really-accurate/?noredirect=on&utm_term=.5693aaa42d65.

Tversky, Amos, and Daniel Kahneman. 1973. "Availability: A Heuristic for Judging Frequency and Probability." *Cognitive Psychology* 5, no. 2 (September): 207–32. https://doi.org/10.1016/0010-0285(73)90033-9.

———. 1974. "Judgment under Uncertainty: Heuristics and Biases." *Science* 185, no. 4157 (1974): 1124–31.

Tversky, Amos, and Daniel Kahneman. 1992. "Advances in Prospect Theory, Cumulative Representation of Uncertainty." *Journal of Risk and Uncertainty* 5 (4): 297–323. https://www.jstor.org/stable/41755005.

Uebersax, John S. 2004. "Genetic Counseling and Cancer Risk Modeling: An Application of Bayes Networks." Research report. Marbella, Spain: Ravenpack International.

Underwood, Sarah. 2017. "Potential and Peril." *Communications of the ACM* 60, no. 6 (May): 17–19. https://doi.org/10.1145/3077231.

USAF. 2014. "Future Capabilities Wargame 2013 Final Report." May 2014. Directorate of Strategic Planning, Future Concepts Division. Washington, DC.

———. 2015. "Air Force Future Operating Concept: A View of the Air Force in 2035." Washington, DC: Department of the Air Force, September 2015. https://www.af.mil/Portals/1/images/airpower/AFFOC.pdf.

———. 2016. "Future Capabilities Wargame 2015 Final Report." February 2016. Strategic Plans, Programs, and Requirements Wargaming Division. Washington DC.

USAF ECCT (Enterprise Capability Collaboration Team). 2016. *Air Superiority 2030 Flight Plan*. Washington, DC: Office of the Secretary of the Air Force (Public Affairs).

USAF Space Command. 2015. "Defense Support Program Satellites." November 23, 2015. https://www.af.mil/About-Us/Fact-Sheets/Display/Article/104611/defense-support-program-satellites/.

Valenzano, Richard, Jonathan Schaeffer, and Nathan Sturtevant. 2013. "Finding Better Candidate Algorithms for Portfolio-Based Planners." *Proceedings of the 23rd International Conference on Automated Planning and Scheduling* (ICAPS '13). https://pdfs.semanticscholar.org/9a2e/e1f3fb731c481c2d61a304c8bcd99307d77d.pdf.

Vallati, Mauro, Lukas Chrpa, Marek Grzes, Thomas L. McCluskey, Mark Roberts, and Scott Sanner. 2015. "The 2014 International Planning Competition: Progress and Trends." *AI Magazine* 36, no. 3 (Fall): 90–98. https://doi.org/10.1609/aimag.v36i3.2571.

Valve. 2012. *Handbook for New Employees*. 1st ed. Bellevue, WA: Valve Press. http://www.valvesoftware.com/company/Valve_Handbook_LowRes.pdf.

van der Malsburg, C. 1973. "Self-Organization of Orientation Sensitive Cells in the Striate Cortex." *Kybernetik* 14, no. 2 (December): 85–100.

Vapnik, Vladimir Naumovich. 1998. *Statistical Learning Theory*. New York: Wiley.

Vaughan, Sandra L., Robert F. Mills, Michael R. Grimaila, Gilbert L. Peterson, and Steven K. Rogers. 2014. "Narratives as a Fundamental Component of Consciousness." In *5th Workshop on Computational Models of Narrative*, edited by Mark A. Finlayson, Jan Christoph Meister, and Emile G. Bruneau, 246–50. http://dx.doi.org/10.4230/OASIcs.CMN.2014.246.

Vermesan, Ovidiu, and Peter Friess. 2013. *Internet of Things: Converging Technologies for Smart Environments and Integrated Ecosystems*. Aalborg, Denmark: River Publishers.

Vicente, Kim J., and Jens Rasmussen. 1990. "The Ecology of Human-Machine Systems II: Mediating 'Direct Perception' in Complex Work Domains." *Ecological Psychology* 2 (3): 207–49.

Vizer, Lisa M, and Andrew Sears. 2015. "Classifying Text-Based Computer Interactions for Health Monitoring." *IEEE Pervasive Computing* 14 (4): 4–71.

von Neumann, John, and Oskar Morgenstern. 1944. *Theory of Games and Economic Behavior*. 1st ed. Princeton, NJ: Princeton University Press.

Vongsy, K., M. T. Eismann, and M. J. Mendenhall. 2015. "Extension of the Linear Chromodynamics Model for Spectral Change Detection in the Presence of Residual Spatial Misregistration." *IEEE Transactions on Geoscience and Remote Sensing* 53(6): 3005–3021.

Vongsy, Karmon, and Michael J. Mendenhall. 2010. "Improved Change Detection through Post Change Classification: A Case Study Using Synthetic Hyperspectral Imagery." In *2010 2nd Workshop on Hyperspectral Image and Signal Processing: Evolution in Remote Sensing*, Reykjavik, 1–4. https://doi.org/10.1109/WHISPERS.2010.5594930.

Vongsy, Karmon, Michael J. Mendenhall, Michael T. Eismann, and Gilbert L. Peterson. 2012. "Removing Parallax-Induced Changes in Hyperspectral Change Detection." In *2012 IEEE International Geoscience and Remote Sensing Symposium*, Munich, 2012–15. https://doi.org/10.1109/IGARSS.2012.6350982.

Vongsy, Karmon, Michael J. Mendenhall, Philip M. Hanna, and Jason Kaufman. 2009. "Change Detection Using Synthetic Hyperspectral Imagery." *In 2009 First Workshop on Hyperspectral Image and Signal Processing: Evolution in Remote Sensing*, Grenoble, France, 1–4. https://doi.org/10.1109/WHISPERS.2009.5289016.

Wallach, Wendell. 2017. "Toward a Ban on Lethal Autonomous Weapons: Surmounting the Obstacles." *Communications of the ACM* 60, no. 5 (April): 28–34. https://doi.org/10.1145/2998579.

Walsh, Toby. 2016. "Turing's Red Flag." *Communications of the ACM* 59, no. 7 (July 2016): 34–37.

Walter, W. Grey. 1950. "An Imitation of Life." *Scientific American*, May 1950.

Wang, Ning, David V. Pynadath, and Susan G. Hill. 2015. "Building Trust in a Human-Robot Team with Automatically Generated Explanations." In *Interservice/Industry Training, Simulation and Education Conference (I/ITSEC)*, Vol. 2015, Orlando, FL, http://www.iitsecdocs.com/volumes/2015.

Wang, Pei. 2006. "From NARS to a Thinking Machine." In *Proceedings of the 2007 Conference on Advances in Artificial General Intelligence: Con-*

cepts, Architectures and Algorithms: Proceedings of the AGI Workshop 2006, 75–93. Amsterdam: IOS Press. http://dl.acm.org/citation.cfm?id=1565455.1565462.

Ware, Colin. 2013. *Information Visualization: Perception for Design*. 3rd ed. Waltham, MA: Morgan Kaufmann.

Waymo. 2018. "Waymo." October 12, 2018. https://waymo.com/.

Waytz, Adam, Joy Heafner, and Nicholas Epley. 2014. "The Mind in the Machine: Anthropomorphism Increases Trust in an Autonomous Vehicle." *Journal of Experimental Social Psychology* 52 (May): 113–17. https://doi.org/10.1016/j.jesp.2014.01.005.

Weber, P., G. Modina-Oliva, C. Simon, and B. Iung. 2012. "Overview on Bayesian Networks Applications for Dependability, Risk Analysis and Maintenance Areas." *Engineering Applications of Artificial Intelligence* 25, no. 4 (June): 671–82. https://doi.org/10.1016/j.engappai.2010.06.002.

Wedell-Wedellsborg, Thomas. 2017. "Are You Solving the Right Problems?" *Harvard Business Review*, February 2017. https://hbr.org/2017/01/are-you-solving-the-right-problems.

Weerdt, Mathijs de, and Brad Clement. 2009. "Introduction to Planning in Multiagent Systems." *Multiagent and Grid Systems* 5:345–55. https://doi.org/10.3233/MGS-2009-0133.

Weizenbaum, Joseph. 1966. "ELIZA—a Computer Program for the Study of Natural Language Communication Between Man and Machine." *Communications of the ACM* 9, no. 1 (January): 36–45. https://doi.org/10.1145/365153.365168.

Wenzhi Liao, Aleksandra Pižurica, Rik Bellens, Sidharta Gautama, and Wilfried Philips. 2015. "Generalized Graph-Based Fusion of Hyperspectral and LiDAR Data Using Morphological Features." *IEEE Geoscience and Remote Sensing Letters* 12, no. 3 (March): 552–56. https://doi.org/10.1109/LGRS.2014.2350263.

Werbos, Paul. 1974. "Beyond Regression: New Tools for Prediction and Analysis in the Behavioral Sciences." Dissertation, Harvard University.

Wernicke, Carl. 1908. "The Symptom-Complex of Aphasia." In *Diseases of the Nervous System*, edited by Archibald Church, 265–324. New York: Appleton.

Weyer, Johannes. 2006. "Modes of Governance of Hybrid Systems. The Mid-Air Collision at Ueberlingen and the Impact of Smart Technology." *Science* 2 (November): 127–49. https://doi.org/10.17877/DE290R-720.

Whalley, Katherine. 2013. "Humans Are on the Grid." *Nature Reviews Neuroscience* 14 (10): 667. https://doi.org/10.1038/nrn3588.

White, Franklin E. 1991. "Data Fusion Lexicon." Fort Belvoir, VA: Data Fusion Subpanel of the Joint Directors of Laboratories, October 1, 1991. https://doi.org/10.21236/ADA529661.

Wickens, Christoper, John Lee, Yili D. Liu, and Sallie Gordon-Becker. 2004. *An Introduction to Human Factors Engineering*. 2nd ed. Upper Saddle River, NJ: Prentice Hall.

Wiener, Earl L. 1989. "Human Factors of Advanced Technology ('Glass Cockpit') Transport Aircraft." Moffett Field, CA: NASA. https://ntrs.nasa.gov/archive/nasa/casi.ntrs.nasa.gov/19890016609.pdf.

Wiener, Norbert. 1948. *Cybernetics, or Control and Communication in the Animal and the Machine*. New York: John Wiley and Sons.

Wilkins, David E. 1988. *Practical Planning: Extending the Classical AI Planning Paradigm*. San Francisco: Morgan Kaufmann.

Winston, Patrick. 1992. *Artificial Intelligence*. 3rd ed. London: Pearson.

Witten, Ian, Elbe Frank, and Mark A. Hall. 2011. *Data Mining: Practical Machine Learning Tools and Techniques*. 3rd ed. Amsterdam: Morgan-Kaufmann.

Wong, Hong Yu. 2010. "The Secret Lives of Emergents." In *Emergence in Science and Philosophy*, edited by Antonella Corradini and Timothy O'Connor, 7. New York: Routledge, 2010.

Woods, David D. 1984. "Visual Momentum: A Concept to Improve the Cognitive Coupling of Person and Computer." *International Journal of Man-Machine Studies* 21 no. 3 (September) 229–44. https://doi.org/10.1016/S0020-7373(84)80043-7.

Woods, David D., and Erik Hollnagel. 2006. *Joint Cognitive Systems: Patterns in Cognitive Systems Engineering*. Boca Raton, FL: CRC, Taylor & Francis, 2006.

Wooldridge, Michael J. 2009. *An Introduction to Multiagent Systems*. Chichester, UK: J. Wiley & Sons, 2009.

Woolley, Brian G., and Gilbert L. Peterson. 2009. "Unified Behavior Framework for Reactive Robot Control." *Journal of Intelligent and Robotic Systems* 55, no. 2 (July): 155–76. https://doi.org/10.1007/s10846-008-9299-1.

Work, Robert. 2015. "Remarks by Defense Deputy Secretary Robert Work at the CNAS Inaugural National Security Forum." Washington, DC, December 14, 2015. https://www.cnas.org/publications/transcript/remarks-by-defense-deputy-secretary-robert-work-at-the-cnas-inaugural-national-security-forum.

Wunker, Stephen. 2012. "Seven Signs of a Successful Incubator." Forbes.com, December 16, 2012. https://www.forbes.com/sites/stephenwunker/2012/12/16/seven-signs-of-a-successful-incubator/.

Xu, Lin, Frank Hutter, Holger H. Hoos, and Kevin Leyton-Brown. 2011. "Hydra-MIP: Automated Algorithm Configuration and Selection for Mixed Integer

Programming." In *Proceedings of the 18th RCRA Workshop on Experimental Evaluation of Algorithms for Solving Problems with Combinatorial Explosion (RCRA 2011)*, 11–15. Barcelona, Spain, 2011. http://citeseerx.ist.psu.edu/viewdoc/download?doi=10.1.1.221.4923&rep=rep1&type=pdf.

Yadron, Danny, and Dan Tynan. 2016. "Tesla Driver Dies in First Fatal Crash While Using Autopilot Mode." *The Guardian.* June 30, 2016, sec. Technology. https://www.theguardian.com/technology/2016/jun/30/tesla-autopilot-death-self-driving-car-elon-musk.

Yeh, Wenchi, and Lawrence W. Barsalou. 2006. "The Situated Nature of Concepts." *American Journal of Psychology* 119, no. 3 (Fall): 349–84.

Yuan, Li. 2018. "Google and Intel Beware: China Is Gunning for Dominance in AI Chips." *Wall Street Journal*, January 4, 2018, sec. Tech. https://www.wsj.com/articles/google-and-intel-beware-china-is-gunning-for-dominance-in-ai-chips-1515060224.

Yuste, Antonio Pérez, and Magdalena Salazar Palma. 2004. "The First Wireless Remote-Control: The Telekino of Torres Quevedo." *IEEE Conference on the History of Electronics* (CHE 2004). Bletchley Park, UK.

Zacharias, Greg L. 1974. "A Digital Autopilot for the Space Shuttle Vehicle." Thesis, Massachusetts Institute of Technology. http://hdl.handle.net/1721.1/28019.

Zacharias, Greg L., J. MacMillan, S. B. Van Hemel, eds. 2008. *Behavioral Modeling and Simulation: From Individuals to Societies.* Washington, DC: The National Academies Press. https://doi.org/10.17226/12169.

Zacharias, Greg, and Mark Maybury. 2010. "Operating Next-Generation Remotely Piloted Aircraft for Irregular Warfare." Washington, DC: US Air Force Scientific Advisory Board, April 2011. https://info.publicintelligence.net/USAF-RemoteIrregularWarfare.pdf.

Zacharias, Greg, and L. Young. 1981. "Influence of Combined Visual and Vestibular Cues on Human Perception and Control of Horizontal Rotation." *Experimental Brain Research* 41 (2): 159–71.

Zadeh, L. A. 1965. "Fuzzy Sets." *Information and Control* 8, no. 3 (June): 338–53. https://doi.org/10.1016/S0019-9958(65)90241-X.

———. 1996. "Fuzzy Logic = Computing with Words." *IEEE Transactions on Fuzzy Systems* 4, no. 2 (May): 103–11. https://doi.org/10.1109/91.493904.

Zajac, Zygmunt. 2014. "Yann LeCun's Answers from the Reddit AMA." FastML (website). May 26, 2014. http://fastml.com/yann-lecuns-answers-from-the-reddit-ama/.

Zheng, Yu. 2015. "Methodologies for Cross-Domain Data Fusion: An Overview." *IEEE Transactions on Big Data* 1 (1): 16–34. https://doi.org/10.1109/TBDATA.2015.2465959.

Ziegler, Bernard, Herbert Praehofer, and Tag Gon Kim. 2000. *Theory of Modeling and Simulation*. 2nd ed. San Diego: Academic Press.

Ziegler, Joshua, Jason Bindewald, and Gilbert Peterson. 2017. "An Introduction to Behavior-Based Robotics." Paper presented at the Seventh Symposium on Educational Advances in Artificial Intelligence (EAAI-2017). San Francisco.

Zigoris, P., J. Siu, O. Wang, and A. T. Hayes. 2003. "Balancing Automated Behavior and Human Control in Multi-Agent Systems: A Case Study in RoboFlag." In *Proceedings of the 2003 American Control Conference, 2003*. Vol.1, 667–71. https://doi.org/10.1109/ACC.2003.1239096.

Złotowski, Jakub, Diane Proudfoot, Kumar Yogeeswaran, and Christoph Bartneck. 2015. "Anthropomorphism: Opportunities and Challenges in Human–Robot Interaction." *International Journal of Social Robotics* 7, no. 3 (June): 347–60. https://doi.org/10.1007/s12369-014-0267-6.